Fuzzy Logic and Probability Applications

ASA-SIAM Series on Statistics and Applied Probability

The ASA-SIAM Series on Statistics and Applied Probability is published jointly by the American Statistical Association and the Society for Industrial and Applied Mathematics. The series consists of a broad spectrum of books on topics in statistics and applied probability. The purpose of the series is to provide inexpensive, quality publications of interest to the intersecting membership of the two societies.

Editorial Board

Fuzzy Logic and Probability Applications

Bridging the Gap

Edited by

Timothy J. Ross

University of New Mexico
Albuquerque, New Mexico

Jane M. Booker

Los Alamos National Laboratory
Los Alamos, New Mexico

W. Jerry Parkinson

Los Alamos National Laboratory
Los Alamos, New Mexico

Society for Industrial and Applied Mathematics
Philadelphia, Pennsylvania

ASA

American Statistical Association
Alexandria, Virginia

The correct bibliographic citation for this book is as follows: Ross, Timothy J., Jane M. Booker, and W. Jerry Parkinson, eds., *Fuzzy Logic and Probability Applications: Bridging the Gap*, ASA-SIAM Series on Statistics and Applied Probability, SIAM, Philadelphia, ASA, Alexandria, VA, 2002.

Library of Congress Cataloging-in-Publication Data

Fuzzy logic and probability applications : bridging the gap / edited by Timothy J. Ross, Jane M. Booker, W. Jerry Parkinson.
    p. cm. — (ASA-SIAM series on statistics and applied probability)
  Includes bibliographical references and index.
  ISBN 0-89871-525-3
    1. Fuzzy logic. 2. Probabilities. 3. Fuzzy logic—Industrial applications. 4. Probabilities—Industrial applications. I. Ross, Timothy J. II. Booker, Jane M. III. Parkinson, W. J. (William Jerry), 1939- IV. Series.

QA9.64 .F894 2002
511.3—dc21

                                                                    2002075756

To the memory of our colleague
and his inspiration for this book,
Dr. Thomas (Tom) R. Bement,
Los Alamos, New Mexico

List of Contributors

Hrishikesh Aradhye
SRI International

Thomas R. Bement (dec.)
Los Alamos National Laboratory

Jane M. Booker
Los Alamos National Laboratory

Kenneth B. Butterfield
Los Alamos National Laboratory

Aly El-Osery
New Mexico Institute of Mining and
Technology

Carlos Ferregut
University of Texas at El Paso

Mo Jamshidi
University of New Mexico

A. Sharif Heger
Los Alamos National Laboratory

Vladik Kreinovich
University of Texas at El Paso

Jonathan L. Lucero
University of New Mexico

Yohans Mendoza
Sirius Images, Inc.

Mary A. Meyer
Los Alamos National Laboratory

William S. Murray
Los Alamos National Laboratory

Roberto A. Osegueda
University of Texas at El Paso

W. Jerry Parkinson
Los Alamos National Laboratory

Timothy J. Ross
University of New Mexico

Kimberly F. Sellers
Carnegie Mellon University

Nozer D. Singpurwalla
George Washington University

Ronald E. Smith
Los Alamos National Laboratory

Contents

Foreword

Probability theory and fuzzy logic are the principal components of an array of methodologies for dealing with problems in which uncertainty and imprecision play important roles. In relation to probability theory, fuzzy logic is a new kid on the block. As such, it has been and continues to be, though to a lesser degree, an object of controversy. The leitmotif of *Fuzzy Logic and Probability Applications: Bridging the Gap* is that fuzzy logic and probability theory are complementary rather than competitive. This is a thesis that I agree with completely. However, in one respect my perspective is more radical. Briefly stated, I believe that it is a fundamental error to base probability theory on bivalent logic. Moreover, it is my conviction that eventually this view will gain wide acceptance. This will happen because with the passage of time it will become increasingly obvious that there is a fundamental conflict between bivalence and reality.

To write a foreword to a book that is aimed at bridging the gap between fuzzy logic and probability theory is a challenge that is hard to meet. But a far greater challenge is to produce a work that illuminates some of the most basic concepts in human cognition—the concepts of randomness, probability, uncertainty, vagueness, possibility, imprecision, and truth. The editors, authors, and publisher of *Fuzzy Logic and Probability Applications* have, in my view, met this challenge.

In *Fuzzy Logic and Probability Applications*, in consonance with this book's central theme, controversial issues relating to fuzzy logic and probability theory are treated with objectivity, authority, and insight. Of particular interest and value is the incisive analysis of the evolution of probability theory and fuzzy logic presented in Chapter 1. One of the basic concepts discussed in this chapter is that of vagueness. In the authors' interpretation, vagueness and fuzziness are almost synonymous. In my view, this is not the case. Basically, vagueness relates to insufficient specificity, as in "I will be back sometime," whereas fuzziness relates to unsharpness of boundaries, as in "I will be back in a few minutes." Thus fuzziness is a property of both predicates and propositions, whereas vagueness is a property of propositions but not of predicates. For example, "tall" is a fuzzy predicate, but "Robert is tall" is fuzzy but not vague.

Complementarity of fuzzy logic and probability theory is rooted in the fact that probability theory is concerned with partial certainty, whereas fuzzy logic is mainly concerned with partial possibility and partial truth. A simple example is that if Robert is half-German, then the proposition "Robert is German" may be viewed as half-true but not uncertain. On the other hand, if it is possible that Robert is German, then the probability that he is German may be 0.5.

Standard probability theory—call it PT—is designed to deal with partial certainty but not with partial possibility or partial truth. This is a serious deficiency of PT since much of human knowledge consists of propositions that in one way or another are partially certain and/or partially possible and/or partially true. For example, in the proposition "Usually,

Robert returns from work at about 6 PM," "usually" is a fuzzy probability and "about 6 PM" is a fuzzy value of time. Propositions of this form may be viewed as descriptors of perceptions. Perceptions are intrinsically imprecise, reflecting the bounded ability of sensory organs and, ultimately, the brain, to resolve detail and store information. More concretely, perceptions are f-granular in the sense that (a) the boundaries of perceived classes are unsharp, and (b) the values of perceived attributes are granulated, with a granule being a clump of values drawn together by indistinguishability, similarity, proximity, or functionality.

f-granularity of perceptions puts them well beyond the reach of conventional meaning-representation methods based on bivalent logic. As a consequence, PT by itself lacks the capability to operate on perception-based information: As an illustration, PT cannot provide an answer to the following query: What is the probability that Robert is home at about t PM, given the perception-based information that (a) usually Robert leaves his office at about 5:30 PM and (b) usually it takes about 30 minutes to reach home?

To add to PT the capability to operate on perception-based information, it is necessary to bridge the gap between fuzzy logic and probability theory. This is the thrust of Chapters 5–8 in *Fuzzy Logic and Probability Applications*. The content of these chapters is a highly important contribution of *Fuzzy Logic and Probability Applications* to the development of a better understanding of how to employ combinations of methods drawn from fuzzy logic and probability theory.

In a related but somewhat different spirit, I have outlined in a recent paper ("Toward a perception-based theory of probabilistic reasoning with imprecise probabilities," *J. Statist. Plann. Inference*, 105 (2002), pp. 233–264) an approach to generalization of PT that adds to PT the capability to deal with perception-based information.

The approach in question involves three stages of generalization: (1) f-generalization, (2) fg-generalization, and (3) nl-generalization. Briefly, these stages are as follows:

(1) f-generalization involves a progression from crisp sets to fuzzy sets in PT, leading to a generalization of PT, which is denoted as PT+. In PT+, probabilities, functions, relations, measures, and everything else are allowed to have fuzzy denotations, that is, be a matter of degree. In particular, probabilities described as low, high, not very high, etc., are interpreted as labels of fuzzy subsets of the unit interval or, equivalently, as possibility distributions of their numerical values.

(2) fg-generalization involves fuzzy granulation of variables, functions, relations, etc., leading to a generalization of PT, which is denoted as PT++. By fuzzy granulation of a variable X we mean a partition of the range of v into fuzzy granules with a granule being a clump of values of X, which are drawn together by indistinguishability, similarity, proximity, or functionality. Membership functions of such granules usually are assumed to be triangular or trapezoidal. Basically, granularity reflects the bounded ability of the human mind to resolve detail and store information.

(3) nl-generalization involves an addition to PT++ of the capability to operate on propositions expressed in a natural language, with the understanding that such propositions serve as descriptors of perceptions. nl-generalization of PT leads to perception-based probability theory, denoted as PTp. By construction, PTp is needed to answer the following query in the Robert example: What is the probability that Robert is home at about t PM?

A serious problem with PT, but not PT+, is the brittleness of crisp definitions. For example, by definition, events A and B are independent iff $P(A, B) = P(A)P(B)$. However, suppose that the equality in question is satisfied to within epsilon in magnitude. In this event, if epsilon is positive, then no matter how small it is, A and B are not independent. What we see is that in PT independence is defined as a crisp concept, but in reality it is a fuzzy concept, implying that independence is a matter of degree. The same applies to the concepts of randomness, stationarity, normality, and almost all other concepts in PT.

The problem in question is closely related to the ancient Greek sorites paradox. More precisely, let C be a crisply defined concept that partitions the space U of objects, $\{u\}$ to which C is applicable into two sets: A^+, the set of objects that satisfy C, and A^-, the set of objects that do not satisfy C. Assume that u has α as a parameter and that $u \in A^+$ iff $\alpha \in A$, where A is a subset of the parameter space, and $u \in A^-$ if α is not in A. Since the boundary between A^+ and A^- is crisply defined, there is a discontinuity such that if $\alpha \in A$, then $\alpha + \alpha \Delta$ may be in A', where $\Delta \alpha$ is an arbitrarily small increment in α and A' is the complement of A. The brittleness of crisply defined concepts is a consequence of this discontinuity.

An even more serious problem is the dilemma of "it is possible but not probable." A simple version of this dilemma is the following. Assume that A is a proper subset of B and that the Lebesgue measure of A is arbitrarily close to the Lebesgue measure of B. Now, what can be said about the probability measure $P(A)$ given the probability measure $P(B)$? The only assertion that can be made is that $P(A)$ lies between 0 and $P(B)$. The uninformativeness of this assertion leads to counterintuitive conclusions. For example, suppose that with probability 0.99 Robert returns from work within one minute of 6 PM. What is the probability that he is home at 6 PM? Using PT, with no additional information or the use of the maximum entropy principle, the answer is "between 0 and 0.99." This simple example is an instance of a basic problem of what to do when we know what is possible but cannot assess the associated probabilities or probability distributions.

These and many other serious problems with PT point to the necessity of generalizing PT by bridging the gap between fuzzy logic and probability theory. *Fuzzy Logic and Probability Applications: Bridging the Gap* is an important move in this direction. The editors, authors, and publisher have produced a book that is a "must-read" for anyone who is interested in solving problems in which uncertainty, imprecision, and partiality of truth play important roles.

Lotfi A. Zadeh
Berkeley, California
April, 2002

Foreword

Writing this foreword after reading Lotfi Zadeh's foreword takes me back more than two decades. When Lotfi was first developing fuzzy logic, we talked about it quite a lot. My most vivid memories of those conversations are about the times when Lotfi presented his ideas at Stanford to my seminar on the foundations of probability. We had some great arguments, with lots of initial misunderstandings, especially on my part. I belonged, in those days, to the school of thought that held that anything scientifically serious could be said within classical logic and probability theory. There were, of course, disagreements about the nature of probability, but not really about the formal properties of probability. The main exception to this last claim was the controversy as to whether a probability measure should be only finitely additive or also countably additive. De Finetti was the leading advocate of the finitely additive viewpoint.

But new ideas and generalizations of probability theory had already been brewing on several fronts. In the early pages of the present book the authors mention the Dempster–Shafer theory of evidence, which essentially requires a generalization to upper and lower probability measures. Moreover, already in the early 1960s, Jack Good, along with de Finetti, a prominent Bayesian, had proposed the use of upper and lower probabilities from a different perspective. In fact, by the late 1970s, just as fuzzy logic was blossoming, many varieties of upper and lower probabilities were being cultivated. Often the ideas could be organized around the concept of a Choquet capacity. For example, Mario Zanotti and I showed that when a pair of upper and lower probability measures is a capacity of infinite order, there exists a probability space and a random relation on this space that generates the pair. On the other hand, in a theory of approximate measurement I introduced at about the same time, the standard pair of upper and lower probability measures is not even a capacity of order two. This is just a sample of the possibilities. But the developing interest in such generalizations of standard probability theory helped create a receptive atmosphere for fuzzy logic.

To continue on this last point, I also want to emphasize that I am impressed by the much greater variety of real applications of fuzzy logic that have been developed than is the case for upper and lower probabilities. I was not surprised by the substantial chapters on aircraft and auto reliability in the present volume, for about six years ago in Germany I attended a conference on uncertainty organized by some smart and sophisticated engineers. They filled my ears with their complaints about the inadequacy of probability theory to provide appropriate methods to analyze an endless array of structural problems generated by reliability and control problems in all parts of engineering. I am not suggesting I fully understand what the final outcome of this direction of work will be, but I am confident that the vigor of the debate, and even more the depth of the new applications of fuzzy logic, constitute a genuinely new turn in the long history of concepts and theories for dealing with uncertainty.

Patrick Suppes
Stanford, California
September, 2002

Preface

This book is designed as a practical guide for scientists and engineers to help them solve problems involving uncertainty and/or probability. Consequently, the book has a practical bent and contains lots of examples. Yet there is enough theory and references to fundamental work to provide firm ground for scientists and engineers.

The point of view of the authors is that probabilists and fuzzy enthusiasts have argued for too long about which philosophy is best. The truth is that both tools have their place in the world of problem solving. In many cases, fuzzy logic is used to solve problems that could be solved with probability because the probability literature is too theoretical and looks impractical to the practitioner. In other cases, fuzzy logic is, indeed, the proper technique to use. Alternatively, some problems are solved using probability because fuzzy techniques appear to be too empirical. Probability is often the correct tool to use. Sometimes both tools can be used together synergistically. This book is intended to help the user choose the best tool for the job.

The distinctive feature of this book is that investigators from these two different fields have combined their talents to provide a text that is useful to multiple communities. This book makes an honest effort to show both the shortcomings and benefits of each technique. It provides examples and insight into the best tool to use for a particular job.

Both fuzzy logic and probability are discussed in other texts. Fuzzy logic texts usually mention probability but are primarily about fuzzy logic. Probability texts seldom even acknowledge fuzzy logic. Because probabilists and fuzzy enthusiasts seldom cooperatively work together, textbooks usually give one method of solving problems or the other, but not both. This book is an exception. The authors *do* work together to mutually benefit both disciplines. In addition, they present examples showing useful combinations of the two techniques.

A familiarity with mathematics through calculus is assumed for this book. The book is intended for the practicing engineer or scientist at the Bachelor of Science level or above. While it is not designed for a specific course, because probability and fuzzy logic are not usually taught together, this book could be used for a course on either subject or for a course on general problem solving. It also has applications to control theory and artificial intelligence, knowledge acquisition/management, and risk/reliability analysis.

Timothy J. Ross

Jane M. Booker

W. Jerry Parkinson

Acknowledgments

Special thanks to the Los Alamos National Laboratory Enhanced Surveillance Program for their support of the research work in Chapters 1, 3, 4, 6, 11, and 14. The Los Alamos National Laboratory is operated by the University of California for the U.S. Department of Energy under contract W-7405-ENG-36.

Also, special thanks to the Los Alamos National Laboratory Educational Research Opportunity Program for their support of the research work in Chapters 1, 2, 8, 12, and 14.

We thank the National Aeronautics and Space Administration University Research Center Office for their partial support of the research in Chapters 1, 5, 7, 9, and 13 under NASA grants NAG2-1480 and NAG2-1196.

The research of Chapter 10 was supported in part by NASA under cooperative agreement NCC5-209; the NSF under grants DUE-9750858 and CDA-9522207; the United Space Alliance under grant NAS 9-20000 (PWO C0C67713A6); the Future Aerospace Science and Technology Program (FAST) Center for Structural Integrity of Aerospace Systems; the Air Force Office of Scientific Research, Air Force Material Command, USAF, under grant F49620-95-1-0518; and the National Security Agency under grant MDA904-98-1-0561. We gratefully acknowledge this support.

Finally, the research of Chapter 9 is an extension of a previous effort appearing in F. S. Wong, T. J. Ross, and A. C. Boissonnade, "Fuzzy sets and survivability analysis of protective structures," in *The Analysis of Fuzzy Information*, Vol. 3, James Bezdek, ed., CRC Press, Boca Raton, FL, 1987, pp. 29–53. The authors are grateful to the investigators (including the senior author) who brought this seminal idea in structural engineering to light. It is our hope that this extension provides renewed effort in this area in the future.

Part I

Fundamentals

Jane M. Booker

I.1 Chapters 1–6

To bridge the gap between probability and fuzzy theories, the first step is to examine and understand the two sides of the gap. The first part of this book consists of six chapters that lay the foundations of both theories and provide the fundamental principles for constructing the bridge.

We (the editors) begin (in Chapter 1) with an introduction to the history of both theories and the stories describing the formulation of the gap between them. It is our intent to represent both "sides," but with the tone of reconciliation. There are cases where applications support one theory more than the other, and these are brought forth in the application chapters in Part II. There are also cases where either probability or fuzzy theory is useful or a hybrid approach combining the two is best, particularly when characterizing different kinds of uncertainties in a complex problem.

Following the philosophical discussion in Chapter 1, Chapters 2 and 3 provide the foundations of fuzzy theory and probability theory, respectively.

Chapter 4 is devoted to Bayesian probability theory. Bayes' theorem provides a powerful structure for bridging the gap between fuzzy and probability theory and lays some of the groundwork for the bridging mechanism. Chapter 5 then completes the building of the bridge in a mathematical and philosophical sense.

Because data and information in today's world often reside within the experience and knowledge of the human mind, Chapter 6 examines the formal use of eliciting and analyzing expert judgment. That topic comprises an entire book in its own right (Meyer and Booker, 2001). The contents of this chapter not only covers both theories but provides applications, making it the perfect transition chapter to the applications chapters in Part II.

I.2   Suggested reading

The reader who has little familiarity with either fuzzy or probability theory should review Chapters 2 and 3, respectively. We highly recommend Chapter 4 for an introductory understanding of how to bridge the gap between the two theories and recommend Chapter 5 to complete this understanding. Because many uncertainties in complex problem solving and mass communication are imbedded in the cognitive processes of the human mind, we suggest a review of the formal methods for handling expert knowledge.

References

M. A. Meyer and J. M. Booker (2001), *Eliciting and Analyzing Expert Judgment: A Practical Guide*, ASA–SIAM Series on Statistics and Applied Probability, SIAM, Philadelphia, ASA, Alexandria, VA.

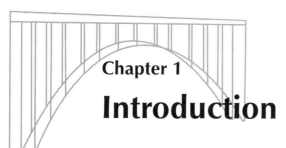

Chapter 1

Introduction

Timothy J. Ross, Jane M. Booker, and W. Jerry Parkinson

1.1   Some history and initial thoughts

The history of humankind and its relationship to technology is well documented. One such technology—information technology—has been impacting human thought over at least the last 400 years.

For example, the 17th century was described by its technology historians as the Age of Experience, in which many developments from direct observation were realized in fundamental astronomy, physics, chemistry, and mathematics despite remaining mysteries in other fundamental concepts (e.g., sphericity of the planet). The 18th century was perceived as the Age of Reason—the Renaissance following the Dark Ages in Europe at the end of the 17th century. From the beginning of the 19th century until about 1950, we had what we could call the Age of Mathematics—here the arithmetical formalisms of the previous century were advanced into more formal calculus-based theories. From 1950 to 1980, we had the Age of Computing—a time where many computational models of physical processes were developed. This period paralleled the developments in digital computing from the ILLIAC through the IBM® 360 to the IBM PC. From 1980 to 1995, we had the Age of Knowledge—a time when, as evidenced in its literature, a great deal of effort was focused on the acquisition and appropriate use of knowledge. Again, we see a parallel development, this time in hardware, between the areas of symbolic computing (LISP machines, Mathematica) and parallel computing (CRAY, Connection-Machine, IBM SP). Finally, 1995 to the present is the Age of Cyberspace—an age of ubiquitous networks and multiple, rapid forms of communication.

Humans cannot be expected to reasonably adjust to today's rapid technological advancements. Today humans are in information overload. The engine behind this can be traced to the individual forces of computer hardware and software; the nascent field of network computing; integration of various engineering fields; new technologies such as GIS (geographic information systems) and GPS (global positioning systems) producing gigabytes of information each hour; intelligent databases; complex simulation models; collaborative engineering; inventive engineering; and new methods to present, distill, and deliver technological education to distant areas—so-called distance education (Arciszewski, 1999).

Superposed with this development in information technology we had a parallel development in the theoretical frameworks for assessing uncertainty in the information. Probability concepts date back to the 1500s, the time of Cardano when gamblers recognized the rules of probability in games of chance and, more important, that avoiding these rules resulted in a sure loss (i.e., the classic coin toss example of "heads you lose, tails I win," referred to as the "Dutch book"). The concepts were still very much in the limelight in 1685, when the Bishop of Wells wrote a paper that discussed a problem in determining the truth of statements made by two witnesses who were both known to be unreliable to the extent that they only tell the truth with probabilities p_1 and p_2, respectively. The Bishop's answer to this was based on his assumption that the two witnesses were independent sources of information (Lindley (1987b)).

Probability theory was initially developed in the 18th century in such landmark treatises as Jacob Bernoulli's *Ars Conjectandi* (1713) and Abraham DeMoiver's *Doctrine of Chances* (1718; 2nd ed., 1738). Later in that century a small number of articles appeared in the periodical literature that would have a profound effect on the field. Most notable of these were Thomas Bayes' *An Essay Towards Solving a Problem in the Doctrine of Chances* (1763) and Pierre Simon Laplace's formulation of the axioms relating to games of chance, *Mémoire sur la probabilité des causes par les evenemens* (1774). Laplace, only 25 years old at the time he began his work in 1772, wrote the first substantial article in mathematical statistics prior to the 19th century. Despite the fact that Laplace, at the same time, was heavily engaged in mathematical astronomy, his memoir was an explosion of ideas that provided the roots for modern decision theory, Bayesian inference with nuisance parameters (historians claim that Laplace did not know of Bayes' earlier work), and the asymptotic approximations of posterior distributions (Stigler (1986)).

By the time of Newton, physicists and mathematicians were formulating different theories of probability (see Chapter 3). The most popular ones remaining today are the relative frequency theory and the subjectivist, or personalistic, theory. The latter theory was initiated by Thomas Bayes (1763), who articulated his very powerful theorem for the assessment of subjective probabilities. The theorem specified that a human's degree of belief could be subjected to an objective, coherent, and measurable mathematical framework within the subjective probability theory. In the early days of the 20th century, Rescher developed a formal framework for a conditional probability theory, and Jan Lukasiewicz developed a multivalued, discrete logic (circa 1930). In the 1960s, Arthur Dempster developed a theory of evidence which, for the first time, included an assessment of ignorance, or the absence of information. In 1965, Zadeh introduced his seminal idea in a continuous-valued logic called fuzzy set theory. In the 1970s, Glenn Shafer extended Dempster's work[1] to produce a complete theory of evidence dealing with information from more than one source, and Lotfi Zadeh illustrated a possibility theory resulting from special cases of fuzzy sets. Later, in the 1980s other investigators showed a strong relationship between evidence theory, probability theory, and possibility theory with the use of what have been called fuzzy measures (Klir and Folger (1988)).

In the over three decades since its inception by Zadeh, fuzzy set theory (and its logical counterpart, fuzzy logic) has undergone tremendous growth. Over ten thousand papers, hundreds of books, almost a dozen journals, and several national and international societies bear witness to this growth. Table 1.1 shows a count of papers containing the word "fuzzy" in the title, as cited by INSPEC and MathSciNet databases. (Data for 2001 are not complete.) To this day, perhaps because the theory is chronologically one of the newest, fuzzy sets and

---

[1] We refer to this extension in the text as the Dempster–Shafer theory.

Table 1.1. *Number of papers with the word "fuzzy" in the title.*

Period	INSPEC	MathSciNet
1970–1979	570	441
1980–1989	2,383	2,463
1990–1999	23,121	5,459
2000–2001	5,940	1,670
Totals	32,014	10,033

*Compiled by Camille Wanat, Head, Engineering Library, University of California at Berkeley, June 21, 2002.

fuzzy logic remain steeped in controversy and debate for a variety of reasons. Although the philosophical and mathematical foundations of fuzzy sets are intuitive, they run counter to the thousands of years of dependence on binary set theory on which our entire Western cultural logic resides (first espoused by Aristotle in ancient Greece). In addition, some have seen fuzzy sets as a competitor to probability theory (the new-kid-on-the-block syndrome) in a variety of settings, such as in competition for precious page space in journals, for classes on campuses, for students in graduate classes, and even for consulting opportunities in industry, to name a few. The statistical societies have even sponsored debates in their own journals (e.g., *Statist. Sci.*, 1 (1986), pp. 335–358) and conferences on topics ranging from the mathematical (e.g., the axioms of subjective probability) to the linguistic (e.g., communicating better with engineers and scientists (Hoadley and Kettering (1990)) in an effort to win back "market share," i.e., to get science and engineering students to take classes in statistics and probability instead of evidence theory, fuzzy logic, soft computing, and other new uncertainty technologies.

However, the debate extends far beyond "market share" competitiveness. The core issues involve the philosophical and theoretical differences between these theories and how these theories are useful for application in today's complex, information-based society. Later in this chapter, there is a lengthy discussion on the running debate between advocates of the two theories concerning the merits of each theory in terms of modeling uncertainty and variability.[2]

It is the premise of this book that this perceived tension between probability theory and fuzzy set theory is precisely the mechanism necessary for scientific advancement. Within this tension are the searches, trials and errors, and shortcomings that all play a part in the evolution of any theory and its applications. In this debate between advocates of these two theories are the iterations necessary to reach a common ground that will one day seem so intuitive and plausible that it will be difficult to reflect or to remember that there ever was a debate at all! Our goal in writing this book is to illustrate how naturally compatible and complementary the two theories are and to help the reader see the power in combining the two theories to address the various forms of uncertainty that plague most complex problems. Some of this compatibility can be reached by examining how probability is interpreted. As will be demonstrated in this book, it is much easier to *bridge the gap* when a subjective or personalistic (e.g., Bayesian-based) interpretation of probability is used (see Chapter 3 for a discussion).

The contributors to this book were chosen for their experience and expertise in both theories but also for their efforts in and understanding of the compatibility of both fuzzy and probability theories. Even though some chapters are more fuzzy oriented and some are more

---

[2]It should be noted that within the probability community there is an equally intense debate over the interpretation of probability (i.e., frequentist versus subjective). To our knowledge, the only attempt ever made in resolving that debate is addressed in Chapter 3.

probability oriented,[3] the goal for each chapter is to provide some insight into establishing the common ground, i.e., *bridging the gap*.

In addition to the motivations for this book as explained above, there is ample evidence in the literature about the need for more information sharing among groups using various theories to assess and quantify uncertainty. For example, Laviolette et al. (1995) claim that "so few statisticians and probabilists have considered the efficacy of fuzzy methods"! Another observation (Bement (1996)) is that "the reason engineers (e.g., Zadeh) had to devise their own theory for handling different kinds of uncertainty was because statisticians failed to respond to the engineering needs." Thus far, most of the debate about FST (fuzzy set theory) and probability has appeared in "fuzzy" journals that "are not frequently read by statisticians." This statement rings true in many other works in the literature.

1.2   The great debate

1.2.1   The debate literature

There have been several organized debates at fuzzy conferences, including a recent one at the annual Joint Statistical Meeting (Anaheim, CA, August 1997); an oral faceoff between two individuals (G. Klir and P. Cheeseman) at the 8th Maximum Entropy Workshop (August 1–5, 1988, St. John's College, Cambridge, U.K.; see Klir (1989)); and there exist several archival collections of written debates. For example, the following citations give an idea of the activity produced in very well written arguments espousing the virtues of various uncertainty methods, although most of the debate centers on Bayesian probability or fuzzy logic:

- *Statist. Sci.*, 2 (1987), pp. 3–44;

- *Comput. Intell.*, 4 (1988), pp. 57–142;

- *IEEE Trans. Fuzzy Systems*, 2 (1994), pp. 1–45;

- *Technometrics*, 37 (1995), pp. 249–292.

In addition, numerous articles have been written outside of organized debates that have been critical of one or the other theory by protagonists of each. These articles include the following:

- Lindley, *Internat. Statist. Rev.*, 50 (1982), pp. 1–26 (with seven commentaries);

- Cheeseman, in the edited volume *Uncertainty in Artificial Intelligence*, North-Holland, Amsterdam, 1986, pp. 85–102;

- Cheeseman, *Comput. Intell.*, 4 (1988), pp. 58–66 (with 22 commentaries);

- Hisdal, *Fuzzy Sets Systems*, 25 (1988), pp. 325–356;

- Kosko, *Internat. J. Gen. Systems*, 17 (1990), pp. 211–240;

- Laviolette and Seaman, *Math. Sci.*, 17 (1992), pp. 26–41;

---

[3]It is not inconceivable that some problems are more fuzzy oriented and some are more probability oriented than others.

- Elkan, in *Proceedings of the American Association for Artificial Intelligence*, MIT Press, Menlo Park, CA, 1993, pp. 698–703 (with numerous commentaries in the subsequent AAAI magazine);

- Zadeh, *IEEE Trans. Circuits Systems*, 45 (1999), pp. 105–119.

The next section takes many of the points made in the various historical debates and organizes them into a few classic paradigms that seem to be at the heart of the philosophical differences between the two theories. While this is merely an attempt to summarize the debates for the purpose of spawning some thinking about the subsequent chapters of this book, it can in no way represent a complete montage of all the fine points made by so many competent scientists. Such a work would be another manuscript in itself.

1.2.2 The issues and controversy

Any attempt to summarize the nearly four decades of debate between the probability and fuzzy communities is a daunting task, and one that is in danger of further criticism for at least two reasons: First, we could not possibly include everyone's arguments (see the brief review of the many organized debates in the previous section); second, we could be precariously close to misrepresenting the opinions of those arguments that we do include here if we have inadvertently taken some arguments out of their originally intended contexts. Therefore, we apologize in advance for any omissions or possible misrepresentations included herein. It is important to mention that many of the opinions and arguments presented in this brief review of the "great debate" should be taken in their historical perspectives. Some of the individuals whose quotes are provided here—quotes perhaps made very early in the evolution of the process—may have changed their minds or changed their perspectives on some of the issues raised over the past 35 years. In addition, in changing views or perspectives, some of these individuals, and more not mentioned here, unwittingly have advanced the knowledge in both fields—fuzzy and probability theories—because it is the debate process, by its very nature, that has forced a positive evolution in bridging the gap between these two very powerful and very useful models of uncertainty and variability.

In what follows, we have organized our review of the great debate into some rather useful, although perhaps arbitrary, epistemological paradigms. We begin with a review of some of the polemical statements that fueled the fires of many of the original debates. Although strident in their character, in retrospect these polemics were well timed in terms of forcing people to look more closely at the side of the debate they were defending. The polemics tended to be one-sided, opposing fuzzy set theory, this being the *new sibling* in the family and requesting equal consideration as a viable theory for assessing uncertainty alongside its more mature and metaphorically larger *brother*, probability theory. Very few of the polemics were aimed in the other direction—probably more as a defensive reaction—and we will mention some of these.

Next, we organize the debate into some paradigms that seemed to characterize much of the disagreement. These are

- philosophical issues of chance, ambiguity, crispness, and vagueness;

- membership functions versus probability density functions;

- Bayes' rule;

- the so-called conjunction fallacy;

- the so-called disjunction contradiction;

- the excluded-middle laws;

- the infusion of fuzziness into probability theory.

We conclude our summary of the great debate with some rather pithy but positive statements.

The early polemics

We give first billing to the father of fuzzy logic, Professor Lotfi Zadeh, whose seminal paper in 1965 obviously started the debate. In a recent paper (Zadeh (1999)), where he discusses the need for developing methods to compute with words, he recounts a few remarks made in the early 1970s by two of his colleagues. These remarks revealed, in a very terse and crude way, a deep-seated proclivity of hard scientists to seriously consider only those things that are numerical in nature. To preface these remarks, Professor Zadeh cited a statement attributed to Lord Kelvin in 1883 about the prevailing 19th century respect for numbers and the utter disrespect for words:

> "I often say that when you can measure what you are speaking about and express it in numbers, you know something about it; but when you cannot measure it, when you cannot express it in numbers, your knowledge is of a meager and unsatisfactory kind: it may be the beginning of knowledge but you have scarcely, in your thoughts, advanced to the state of science, whatever the matter may be."

Zadeh goes on to recall that, in 1972, Rudolph Kalman, remarking on Zadeh's first exposition on a linguistic variable, had this to say:

> "Is Professor Zadeh presenting important ideas or is he indulging in wishful thinking? No doubt Professor Zadeh's enthusiasm for fuzziness has been reinforced by the prevailing climate in the U.S.—one of unprecedented permissiveness. 'Fuzzification' is a kind of scientific permissiveness; it tends to result in socially appealing slogans unaccompanied by the discipline of hard scientific work and patient observation."

In a similar vein, in 1975 Zadeh's colleague at Berkeley, Professor William Kahan, offered his assessment:

> "Fuzzy theory is wrong, wrong, and pernicious. I cannot think of any problem that could not be solved better by ordinary logic. What we need is more logical thinking, not less. The danger of fuzzy theory is that it will encourage the sort of imprecise thinking that brought us so much trouble."

These statements, and others similar to them, set the stage for the great debate that, although continuing today, has calmed in its rhetoric in recent years.

In 1988, Peter Cheeseman excited numerous investigators in the fields of fuzzy set theory, logic, Dempster–Shafer evidence theory, and other theories with his work "An Inquiry into Computer Understanding." In this paper, he made the claim that both fuzzy set theory and Dempster–Shafer evidence theory violated *context dependency*, a required property of any method-assessing beliefs. This statement, and many others included in the paper, such as his misstatements about possibility distributions (Ruspini (1988)), incited such a debate that 22 commentaries followed and were published.

In a series of papers dealing with the expression of uncertainty within artificial intelligence, Dennis Lindley (1982, 1987a, b) perhaps expressed the most vociferous challenge to fuzzy set theory—or any other non-Bayesian theory—with his comments about the inevitability of probability:

> *"The only satisfactory description of uncertainty is probability.* By this is meant that every uncertainty statement must be in the form of a probability; that several uncertainties must be combined using the rules of probability, and that the calculus of probabilities is adequate to handle *all* situations involving uncertainty. In particular, alternative descriptions of uncertainty are unnecessary. These include the procedures of classical statistics; rules of combination... possibility statements in fuzzy logic... use of upper and lower probabilities... and belief functions."

In a single paragraph, Lindley's proclamations were sufficient to not only extend indefinitely the debate between fuzzy and probability, but also to prolong numerous other debates, such as those continuing for the past 100 years between frequentists and Bayesians.

Other statements were made that, although inflammatory, appeared to have less substance. Hisdal (1988a) stated "the fuzzy set group is... in the position of having a solution for which it has not yet found a problem" and "a theory whose formulas must be replaced by other ad hoc ones whenever it does not agree with experiment is not a finished theory." A rather unexplained quote from her, "Fuzziness $\neq$ randomness... this is a very strong, and... also a very surprising assertion," was followed by an even more enigmatic statement: "Fuzzy set theory has mostly assumed that some mystic agent is at work, making its fuzzy decisions according to some undefined procedure." Moreover, in (Hisdal (1988b)), she states of fuzzy set theory that "This seems to imply the belief, that human thinking is based on inexact, fuzzily-defined concepts. As I have heard a colleague express it, the theory of fuzzy sets is no theory at all, it is more like a collection of cooking recipes."

Laviolette and Seaman (1992) remarked that "fuzzy set theory represents a higher level of abstraction relative to probability theory" and questioned whether laws exist to govern the combination of membership values. In a rather curious metaphor explaining the relationship of the axiomatic differences in the two theories, specifically the fuzzy property of supersubsethood as articulated by Bart Kosko in 1990, they begin with "The foundation of probability is both operationally and axiomatically sound. Supersethood alone need not make one theory more broadly applicable than another in any practical sense. If we wish to design and construct a building, we need only Newtonian mechanics and its consequences. The fact that quantum mechanics includes Newtonian mechanics as a subset theory is of no practical consequence." Then, in speaking about the operational effectiveness of the two theories, they further suggest "We have proposed that fuzzy methods be judged by their sensitivity to changes in an associated probabilistic model. We take for granted that fuzzy methods are sub-optimal with respect to probabilistic methods. It is important that a method for judging the efficacy of fuzzy methods be developed and employed, given the widespread interest in FST." Finally, they state "... in our opinion, this operational deficiency remains the chief disadvantage of fuzzy representations of uncertainty."

From the other side of the argument, Kosko (1994) states that "probability is a very special case of fuzziness." In referring to the excluded-middle laws, he states that it "forces us to draw hard lines between things and non-things. We cannot do that in the real world. Zoom in close enough and that breaks down." (See section 2.1.2 in Chapter 2 on excluded-middle laws for an expansion of these ideas.) Kosko further states that "Our math and world view might be different today if modern math had taken root in the A-AND-not-A views of Eastern culture instead of the A-OR-not-A view of ancient Greece." To dismiss this as *unfortunate deconstructionism* is just to name call and to ignore historical fact. For a long time the probability view had a monopoly on uncertainty, but now "fuzzy theory challenges the probability monopoly... the probability monopoly is over."

In his 1990 paper "Fuzziness vs. Probability," Kosko addresses everything from randomness, to conditional probability, to Bayesian subjective probability. He very eloquently shows that his subsethood theorem is derived from first principles and, in commenting on the lack of a derivable expression for conditional probability, remarks that this is the "difference between showing and telling." He remarked that his "subsethood theorem suggests that randomness is a working fiction akin to the luminiferous ether of nineteenth-century physics—the phlogiston of thought." Noting that his derivation of the subsethood theorem had nothing to do with randomness, he states "The identification of relative frequency with probability is cultural, not logical. That may take some getting used to after hundreds of years of casting gambling intuitions as matters of probability and a century of building probability into the description of the universe. It is ironic that to date every assumption of probability—at least in the relative frequency sense of science, engineering, gambling, and daily life—has actually been an invocation of fuzziness." Giving equal attention to the subjective probability groups, he claims that "Bayesianism is a polemical doctrine." To justify this, he shows that Bayes' rule also stems from the subsethood theorem.[4] Additionally, in commenting on Lindley's argument (Lindley (1987a)) that only probability theory is a coherent theory and all other characterizations of uncertainty are incoherent, Kosko states "this polemic evaporates in the face of" the subsethood theorem and "ironically, rather than establish the primacy of axiomatic probability, Lindley seems to argue that it is fuzziness in disguise." Kosko finishes his own series of polemics by euphemistically pointing out that perhaps 100 years from now, "no one... will believe that there was a time when a concept as simple, as intuitive, as expressive as a fuzzy set met with such impassioned denial."

Kosko's prediction reflects the nature of change in human cognition and perhaps even philosophy. Whether that process can be best described as fuzzy or probabilistic in nature we leave as a mental exercise.

Philosophical issues: Chance, ambiguity, and crispness versus vagueness

It is sometimes useful to first sit back and ask some fundamental questions about what it is we are trying to do when we attempt to first characterize various forms of uncertainty, and then to posit mathematical forms for quantifying them. Philosophers excel at these questions. The history of humankind's pondering of this matter is rich, and won't be replicated here, but the ideas of a few individuals who have thought about such notions as chance, ambiguity, crispness, and vagueness are cited here. While we could look at a time in Western culture as far back as that of Socrates, Descartes, or Aristotle, or revisit the writings of Zen Buddhism, we shall skip all that and move directly to the 20th century, where a few individuals spoke profoundly on these matters of uncertainty.

Max Black, in writing his 1937 essay "Vagueness: An exercise in logical analysis," first cites remarks made by the ancient philosopher Plato about uncertainty in geometry, then embellishes on the writings of Bertrand Russell (1923) who emphasized that "all traditional logic habitually assumes that precise symbols are being employed." With these great thoughts prefacing his own arguments, he proceeded to his own, now famous quote:

---

[4]While subsethood by itself may not guarantee performance enhancement in practical applications, enhancements to existing solutions from the use of fuzzy systems is illustrated in the applications chapters of Part II of this book.

"It is a paradox, whose importance familiarity fails to diminish, that the most highly developed and useful scientific theories are ostensibly expressed in terms of objects never encountered in experience. The line traced by a draftsman, no matter how accurate, is seen beneath the microscope as a kind of corrugated trench, far removed from the ideal line of pure geometry. And the 'point-planet' of astronomy, the 'perfect gas' of thermodynamics, or the 'pure-species' of genetics are equally remote from exact realization. Indeed the unintelligibility at the atomic or subatomic level of the notion of a rigidly demarcated boundary shows that such objects not merely are not but could not be encountered. While the mathematician constructs a theory in terms of 'perfect' objects, the experimental scientist observes objects of which the properties demanded by theory are and can, in the very nature of measurement, be only approximately true."

More recently, in support of Black's work, Quine (1981) states

"Diminish a table, conceptually, molecule by molecule: when is a table not a table? No stipulations will avail us here, however arbitrary.... If the term 'table' is to be reconciled with bivalence, we must posit an exact demarcation, exact to the last molecule, even though we cannot specify it. We must hold that there are physical objects, coincident except for one molecule, such that one is a table and the other is not."[5]

Bruno de Finetti, in his landmark 1974 book *Theory of Probability*, quickly gets his readers attention by proclaiming "Probability does not exist; it is a subjective description of a person's uncertainty. We should be normative about uncertainty and not descriptive." He further emphasizes that the frequentist view of probability (objectivist view) "requires individual trials to be equally probable and stochastically independent." In discussing the difference between possibility and probability he states "The logic of certainty furnishes us with the range of possibility (and the possible has no gradations); probability is an additional notion that one applies within the range of possibility, thus giving rise to gradations ('more or less' probable) that are meaningless in the logic of uncertainty." In his book, de Finetti warns us that: "The calculus of probability can say absolutely nothing about reality"; and in referring to the dangers implicit in attempts to confuse certainty with high probability, he states, "We have to stress this point because these attempts assume many forms and are always dangerous. In one sentence: to make a mistake of this kind leaves one inevitably faced with all sorts of fallacious arguments and contradictions whenever an attempt is made to state, on the basis of probabilistic considerations, that something must occur, or that its occurrence confirms or disproves some probabilistic assumptions."

In a discussion about the use of such vague terms as "very probable," "practically certain," or "almost impossible," de Finetti states

"The field of probability and statistics is then transformed into a Tower of Babel, in which only the most naïve amateur claims to understand what he says and hears, and this because, in a language devoid of convention, the fundamental distinctions between what is certain and what is not, and between what is impossible and what is not, are abolished. Certainty and impossibility then become confused with high or low degrees of a subjective probability, which is itself denied precisely by this falsification of the language. On the contrary,

---

[5]Historically, a sandpile paradox analogous to Quine's table was discussed much earlier, perhaps in ancient Greek culture.

the preservation of a clear, terse distinction between certainty and uncertainty, impossibility and possibility, is the unique and essential precondition for making meaningful statements (which could be either right or wrong), whereas the alternative transforms every sentence into a nonsense."

Probability density functions versus membership functions

The early debate in the literature surrounding the meaning of probability and fuzzy logic also dealt with the functions in each theory; i.e., probability density functions (PDFs) and membership functions (MFs). The early claim by probabilists was that membership functions were simply probability density functions couched in a different context, especially in the modeling of subjective probability. The confusion perhaps stemmed from both the similarity between the functions and the similarity between their applications—to address questions about uncertainty. However, this confusion still prevails to a limited extent even today. For example, in the paper (Laviolette et al. (1995)), the authors provide a quote from the literature that intimated that membership functions are "probabilities in disguise." When one looks at the ontological basis for both these functions, it is seen that the only common feature of an MF and a PDF is that both are nonnegative functions. Moreover, PDFs do not represent probabilities and, while the maximum membership value allowed by convention is unity, there is no maximum value for a PDF. On the other hand, a cumulative probability distribution function (CDF), the integral of the PDF, is a nonnegative functions defined on the unit interval [0, 1] just as an MF. A CDF is a monotonically increasing function, and an MF is not. The integral of a PDF (the derivative of the CDF) must equal unity, and the area under an MF need not equal unity.[6] Based on these fundamental differences, it is not so obvious how either function could be a disguise for the other, as claimed by some. However, we recognize that the attempts to overcome, ignore, or mask the differences may be the result of wishful thinking, either from wanting to show that one theory is superfluous because its features can be captured by the other, or wanting to bridge the gap between the two.

One area in which PDFs and MFs have been equated is that of Bayesian probability analysis (Laviolette (1995), Cheeseman (1988)). In this case, the likelihood function in Bayes' rule, the function containing new information and used to update the prior probabilities, need not be a PDF. In Bayes' rule, there is a normalization taking place that ensures the posterior distribution (updated prior) will be a PDF, even if the likelihood function is not a PDF. In many applications of Bayes' rule, this likelihood function measures a degree of belief and, by convention, attains a value of unity. Geometrically, the likelihood function can be used in the same context as an MF. In fact, Laviolette uses this similarity to point out that fuzzy control can be approached in the same way as probabilistic control (see Chapter 8 for a more definitive discussion of this claim), and Cheeseman uses this similarity to mistakenly imply that membership functions are simply likelihood functions. But the bottom line is that subjective probabilities—those arising from beliefs rather than from relative frequencies—still must adhere to the properties mentioned above for PDFs and CDFs. They cannot be equated with MFs without making extreme assumptions to account for the differences. Another fundamental difference is that MFs measure degrees of belongingness. They do not measure likelihood (except in a similarity sense, i.e., likeliness) or degrees of belief or frequencies, or perceptions of chance, and the area under their curve is not constrained to a value of unity. They measure something less measurable—set membership in ambiguous

---

[6]While the area under an MF is not a constraint on these functions, it is a useful metric in some applications involving defuzzification methods (see Chapter 2).

or vague sets. Although both membership values and probabilities map to the unit interval [0, 1], they are axiomatically different (see the discussion in Chapter 5).

James Bezdek gives some excellent examples of the differences between membership values and probabilities (Bezdek (1993, 1994a, 1994b)). He summarizes much of the debate by pointing out that the probability-only advocates base their arguments on one of two philosophical themes: (i) nonstatistical uncertainty does not exist, or (ii) maybe nonrandom uncertainty does exist, but a probability model still is the best choice for modeling all uncertainties. Theme (i) engenders a debate even within the statistical community because there is a nice variety of choices on which to base a probabilistic approach: relative frequency, subjectivity, or the axiomatic approach (von Mises (1957)). In theme (ii), it has been argued by many (e.g., Klir (1989, 1994)) that nonstatistical forms of uncertainty do exist and that numerous theories address them in very useful ways.

Bezdek (1993) gives the following example that, in turn, has spawned numerous other examples. Suppose you are very thirsty and have the chance to drink the liquid from one of two different bottles, A and B. You must decide from which bottle to drink using the following information: bottle A has 0.91 membership in the set of potable liquids, whereas bottle B has a probability of 0.91 of being potable. For the sake of illustration, we will define this probability as a frequentist interpretation. Most people would choose bottle A since its contents are at least "reasonably similar" to a potable liquid and because bottle B has a 9% chance of being unsavory, or even deadly. Moreover, if both bottles could be tested for potability, the membership value for bottle A would remain unchanged, while the probability for bottle B being potable would become either 0 or 1—it is either potable or not potable after a test.[7] These two different kinds of information inherently possess philosophically different notions: fuzzy memberships represent similarities of objects to imprecisely defined properties, and probabilities convey assessments of relative frequencies.

Some probabilists have claimed that natural language is precise, indeed, and that there is no need to use a method like fuzzy logic to model linguistic imprecision. An example dealing with imprecision in natural language is also given by Bezdek (1993). Suppose that as you approach a red light, you must advise a driving student when to apply the brakes. Would you say "begin braking 74 feet from the crosswalk," or would your advice be for the student to "apply the brakes pretty soon"? As an exercise, think of how you would address this uncertainty with probability theory.[8] In this case, precision in the instruction is of little value. The latter instruction would be used by humans since the first command is too precise to be implemented in real time. Sometimes, striving for precision can be expensive, or adds little or no useful information, or both. Many times, such uncertainty cannot be presumed to be, or adequately assessed as, an element of chance.

Bayes' rule

Cheeseman (1988) gives an eloquent discussion of the utility of Bayesian probability in reasoning with subjective knowledge. He and Laviolette (1985) suggest that there is no need to use fuzzy set theory to model subjective knowledge since Bayesian methods are more than adequate for this. However, even the subjective interpretation of probability still has the restrictions of that theory. For example, there is no apparent difference in the way one would model ignorance that arises from lack of knowledge or ignorance that arises from

---

[7]We could argue that the subjective probability would also remain unchanged; however, this is more of a frequentist interpretation.

[8]Using probabilities, we could estimate the required distance and then instruct the student to brake where he/she maximized his/her probability assessment of stopping in time.

only knowing something about a random process (McDermott (1988)). Moreover, others still question the utility of Bayesian models in epistemology (Dempster (1988), Dubois and Prade (1988)).

Shafer (1987) makes the following points about Bayes' formula $P(A|U_0) = P(U_0|A)P(A)/P(U_0)$:

> "[It] can serve as the symbolic expression of a rule. This rule, the Bayesian rule of conditioning, says that when new knowledge or evidence tells us that the correct answer to the question considered by U is in the subset U_0, we should change our probability for another subset A from $P(A)$ to the new probability given by this quotient, above. Strictly speaking, this rule is valid only when the receipt of the information, U_0, is itself a part of our probability model—i.e., it was foreseen by the protocol. In practice, however, the rule is used more broadly than this in subjective probability judgment. This broader use is reasonable, but it must be emphasized that the calculation is only an argument—an argument that is imperfect because it involves a comparison of the actual evidential situation with a situation where U_0 was specified by a protocol, as one of the possibilities for the set of still possible outcomes at some point in the game of chance. The protocol is the specification, at each step, of what may happen next, i.e., the rules of the game of chance."

Alleged conjunction fallacy

Conjunction in fuzzy logic is modeled with the minimum operator as the smaller of the membership values between two sets. Woodall (1997) and Osherson and Smith (1981) cite the conjunction fallacy as one of the frailties of fuzzy logic. Others, such as McNeil and Freiberger (1993), have called this an intriguing cognitive puzzle that is not a frailty of fuzzy logic but rather a strength, as it exposes with human reasoning the reality of conjunctive processes. Basically, the cognitive puzzle can be illustrated with the following example: The intersection between the set "pet" and the set "fish" produces a membership value of the conjunction "pet fish" that is the smaller of the membership values for "pet" and for "fish" for a particular species of fish in the universe of vertebrates. Osherson and Smith considered that guppy, as a specific species, should have a higher value in the conjunctive set "pet fish" than it does in either "pet" or "fish"; it is not a common pet, nor a common fish, they argued, but it is certainly a strong prototype of the concept "pet fish." In explaining the minimum operator as a preferred operator for the conjunction of fuzzy sets, Zadeh (and others, e.g., Oden (1984) and Rosch (1973)) pointed out that humans *normalize* in their thinking of concepts, whereas strict logical operations don't involve such normalization. Being in the intersection of set "pet" and set "fish," i.e., in the set "pets and fish," is different from being in the set "pet fish." A linguistic modification (pet modifies fish) does *not* equate to a logical conjunction of concepts (sets). Other, more easily understood examples are "red hair" or "ice cube," where the terms "red" and "cube" are altered significantly from their usual meaning. In thinking about *pet–fish* people normalize; they compare a guppy to other pets and other fish and the guppy has a low membership, but when they compare it only to other *pet–fish*, it grades higher. Such cognitive puzzles, which lead to faulty human reasoning, have long been discussed in the literature (Tversky and Kahneman (1983)).

Normalization is typical in many uncertainty operations. A conditional probability is a normalization and, as we have seen in the previous section, Bayes' rule also involves normalization. For example, the probability of getting a 3 on one toss of a die is $\frac{1}{6}$, but the probability changes to $\frac{1}{3}$ if the result is conditioned (i.e., normalized) on the outcome being an odd number (i.e., 1, 3, or 5). In their arguments about "pet fish," Osherson and Smith implicitly normalized for their example of guppies.

Alleged disjunction contradiction

Another allegation about fuzzy logic suggested by some probabilists in the literature (Hisdal (1988b), Laviolette et al. (1985), Lindley (1987a)) has been termed the "disjunctive contradiction." In describing a fuzzy set modeling the heights of people, Hisdal (1988b) complains that the logical union between the fuzzy sets *medium height* and *tall height* should *not* be less than unity between the prototypical values of medium and tall. She calls this the "sharp depression" in the grade of membership curve for two adjacent linguistic labels (see Figure 1.1).

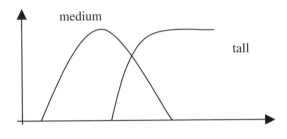

Figure 1.1.

In other words, Hisdal's question is, how could any person's membership in the set "medium OR tall" (which, for fuzzy operations, involves the maximum operator) be less than the largest membership of either if that person's height was between the prototype heights representing medium height and tall? The answer is quite simple. As the figure shows, a "tall" person can have a low membership in the set "medium." The set "tall" is not a subset of the set "medium," but there is an overlap—there exists a common ground in which members of both sets coexist with lower membership. Hisdal confused "medium or tall" with the set "at least medium." Hence there is no inconsistency in the use of the maximum operator here for disjunction, as stated by Hisdal, but rather a misinterpretation of the meaning of disjunction—or logical union. Finally, the argument of Osherson and Smith (1981) on this issue confuses the operations of "sum" and "max" in their arguments on the union operator; the former is an operator on functions and the latter is an operator on sets.

Those controversial excluded-middle laws

For crisp sets, Lindley (1971) showed that the excluded-middle laws result from the three basic laws of probability theory (convexity, addition, and multiplication) when he surmised that an event, E, and its negation, not-E, are mutually exclusive and their union exhaustive. He implicitly assumes that event occurrence is binary—it does or does not occur. Gaines (1978) showed that a probability theory results only after the law of the excluded-middle is added as an axiom to a general uncertainty logic. However, in either case, the critical assumption is that there is no provision for the common ground; as termed by Osherson and Smith, "there can be no apple that is not an apple." Lindley is convinced—he claims that "uncertainties are constrained by the laws of probability" and thereby ignores vagueness, fuzziness, ambiguity, and linguistic imprecision. The laws of probability theory work wonderfully with sets whose boundaries are clear and unambiguous, but a theory is also needed for sets with fuzzy boundaries.

On this same matter, Osherson and Smith (1981), and certainly many others at that time, argued that the "apple that is not an apple" concept is logically empty. Hisdal (1988b) erroneously termed this the *complementation paradox*. In fuzzy set theory, locutions of the type "tomatoes are both fruit and not fruit" can be supported, but in classical Aristotelian logic they cannot. Basically, how can the complement of a fuzzy set, where the set itself does not enjoy full membership over its support, be null throughout that support? The only way to accept the excluded-middle is to accept only "all or nothing" memberships. For example, if the concept "tall" is vague, then so is the concept "not tall." In this case, why would the conjunction "tall *and* not tall" still not be ambiguous to some degree (i.e., not null everywhere)? Alternatively, why wouldn't the disjunction "tall *or* not tall" have less than full membership everywhere?

Classical sets deal with sets whose boundaries are crisp, but very few concepts yield crisp boundaries—in reality, they don't exist (see Black (1937)). As espoused by Howard Gardner (1985), the greatest proponents of crispness—Socrates, Locke, Wittgenstein—all had labored under the deepest illusion!

There are at least two reasons why the law of the excluded-middle might be inappropriate for some problems. First, people may have a high degree of belief about a number of possibilities in a problem. A proposition should not be "crowded out" just because it has a large number of competing possibilities. The difficulties people have in expressing beliefs consistent with the axioms of probability logic are sometimes manifested in the rigidity of the law of the excluded-middle (Wallsten and Budescu (1983)). Second, the law of the excluded-middle results in an inverse relationship between the informational content of a proposition and it probability. For example, in a universe of n singletons, as more and more evidence becomes available on each of the singletons, the relative amount of evidence on any one diminishes (Blockley (1983)). This makes the excluded-middle laws inappropriate as a measure of modeling uncertainty in many situations.

Since the law of the excluded-middle (and its dual, the law of contradiction) is responsible for spawning a theory of probability from a general uncertainty theory (see Chapter 5), it is interesting to consider examples where it proves cumbersome and counterintuitive. David Blockley, in civil engineering, and Wallsen and Budescu, both in business and management, all argued in 1983 the points made in the previous paragraph.

Consider a hypothetical example involving several alternatives. A patient goes to two doctors to get opinions on persistent head pain. Both doctors consider four possible causes: stress headache (H), concussion (Co), meningitis (M), or a tumor (T). Prior to asking the patient to undergo more extensive and expensive tests, both doctors conduct a routine office examination, at different times and locations, and come to the same conclusions: they give little weight to meningitis but considerable weight to the other three possible causes. When pressed to give numerical values for their assessments (their degrees of belief) they give the following numbers: $w(T) = 0.8$, $w(H) = 0.8$, $w(Co) = 0.8$, $w(M) = 0.1$. Of course, a probabilist would normalize these to obtain probabilities and to get their sum to equal unity; i.e., $p(T) = 0.32$, $p(H) = 0.32$, $p(Co) = 0.32$, $p(M) = 0.04$. The problem arises when the probabilist informs the doctor that the probability that the patient's pain is due to something "other than a tumor" is $0.68 = 1 - p(T)$, which is a number higher than the normalized value for any of the other three possible causes! This runs counter to the beliefs of the doctors, who feel that a concussion and a headache are also strong possibilities based on their "limited" diagnosis. Such is the requirement imposed by the excluded-middle laws. Based on this information, should the patient spend more money on tests? If the doctors' original weights were assigned as possibilities, these weights would not need any normalization.

On the other hand, Bayesianists would claim that the original weights were simple likelihood values obtained by diagnosis to be used to modify prior probabilities of the historical evidence of the four possible causes of the patient's head pain. For example, suppose that the prior probabilities ϕ_i are $\phi(H) = 0.8$, $\phi(T) = 0.05$, $\phi(M) = 0.01$, and $\phi(Co) = 0.30$; then from Bayes' formula, we can predict for each potential cause of the patient's head pain (HP), C_i,

$$p(C_i|HP) = \frac{w(HP|C_i)\phi(C_i)}{\sum_{i=1}^{4} w(HP|C_i)\phi(C_i)},$$

where $w(HP|C_i)$ is the likelihood value for the ith possible cause of the head pain, as given in the previous paragraph. The resulting posterior probabilities (updated priors) would be calculated as

$$w(C_1|HP) = w(H|HP) = 0.695,$$
$$w(C_2|HP) = w(T|HP) = 0.043,$$
$$w(C_3|HP) = w(C|HP) = 0.261,$$
$$w(C_4|HP) = w(M|HP) = 0.001.$$

Now the probability that the head pain is due to a tumor is 0.043, slightly higher than the historical evidence due to the doctors' rather high weight assigned after a rudimentary diagnostic exam. However, it remains that the probability (now updated) that it is a cause other than a tumor is 0.957, a number much higher than the rest of the potential causes. Therefore, whether one takes the philosophy of a Bayesianist or a frequentist, the excluded-middle law can be counterintuitive to people's initial judgments; in this case, the doctors' original assessments that three of the four possible causes all have a high likelihood hold. Since the evidence (or degrees of belief, or likelihood values, etc.) comes from humans, it is *not* a justification for using probability theory, as is often claimed by Lindley and others, to suggest that humans must be forced to think in "coherent" terms (the term "coherent" borrowed by probabilists to describe the additivity axiom of probability theory). Meyer and Booker (2001) in their formal elicitation methods advocate just the opposite approach. The highest quality of expert information comes from allowing the experts to express their beliefs free of such probabilistic constraints. Of course, we expect (or perhaps hope) that humans exhibit coherence as a linguistic feature of reasoning. However, studies (Meyer and Booker (2001)) have too often shown this is not the case, and we should not force them to obey the coherence property of probability theory. More to the point, we also should not force them to obey a set of axiomatic structures that run counter to their thinking or problem solving because such restrictions induce cognitive and motivational biases that both degrades the quality and changes the nature and content of the information given.

Probability theory needs an infusion of fuzzy logic

In a 1998 email communication to colleagues at the University of California at Berkeley, Lotfi Zadeh stated "probability theory needs an infusion of fuzzy logic to enhance its ability to deal with real-world problems." Earlier, in their 1988 series of commentaries on Peter Cheeseman's essay on his controversial *computer understanding*, Enrique Ruspini, Ron

Yager, and Lotfi Zadeh all made suggestions (*Comput. Intell.*, 4 (1988), pp. 57–142) about how powerful probability theory would be if it were able to consider notions of overlapping sets—however, then it wouldn't be the probability theory that we all know (or think we know) today. In a 1996 paper, Zadeh alludes to the fact that the historically main contribution—but not the only one—of fuzzy logic will be its ability to help us develop a new field, computing with words (CW). Here he states that a frequently asked question is, "What can be done with fuzzy logic that cannot be done equally well with other methodologies, e.g., predicate logic, probability theory, neural network theory, Bayesian network, and classical control?" One such answer involves the nascent field of CW.

In 1999, Zadeh stated that information granulation is associated with an increasing order of precision: from interval to fuzzy to evidence to random, the granulation goes from coarse to fine. He cites the successes of precision on monumental scientific tasks involving the atomic bomb, moon landings, supercomputers, telescopes, and carbon dating of rock. He also cites failures on more simple tasks: we cannot build robots that can move with the agility of animals; we cannot automate driving tasks in heavy traffic; we cannot automate language translation; we cannot summarize nontrivial stories automatically; we cannot model the behavior of economic systems; and we cannot construct machines that can compete with children in the performance of simple physical and cognitive tasks. Hoadley and Kettering (1990) point out that follow-up studies to the Space Shuttle *Challenger* disaster revealed that "relevant statistical data had been looked at incorrectly. . . probabilistic-risk assessment methods were needed to support NASA, and. . . its staff lacked specialists and engineers trained in the statistical sciences." They admit, however, that statistical science does not seem to deal well with the probabilistic risk assessments of complex engineered systems, because "there are too little data in the traditional sense."

The subject of specifically how fuzzy set theory can actually enhance the effectiveness of probability theory when dealing with some problems has been articulated recently by Zadeh (1995, 1996). In these works, Zadeh points out that probability theory is very useful in dealing with the uncertainty inherent in measurements or objects that can be measured; however, it is not very useful in dealing with the uncertainty imbedded in perceptions by humans. The former involves crisp sets, while the latter involves fuzzy sets. Examples of the latter are the following:

1. What is the probability that your tax return will be audited?

2. What is the probability that tomorrow will be a warm day?

3. Team A and Team B played each other 10 times. Team A won the first seven times, and Team B won the last three times. What is the probability that Team A will win the next game between the two teams?

4. Most young women are healthy. Mary is young. What is the probability that Mary is healthy?

In question 1, the difficulty arises from the basic property of conditional probability—namely, given $p(x)$, all that can be said is that the value of $p(x|y)$ is between 0 and 1. Thus if we know that 1% of all tax returns are audited, this tells us little about the probability that your tax return will be audited, except in a bounding sense that your return is one of the many from which the audit sample is taken. You may have an estimate of how often you have been audited, which could be used in a Bayesian probability model, and you can always provide your subjective belief as a probability for an estimate. If we have more detailed information about you, e.g., income, age, residence, occupation, etc., a better estimate might be possible from IRS data based upon the fraction of returns that are audited in a certain category, but this detail will never reach the level of a single individual. Probability theory alone cannot directly handle this question at the appropriate level.

In question 2, the difficulty is that the warmth is a matter of degree. The event "a warm day" cannot be defined precisely; it is a fuzzy event based on personal preference.

In question 3, the difficulty is that we are dealing with a time series drawn from a nonstationary process. In such cases, probabilities do not exist. In fact, that statement is the familiar quote attributed to de Finetti (1974) in his discussion about the origins of subjective probabilities, as enunciated earlier in this debate summary.

In question 4, we have a point of common sense reasoning, and probability theory is inadequate to handle the uncertainty in concepts such as "most young women" or "healthy."

To one degree or another, all four questions involve the same theme: the answers to these questions are not numbers; they are linguistic descriptions of fuzzy perceptions of probabilities. As pointed out by Zadeh (1999), these kinds of questions can be addressed by an enhanced version of probability theory—a theory enhanced by a generalization that accounts for both fuzzification and granulation. Here fuzzy granulation reflects the finite cognitive ability of humans to resolve detail and store information. Classical probability theory is much less effective in those fields in which dependencies between variables are not well defined; the knowledge of probabilities is imprecise and/or incomplete; the systems are not mechanistic; and human reasoning, perception, and emotion play an important role. This is the case in many fields ranging from economics to weather forecasting.

Our summary of the debate

In 1997, in discussing the choices available to analysts who are modeling uncertainty in their systems, Nguyen had this to say:

> "Of course, we are free to model mathematically the way we wish, but unless the modeling is useful for applications, the modeling problem may be simply a mathematical game. In the axiomatic theory of probability we do not allow all possible subsets of a sample space to be qualified as events. Instead we take a sigma-field of subsets of the sample space to be the collection of events of interest. It is not possible to extend the sigma-additive set-function to the whole power set of the reals. Acknowledging that randomness and fuzziness are different types of uncertainties, there is no compelling reason to believe that their associated calculi are the same."

The arguments of Lindley (1982) have been countered on mathematical grounds by Goodman, Nguyen, and Rogers (1991). "Saying that fuzzy sets theory is complementary rather than competitive does not presume deficiencies in probability."

Finally, rather than debate what is the correct set of axioms to use (i.e., which logic structure) for a given problem involving uncertainty, one should look closely at the problem, determine which propositions are vague or imprecise and which ones are statistically independent or mutually exclusive, determine which ones are consistent with human cognition, and use these considerations to apply a proper uncertainty logic, with or without the law of the excluded-middle. By examining a problem so closely as to determine these relationships, one finds out more about the structure of the problem in the first place. For example the assumption of a strong truth-functionality (for fuzzy logic) could be viewed as a computational device that simplifies calculations, and the resulting solutions would be presented as ranges of values that most certainly bound the true answer if the assumption is not reasonable. A choice of whether fuzzy logic is appropriate is, after all, a question of balancing the model with the nature of the uncertainty contained within it. Problems without an underlying physical model, problems involving a complicated weave of technical, social, political, and economic factors, and problems with incomplete, ill-defined, and inconsistent information where conditional probabilities cannot be supplied or rationally formulated perhaps are candidates for fuzzy set applications. It has become apparent that the engineering profession is overwhelmed by these sorts of problems.

The argument, then, appears to be focused on the question, "Which theory is more practical in assessing the various forms of uncertainty that accompany any problem?" The major intent of this book, therefore, is to shed light on this question by comparing just two of the theories, fuzzy sets and probability, for various applications in different fields. While we know that this book will not end the philosophical debate (which could be waged for many more decades), will not convert either the protagonists or antagonists of a particular theory, and will not lessen the importance or usefulness of any of the theories, it is our hope that its contents will serve to educate the various audiences about the possibility that the complementary use of these two theories in addressing uncertainties in many different kinds of problems can be very powerful indeed. Hence we hope that this book will help "bridge the gap" between probability theory and other nonprobabilistic alternatives in general, and fuzzy set theory in particular.

1.3   Fuzzy logic and probability: The best of both worlds

In contrast to the literature cited in section 1.2, there is a growing list of recent papers illustrating the power of combining the theories. What follows are very brief, very incomplete descriptions of a few of the works readily available to the editors of this book:

- *Cooper and Ross* (1998) *on subjective knowledge in system safety.* The authors describe a series of mathematical developments combining probability operations and fuzzy operations in the area of system reliability. They provide sort of a shopping cart of potential ideas in combining the two methods to assess both modeling and parametric uncertainty within the context of assessing the safety and reliability of manmade systems.

- *Ross et al.* (1999) *on system reliability.* In this work, the authors discuss a way to fuse probabilities, using fuzzy norms, to address the fact that real systems contain components whose interactions are between the extremes of completely independent and completely dependent. The fuzzy norms are based on Frank's *t*-norms, which can be shown to contain, as a family of norms, the probabilistic and fuzzy norms. The paper further shows that memberships and probability density functions can be combined using the same norms, thus paving the way for an algebraic approach to fuse both (random and fuzzy) kinds of information.

- *Smith et al.* (1997, 1998) *on uncertainty estimation.* In a series of two papers, Smith et al. used the graphical methods typical in fuzzy control applications to develop probabilistic density functions of the uncertainties inherent in the working parts of a system to estimate the systems reliability. In this case, the output MFs in the inference of a fuzzy control method were used to develop the density functions.

- *Zadeh* (1999) *on computing with words.* In this paper, Zadeh shows that a question of the type "What is the probability of drawing a small ball from an urn of balls of various sizes?" can be addressed with a combination of probability and fuzzy set theory. In this work there is a distinction drawn between being able to measure and manipulate numbers (as done with probabilities) and being able to measure and manipulate perceptions (linguistic data).

- *Nikolaidis* (1997–1999) *on designing under uncertainty*. In a sequence of papers (Nikolaidis et al. (1999) and Maglaras et al. (1997)), a contrast is established between probability theory and possibility theory in the area of structural design. In this case, structural failure and survival (safety) usually are taken to be complementary states, even though the transition between the two is gradual for most systems. In designs under uncertainty, it is important to understand how each of the two methods maximizes the expression of safety. In terms of design optimization against failure, the probabilistic optimization tries to reduce the probabilities of failure of the modes that are easiest to control in order to minimize the system failure probability, whereas fuzzy set optimization simply tries to equalize the possibilities of failure of all failure modes in minimizing the system failure possibility. They show that fuzzy set methods are not necessarily more conservative than probabilistic methods when assessing system failure. They also show that fuzzy set methods yield safer designs than probabilistic designs when there is limited data.

- *Rousseeuw* (1995) *on fuzzy and probabilistic clustering*. This work is important because it shows an application in classification where a fuzzy clustering method is actually a collaboration of two approaches: fuzzy theory and statistics. The usefulness of fuzzy clustering is apparent in that it helps the classification process avoid convergence only to local minima. In a completely fuzzy approach, each cluster center is influenced also by the many objects that essentially belong to other clusters. This bias may keep the fuzzy method from finding the true clusters. This problem is overcome by the use of objective functions, which are hybrids of purely fuzzy and purely probabilistic formalisms and which produce a high contrast clustering method capable of finding the true clusters in data.

- *Singpurwalla et al.* (2000) *on membership and likelihood functions*. In this published laboratory report, Singpurwalla et al. explore how probability theory and fuzzy MFs can be made to work in concert so that uncertainty of both outcomes and imprecision can be treated in a unified and coherent manner. In this work, the authors show that the MF can be interpreted as a likelihood if the fuzzy set takes the role of an observation and the elements of the fuzzy set take the role of the hypothesis. They show further that the early work of Zadeh (1968) in defining the probability of a fuzzy event as being equal to the expected value of the event's membership function is proportional to the measure they develop, which is a function of a prior probability of the classification of the event.

As indicated earlier, technology is moving so fast, in fact, that the systems on which we rely for our support and our discourse are becoming more, not less, complex. Decision-making has become so much guesswork under time constraints that analysts rarely have time to develop well-conceived models of today's systems. Complexity in the real world generally arises from uncertainty in various forms. Complexity and uncertainty are features of problems that have been addressed since humans began thinking abstractly; they are ubiquitous features that imbue most social, technical, and economic problems faced by the human race. Why is it then that computers, which have been designed by humans, after all, are not capable of addressing complex issues, that is, issues characterized by vagueness, ambiguity, imprecision, and other forms of uncertainty? How can humans reason about real systems when the complete description of a real system often requires more detailed data than a human could ever hope to recognize simultaneously and assimilate with understanding? It is because humans have the capacity to reason approximately, a capability that computers currently do not have. In reasoning about such systems, humans simply approximate behavior, thereby maintaining only a generic understanding about the problem. Fortunately, for humans' ability to understand complex systems, this generality and ambiguity is sufficient.

When we learn more and more about a system, its complexity decreases and our understanding increases. As complexity decreases, the precision afforded by computational methods become more useful in modeling the system. For systems with little complexity, and hence little uncertainty, closed-form mathematical expressions provide precise descriptions of the systems. For systems that are a little more complex, but for which significant data exists, model-free methods, such as artificial neural networks, provide a powerful and robust means to reduce some uncertainty through learning based on patterns in the available data. Finally, for the most complex systems, where little numerical data exists and where only ambiguous or imprecise information may be available, probabilistic or fuzzy reasoning can provide a way to understand system behavior by allowing us to interpolate approximately between observed input and output situations and thereby measuring, in some way, the uncertainty and variability. The imprecision in fuzzy models and the variability in probabilistic predictions generally can be quite high. The point, however, is to match the model type with the character of the uncertainty exhibited in the problem. In situations where precision is apparent, for example, fuzzy systems are not as efficient as more precise algorithms in providing us with the best understanding of the problem. On the other hand, fuzzy systems can focus on modeling problems with imprecise or ambiguous information. In systems where the model is understood but where variability in parameters of the model can be quite high, a probabilistic model can enrich understanding. Knowledge in using the tools illustrated in the subsequent chapters of this book will allow readers to address the vast majority of problems that are characterized either by their complexity or by their lack of a requirement for precision.

Our philosophy in preparing this book for a world of complex problems is to set a tone of conciliation between fuzzy theory and probability. We admit that there are differences between them. We do not propose that one is better than the other. We agree that one is more mature than the other. We admit that some applications are better served by one or the other. However, we contend that some applications can benefit from a hybrid approach of the two. Our intent is to help probabilists understand fuzzy set theory and fuzzy logic and to help fuzzy advocates see that probability theory, in conjunction with fuzzy logic, can be very powerful for some types of problems. Another objective of this book is to help the user select the best tool for the problem at hand. Our intended audience is statisticians, probabilists, and engineers.

1.4   Organization of the book

The book is organized into two major divisions. The chapters in Part I cover fundamental topics of probability, fuzzy set theory, uncertainty, and expert judgment elicitation. In Part I, we move from specific explanations of probability theory to more general descriptions of various forms of uncertainty, culminating in the elicitation of human evaluations and assessments of these uncertainties. Part II includes applications and case studies that illustrate the uses of fuzzy theory and probability and hybrid approaches.

Chapter 2 covers fuzzy set theory, fuzzy logic, and fuzzy systems. The origins of membership sets and functions are included, along with a discussion on the law of the excluded-middle. Fuzzy relationships and operations are illustrated with examples. Fuzzy logic and classical logic are compared.

Chapter 3 covers the foundations of probability theory and includes a discussion on the different interpretations of probability. Topics relating to probability distribution functions are presented as they relate to uncertainty and to hybrid fuzzy/probability approaches. The last section of this chapter provides a transition from probability to fuzzy theory by discussing the concepts of data, knowledge, and information.

One of the probability theories from Chapter 3 has Bayes' theorem at its core. Details regarding methods based on this approach to probability are presented in Chapter 4. Interpretations, applications, and issues associated with the Bayesian-based methods are presented. One reason for the special emphasis on these methods is the link it provides to fuzzy MFs, as described in the last section of the chapter.

Chapter 5 considers the uses of fuzzy set theory and probability theory. As such, it covers and compares the topics of vagueness, imprecision, chance, uncertainty, and ambiguity as they relate to both theories. Historical development of these topics is discussed, leading up to recent research work in possibility theory.

The process of defining and understanding the various kinds of uncertainty crosses into another subjective arena: the elicitation of expert knowledge. Chapter 6 provides guidelines for eliciting expert judgment in both realms of probability and fuzzy logic. While this chapter provides many of the elicitation fundamentals, it does so using two different applications for illustration.

Part II of the book, on applications and case studies, begins with Chapter 7 on image enhancement processing. This chapter compares image enhancement via the modification of the PDF of the gray levels with the new techniques that involve the use of knowledge-based (fuzzy expert) systems capable of mimicking the behavior of a human expert. This chapter includes some color images.

Chapter 8 provides a brief, basic introduction to engineering process control. It begins with classical fuzzy process control and introduces the relatively new idea of probabilistic process control. Two different scenarios of a problem are examined using classical process control, proportional-integral-derivative (PID) control, and probabilistic control techniques.

Chapter 9 presents a structural safety analysis study for multistory framed structures using a combined fuzzy and probability approach. The current probabilistic approach in treating system and human uncertainties and its inadequacy is discussed. The alternative approach of using fuzzy sets to model and analyze these uncertainties is proposed and illustrated with examples. Fuzzy set models for the treatment of some uncertainties in reliability assessment complement probabilistic models and are readily incorporated into the current analysis procedure for the safety assessment of structures.

Reliability is usually considered a probabilistic concept. Chapter 10 presents a case study of the reliability of aircraft structural integrity. The usefulness of fuzzy methods is demonstrated when insufficient test data are available and when reliance is necessary on the use of expert judgment.

Chapter 11 presents another case study in reliability. It begins with a complex reliability problem in the automotive industry, carries it through using Bayesian methods, and finally covers how certain aspects of the problem would benefit from the use of MFs. As in Chapter 10, here the use of expert judgment is prevalent when it is not feasible or practical to obtain sufficient test data for traditional reliability calculations.

Statistical process control (SPC) is a long-standing set of probability-based techniques for monitoring the behavior of a manufacturing process. Chapter 12 describes an application of fuzzy set theory for exposure control in beryllium part manufacturing by adapting and comparing fuzzy methods to the classical SPC. This very lengthy chapter considers the topic of fuzzy SPC as sufficiently important and novel (as introduced in the third author's Ph.D. dissertation); the chapter is essentially a tutorial on standard SPC extended to incorporate fuzzy information and fuzzy metrics.

Combining, propagating, and accounting for uncertainties is considered for fault tree logic models in Chapter 13. This scheme is based on classical (Boolean) logic. Now this method is expanded to include other logics. Three of the many proposed logics in the literature considered here are Lukasiewicz, Boolean, and fuzzy. This is a step towards answering the questions "Which is the most appropriate logic for the given situation?" and "How can uncertainty and imprecision be addressed?"

Chapter 14 investigates the problem of using fuzzy techniques to quantify information from experts when test data or a model of the system is unavailable. MFs are useful when the uncertainty associated with this expertise is expressed as rules. Two examples illustrate how a fuzzy-probability hybrid technique can be used to develop uncertainty distributions. One example is for cutting tool wear and the other is for reliability of a simple series system.

Chapter 15 introduces the use of Bayesian belief networks as an effective tool to show flow of information and to represent and propagate uncertainty based on a mathematically sound platform. Several illustrations are presented in detecting, isolating, and accommodating sensor faults. A probabilistic representation of sensor errors and faults is used for the construction of the Bayesian network. Fuzzy logic forms the basis for a second approach to sensor networks.

These chapters represent the collective experience of the authors' and editors' efforts and research to bridge the gap between the theories of fuzzy logic and probability. As stated above, our goal is to continue to develop, support, and promote these techniques for solving the complex problems of modern information technology.

References

T. Arciszewski (1999), *History of Information Technology*, discussion at the American Society of Civil Engineers Committee on Artificial Intelligence and Expert Systems Workshop, George Mason University, Fairfax, VA.

T. Bayes (1763), An essay towards solving a problem in the doctrine of chances, *Philos. Trans. Roy. Soc.*, 53, pp. 370–418.

T. Bement (1996), private communication.

J. Bezdek (1993), Fuzzy models—what are they and why? *IEEE Trans. Fuzzy Systems*, 1, pp. 1–6.

J. Bezdek (1994a), Fuzziness vs. probability: Again(!?), *IEEE Trans. Fuzzy Systems*, 2, pp. 1–3.

J. Bezdek (1994b), The thirsty traveler visits Gamont: A rejoinder to "Comments on fuzzy sets—what are they and why?" *IEEE Trans. Fuzzy Systems*, 2, pp. 43–45.

M. Black (1937), Vagueness: An exercise in logical analysis, *Internat. J. Gen. Systems*, 17, pp. 107–128.

D. BLOCKLEY (1983), Comments on "Model uncertainty in structural reliability," by Ove Ditlevsen, *J. Structural Safety*, 1, pp. 233–235.

P. CHEESEMAN (1986), Probabilistic versus fuzzy reasoning, in *Uncertainty in Artificial Intelligence*, L. Kanal and J. Lemmer, eds., North-Holland, Amsterdam, pp. 85–102.

P. CHEESEMAN (1988), An inquiry into computer understanding, *Comput. Intell.*, 4, pp. 58–66.

J. A. COOPER AND T. ROSS (1998), The application of new mathematical structures to safety analysis, in *Proceedings of the IEEE Conference on Fuzzy Systems (FUZZ-IEEE '98)*, Vol. 1, IEEE, Piscataway, NJ, pp. 692–697.

B. DE FINETTI (1974), *Theory of Probability*, John Wiley, New York.

A. DEMPSTER (1988), Comments on "An inquiry into computer understanding" by Peter Cheeseman, *Comput. Intell.*, 4, pp. 72–73.

D. DUBOIS AND H. PRADE (1988), Comments on "An inquiry into computer understanding" by Peter Cheeseman, *Comput. Intell.*, 4, pp. 73–76.

C. ELKAN (1993), The paradoxical success of fuzzy logic, in *Proceedings of the American Association for Artificial Intelligence*, MIT Press, Menlo Park, CA, pp. 698–703.

B. GAINES (1978), Fuzzy and probability uncertainty logics, *Inform. and Control*, 38, pp. 154–169.

H. GARDNER (1985), *The Mind's New Science: A History of the Cognitive Revolution*, Basic Books, New York.

I. GOODMAN, H. NGUYEN, AND G. ROGERS (1991), On scoring approach to the admissibility of uncertainty measures in expert systems, *J. Math. Anal. Appl.*, 159, pp. 550–594.

E. HISDAL (1988A), Are grades of membership probabilities? *Fuzzy Sets Systems*, 25, pp. 325–348.

E. HISDAL (1988B), The philosophical issues raised by fuzzy set theory, *Fuzzy Sets Systems*, 25, pp. 349–356.

A. HOADLEY AND J. KETTERING (1990), Communication between statisticians and engineer/physical scientists, *Technometrics*, 32, pp. 243–274 (includes 11 commentaries).

G. KLIR (1989), Is there more to uncertainty than some probability theorists might have us believe? *Internat. J. Gen. Systems*, 15, pp. 347–378.

G. KLIR (1994), On the alleged superiority of probabilistic representation of uncertainty, *IEEE Trans. Fuzzy Systems*, 2, pp. 27–31.

G. KLIR AND T. FOLGER (1988), *Fuzzy Sets, Uncertainty, and Information*, Prentice–Hall, Englewood Cliffs, NJ.

B. KOSKO (1990), Fuzziness vs. probability, *Internat. J. Gen. Systems*, 17, pp. 211–240.

B. KOSKO (1994), The probability monopoly, *IEEE Trans. Fuzzy Systems*, 2, pp. 32–33.

M. LAVIOLETTE AND J. SEAMAN (1992), Evaluating fuzzy representations of uncertainty, *Math. Sci.*, 17, pp. 26–41.

M. LAVIOLETTE, J. SEAMAN, J. BARRETT, AND W. WOODALL (1995), A probabilistic and statistical view of fuzzy methods, *Technometrics*, 37, pp. 249–292 (with seven commentaries).

D. LINDLEY (1971), *Making Decisions*, 2nd ed., John Wiley, New York.

D. LINDLEY (1982), Scoring rules and the inevitability of probability, *Internat. Statist. Rev.*, 50, pp. 1–26 (with seven commentaries).

D. LINDLEY (1987A), The probability approach to the treatment of uncertainty in artificial intelligence and expert systems, *Statist. Sci.*, 2, pp. 17–24.

D. LINDLEY (1987B), Comment: A tale of two wells, *Statist. Sci.*, 2, pp. 38–40.

G. MAGLARAS, E. NIKOLAIDIS, R. HAFTKA, AND H. CUDNEY (1997), Analytical-experimental comparison of probabilistic methods and fuzzy set based methods for designing under uncertainty, *Structural Optim.*, 13, pp. 69–80.

D. MCDERMOTT (1988), Comments on Cheeseman: Why plausible reasoning? Bayesian inference, *Comput. Intell.*, 4, pp. 91–92.

D. MCNEIL AND P. FRIEBERGER (1993), *Fuzzy Logic*, Simon and Schuster, New York.

M. A. MEYER AND J. M. BOOKER (2001), *Eliciting and Analyzing Expert Judgment: A Practical Guide*, ASA–SIAM Series on Statistics and Applied Probability, SIAM, Philadelphia, ASA, Alexandria, VA.

H. NGUYEN (1997), Fuzzy sets and probability, *Fuzzy Sets Systems*, 90, pp. 129–132.

E. NIKOLAIDIS, H. CUDNEY, R. ROSCA, AND R. HAFTKA (1999), Comparison of probabilistic and fuzzy set-based methods for designing under uncertainty, in *Proceedings of the 40th AIAA/ASME/ASCE/AHS/ASC Structures, Structural Dynamics, and Materials Conference, St. Louis, MO*, Vol. 4, American Institute of Aeronautics and Astronautics, Reston, VA, pp. 2860–2874.

G. ODEN (1994), Integration of fuzzy linguistic information in language comprehension, *Fuzzy Sets Systems*, 14, pp. 29–41.

D. OSHERSON AND E. SMITH (1981), On the adequacy of prototype theory as a theory of concepts, *Cognition*, 9, pp. 35–58.

W. QUINE (1981), *Theories and Things*, Harvard University Press, Cambridge, MA.

E. ROSCH (1973), Natural categories, *Cognitive Psychology*, 4, pp. 328–350.

T. ROSS, C. FERREGUT, R. OSEGUEDA, AND V. KREINOVICH (1999), System reliability: A case when fuzzy logic enhances probability theory's ability to deal with real-world problems, in *Proceedings of the 18th International Conference of the North American Fuzzy Information Society*, R. Dave and T. Sudkamp, eds., IEEE, Piscataway, NJ, pp. 81–84.

P. ROUSSEEUW (1995), Discussion: Fuzzy clustering at the intersection, *Technometrics*, 37, pp. 283–286.

E. RUSPINI (1988), Intelligent understanding and wishful thinking: The dangers of episte-mological alchemy, *Comput. Intell.*, 4, pp. 105–117.

B. RUSSELL (1923), Vagueness, *Australasian J. Philos.*, 1, pp. 85–88.

G. SHAFER (1987), Belief functions and possibility measures, in *Analysis of Fuzzy Informa-tion*, Vol. 1, J. Bezdek, ed., CRC Press, Boca Raton, FL, pp. 51–84.

N. SINGPURWALLA, T. BEMENT, AND J. BOOKER (2000), *Membership Functions and Prob-ability Measures of Fuzzy Sets*, Report LA-UR 00-3660, Los Alamos National Laboratory, Los Alamos, NM.

R. SMITH, T. BEMENT, W. PARKINSON, F. MORTENSEN, S. BECKER, AND M. MEYER (1997), The use of fuzzy control system techniques to develop uncertainty distributions, in *Pro-ceedings of the Joint Statistical Meetings, Anaheim, CA*, American Statistical Association, Alexandria, VA.

R. E. SMITH, J. M. BOOKER, T. R. BEMENT, M. A. MEYER, W. J. PARKINSON, AND M. JAMSHIDI (1998), The use of fuzzy control system methods for characterizing expert judgment uncer-tainty distributions, in *Proceedings of the 4th International Conference on Probabilistic Safety Assessment and Management (PSAM 4)*, Vol. 1: Probabilistic Safety Assessment and Management, A. Mosleh and R. A. Bari, eds., Springer, London, pp. 499–502.

S. STIGLER (1986), Laplace's 1774 memoir on inverse probability, *Statist. Sci.*, 1, pp. 359–378.

A. TVERSKY AND D. KAHNEMAN (1983), Probability, representativeness, and the conjunction fallacy, *Psychological Rev.*, 90, pp. 293–315.

R. VON MISES (1957), *Probability, Statistics, and Truth*, Dover, New York.

T. WALLSTEN AND D. BUDESCU (1983), Encoding subjective probabilities: A psychological and psychometric review, *Management Sci.*, 29, pp. 151–173.

W. WOODALL (1997), An overview of comparisons between fuzzy and statistical methods, (extended abstract 154) in *Proceedings of the Joint Statistical Meetings, Anaheim, CA*, Amer-ican Statistical Association, Alexandria, VA, p. 174.

L. ZADEH (1965), Fuzzy sets, *Inform. and Control*, 8, pp. 338–353.

L. ZADEH (1995), Discussion: Probability theory and fuzzy logic are complementary rather than competitive, *Technometrics*, 37, pp. 271–276.

L. ZADEH (1996), Fuzzy logic = computing with words, *IEEE Trans. Fuzzy Systems*, 4, pp. 103–111.

L. ZADEH (1998), *PT Needs FL (Part 2)*, email communication to the BISC (Berkeley Initiative for Soft Computing) Group, Berkeley, CA, November 20.

L. ZADEH (1999), From computing with numbers to computing with words—from manip-ulation of measurements to manipulation of perceptions, *IEEE Trans. Circuits Systems*, 45, pp. 105–119.

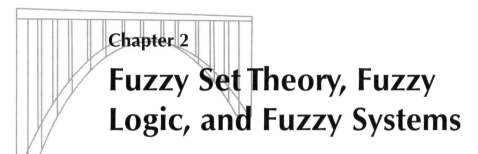

Chapter 2

Fuzzy Set Theory, Fuzzy Logic, and Fuzzy Systems

Timothy J. Ross and W. Jerry Parkinson

2.1   Introduction

Making decisions about processes that contain nonrandom uncertainty, such as the uncertainty in natural language, with the use of classical theories has been shown to be less than perfect. Lotfi Zadeh suggested that *set membership* is the key to decision-making when faced with linguistic and nonrandom uncertainty. In fact, Dr. Zadeh stated in his seminal paper of 1965:

> "The notion of a fuzzy set provides a convenient point of departure for the construction of a conceptual framework which parallels in many respects the framework used in the case of ordinary sets, but is more general than the latter and, potentially, may prove to have a much wider scope of applicability, particularly in the fields of pattern classification and information processing. Essentially, such a framework provides a natural way of dealing with problems in which the source of imprecision is the absence of sharply defined criteria of class membership rather than the presence of random variables."

Suppose we are interested in the height of people. We can easily assess whether someone is over 6 feet tall. In a binary sense, this person either is or is not, based on the accuracy, or imprecision, of our measuring device. For example, if "tall" is a set defined as heights equal to or greater than 6 feet, a computer would not recognize an individual of height 5′11.999″ as being a member of the set "tall." But how do we assess the uncertainty in the following question: Is the person *nearly* 6 feet tall? The uncertainty in this case is due to the vagueness, or ambiguity, of the adjective *nearly*. A 5′11″ person clearly could be a member of the set "nearly 6 feet tall" people. In the first situation, the uncertainty of whether a person's height, which is unknown, is 6 feet or not is binary; it either is or is not, and we can produce a probability assessment of that prospect based on height data from many people. But the uncertainty of whether a person is *nearly* 6 feet tall is nonrandom. The degree to

29

which the person approaches a height of 6 feet is fuzzy. In reality, "tallness" is a matter of degree and is relative. Among peoples of some tribes in Africa, a height of 6 feet for a male is considered short. Therefore, 6 feet can be tall in one context and short in another. In the real (fuzzy) world, the set of tall people can overlap with the set of not-tall people; such an overlap is null in the world of binary logic.

This notion of set membership, then, is central to the representation of objects within a universe and to sets defined in the universe. Classical sets contain objects that satisfy precise properties of membership; fuzzy sets contain objects that satisfy imprecise properties of membership, i.e., membership of an object in a fuzzy set can be approximate. For example, the set of heights *from 5 to 7 feet* is crisp; the set of heights in the region *around 6 feet* is fuzzy. To elaborate, suppose we have an exhaustive collection of individual elements (singletons) x that make up a universe of information (discourse) X. Further, various combinations of these individual elements make up sets, say, A, in the universe. For crisp sets an element x in the universe X either is a member of some crisp set A or is not. This binary issue of membership can be represented mathematically with the indicator function

$$\chi_A(x) = \begin{cases} 1, & x \in A, \\ 0, & x \notin A, \end{cases} \tag{2.1}$$

where $\chi_A(A)$ indicates an unambiguous membership of element x in set A, and $\in$ and $\notin$ denote contained-in and not contained-in, respectively. For our example of the universe of heights of people, suppose set A is the crisp set of all people with $5.0 \le x \le 7.0$ feet, shown in Figure 2.1(a). A particular individual, x_1, has a height of 6.0 feet. The membership of this individual in crisp set A is equal to 1, or full membership, given symbolically as $\chi_A(x_1) = 1$. Another individual, say, x_2, has a height of 4.99 feet. The membership of this individual in set A is equal to 0, or no membership, and hence $\chi_A(x_2) = 0$, also seen in Figure 2.1(a). In these cases, the membership in a set is binary: an element either is a member of a set or is not.

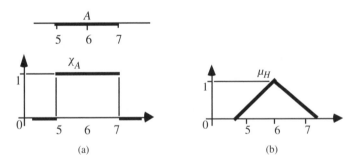

(a)   (b)

Figure 2.1. *Height membership functions for* (a) *a crisp set* (A) *and* (b) *a fuzzy set* (H).

Zadeh extended the notion of binary membership to accommodate various "degrees of membership" on the real continuous interval [0, 1], where the endpoints of 0 and 1 conform to no membership and full membership, respectively, just as the indicator function does for crisp sets, but where the infinite number of values in between the endpoints can represent various degrees of membership for an element x in some set in the universe. The sets in the universe X that can accommodate "degrees of membership" were termed by Zadeh as "fuzzy sets."

Continuing further with the example on heights, consider a set H consisting of heights *near 6 feet*. Since the property *near 6 feet* is fuzzy, there is not a unique membership function for H. Rather, the analyst must decide what the membership function, denoted μ_H, should look like. Plausible properties of this function might be (i) normality ($\mu_H(6) = 1$), (ii) monotonicity (the closer x is to 6, the closer μ_H is to 1), and (iii) symmetry (numbers equidistant from 6 should have the same value of μ_H) (Bezdek (1993)). Such a membership function is illustrated in Figure 2.1(b). A key difference between crisp and fuzzy sets is their membership function; a crisp set has a unique membership function, whereas a fuzzy set can have an infinite number of membership functions to represent it. For fuzzy sets, the uniqueness is sacrificed, but flexibility is gained because the membership function can be adjusted to maximize the utility for a particular application.

James Bezdek provided one of the most lucid comparisons between crisp and fuzzy sets that warrants repeating here (Bezdek (1993)). Crisp sets of real objects are equivalent to, and isomorphically described by, a unique membership function, such as χ_A in Figure 2.1(a). However, there is no set-theoretic equivalent of "real objects" corresponding to χ_A. Fuzzy sets are *always functions*, which map a universe of objects, say, X, onto the unit interval [0, 1]; that is, the fuzzy set H is the *function* μ_H that carries X into [0, 1]. Hence *every* function that maps X onto [0, 1] is a fuzzy set. While this is true in a formal mathematical sense, many functions that qualify on the basis of this definition cannot be suitable fuzzy sets. However, they *become* fuzzy sets when—and only when—they match some intuitively plausible semantic description of imprecise properties of the objects in X.

The membership function embodies the mathematical representation of membership in a set, and the notation used throughout this text for a fuzzy set is a set symbol with a "tilde" underscore, say, $\underset{\sim}{A}$, where the functional mapping is given by

$$\mu_{\underset{\sim}{A}}(x) \in [0, 1],\tag{2.2}$$

where $\mu_{\underset{\sim}{A}}(x)$ gives the degree of membership of element x in fuzzy set $\underset{\sim}{A}$. Therefore, $\mu_{\underset{\sim}{A}}(x)$ is a value on the unit interval which measures the degree to which element x belongs to fuzzy set $\underset{\sim}{A}$; equivalently, $\mu_{\underset{\sim}{A}}(x) = \text{Degree}(x \in \underset{\sim}{A})$.

To summarize, there is a clear distinction between fuzziness and randomness of an event. *Fuzziness describes the ambiguity of an event, whereas randomness describes the likelihood of occurrence of the event.* The event will occur or will not occur; but is the description of the event unambiguous enough to quantify its occurrence or nonoccurrence?

2.1.1 Fuzzy sets

In classical sets, the transition for an element in the universe between membership and nonmembership in a given set is abrupt and well defined (hence termed "crisp"). For an element in a universe that contains fuzzy sets, this transition can be gradual and, among various degrees of membership, can be thought of as conforming to the fact that the boundaries of the fuzzy sets are vague and ambiguous. Hence membership in this set of an element from the universe is measured by a function which attempts to describe vagueness and ambiguity.

A fuzzy set, then, is a set containing elements which have varying degrees of membership in the set. This idea is contrasted with classical, or crisp, sets because members of a crisp set would not be members unless their membership was full or complete in that set (i.e., their membership is assigned a value of 1). Elements in a fuzzy set also can be members of other fuzzy sets on the same universe because their membership can be a value other than unity.

Elements of a fuzzy set are mapped to a universe of "membership values" using a function-theoretic form. As shown in (2.2), fuzzy sets are denoted in this text by a set

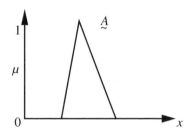

Figure 2.2. *Membership function for fuzzy set A.*

symbol with a tilde understrike; so, for example, A would be the "fuzzy set" A. This function maps elements of a fuzzy set A to a real numbered value on the interval 0 to 1. If an element in the universe, say, x, is a member of fuzzy set A, then this mapping is given by (2.2), or $\mu_A(x) \in [0, 1]$. This mapping is shown in Figure 2.2 for a fuzzy set.

A notation convention for fuzzy sets when the universe of discourse, X, is discrete and finite, is given below for a fuzzy set A:

$$A = \frac{\mu_A(x_1)}{x_1} + \frac{\mu_A(x_2)}{x_2} + \cdots + = \sum_i \frac{\mu_A(x_i)}{x_i}. \tag{2.3}$$

In the notation the horizontal bar is not a quotient, but rather a delimiter. In addition, the numerator in each individual expression is the membership value in set A associated with the element of the universe indicated in the denominator of each expression. The summation symbol is not for algebraic summation, but rather denotes the collection or aggregation of each element; hence the "+" signs are not the algebraic "add" but rather are a function-theoretic union.

2.1.2   Fuzzy set operations

Define three fuzzy sets A, B, and C in the universe X. For a given element x of the universe, the following function-theoretic operations for the set-theoretic operations of union, intersection, and complement are defined for A, B, and C on X:

$$\text{union:} \qquad \mu_{A \cup B}(x) = \mu_A(x) \vee \mu_B(x), \tag{2.4}$$

$$\text{intersection:} \quad \mu_{A \cap B}(x) = \mu_A(x) \wedge \mu_B(x), \tag{2.5}$$

$$\text{complement:} \quad \mu_{\bar{A}}(x) = 1 - \mu_A(x), \tag{2.6}$$

where the symbols $\cup$, $\cap$, $\vee$, $\wedge$, and $^-$ denote the operations of union, intersection, max, min, and complementation, respectively. Venn diagrams for these operations, extended to consider fuzzy sets, are shown in Figures 2.3–2.5.

Any fuzzy set A defined on a universe X is a subset of that universe. Also, by definition and just as with classical sets, the membership value of any element x in the null set $\emptyset$ is 0 and the membership value of any element x in the whole set X is 1. Note that the null set and the whole set are not fuzzy sets in this context (no tilde underscore). The appropriate notation for these ideas is as follows:

$$A \subseteq X \rightarrow \mu_A(x) \leq \mu_x(x) \tag{2.7a}$$

$$\text{for all } x \in X, \ \mu_{\emptyset}(x) = 0 \tag{2.7b}$$

$$\text{for all } x \in X, \ \mu_x(x) = 1. \tag{2.7c}$$

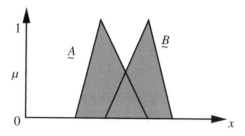

Figure 2.3. *Union of fuzzy sets $\underset{\sim}{A}$ and $\underset{\sim}{B}$.*

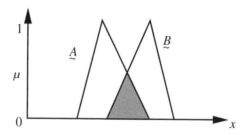

Figure 2.4. *Intersection of fuzzy sets $\underset{\sim}{A}$ and $\underset{\sim}{B}$.*

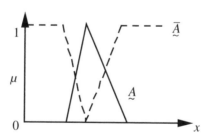

Figure 2.5. *Complement of fuzzy set $\underset{\sim}{A}$.*

The collection of all fuzzy sets and fuzzy subsets on X is denoted as the fuzzy power set Power(X). Because all fuzzy sets can overlap, the cardinality of the fuzzy power set is infinite.

DeMorgan's laws for classical sets also hold for fuzzy sets, as denoted by

$$\overline{(\underset{\sim}{A} \cap \underset{\sim}{B})} = \bar{\underset{\sim}{A}} \cup \bar{\underset{\sim}{B}}, \tag{2.8a}$$

$$\overline{(\underset{\sim}{A} \cup \underset{\sim}{B})} = \bar{\underset{\sim}{A}} \cap \bar{\underset{\sim}{B}}. \tag{2.8b}$$

As indicated in Ross (1995), all other operations on classical sets also hold for fuzzy sets, except for the excluded-middle laws. These two laws do not hold for fuzzy sets because of the fact that, since fuzzy sets can overlap, a set and its complement can also overlap. The *excluded-middle laws*, extended for fuzzy sets, are expressed by

$$\underset{\sim}{A} \cup \bar{\underset{\sim}{A}} \neq X, \tag{2.9a}$$

$$\underset{\sim}{A} \cap \bar{\underset{\sim}{A}} \neq \emptyset. \tag{2.9b}$$

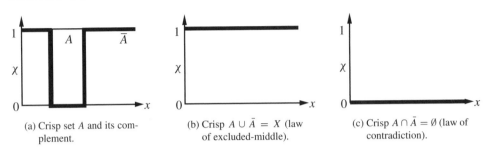

(a) Crisp set A and its com-   (b) Crisp $A \cup \bar{A} = X$ (law   (c) Crisp $A \cap \bar{A} = \emptyset$ (law of
plement.   of excluded-middle).   contradiction).

Figure 2.6. *Excluded-middle laws for crisp sets.*

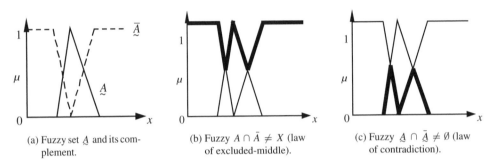

(a) Fuzzy set $\underset{\sim}{A}$ and its com-   (b) Fuzzy $A \cap \bar{A} \neq X$ (law   (c) Fuzzy $\underset{\sim}{A} \cap \underset{\sim}{\bar{A}} \neq \emptyset$ (law
plement.   of excluded-middle).   of contradiction).

Figure 2.7. *Excluded-middle laws for fuzzy sets.*

Extended Venn diagrams comparing the *excluded-middle laws* for classical (crisp) sets and fuzzy sets are shown in Figures 2.6 and 2.7, respectively.

2.2 Fuzzy relations

Fuzzy relations are developed by allowing the relationship between elements of two or more sets to take on an infinite number of degrees of relationship between the extremes of "completely related" and "not related," which are the only degrees of relationship possible in crisp relations. In this sense, fuzzy relations are to crisp relations as fuzzy sets are to crisp sets; crisp sets and relations are more constrained realizations of fuzzy sets and relations.

Fuzzy relations map elements of one universe, say, X, to those of another universe, say, Y, through the Cartesian product of the two universes. However, the "strength" of the relation between ordered pairs of the two universes is not measured with the indicator function (as in the case of crisp relations), but rather with a membership function expressing various "degrees" of strength of the relation on the unit interval [0, 1]. Hence, a fuzzy relation $\underset{\sim}{R}$ is a mapping from the Cartesian space $X \times Y$ to the interval [0, 1], where the strength of the mapping is expressed by the membership function of the relation for ordered pairs from the two universes, or $\mu_{\underset{\sim}{R}}(x, y)$.

2.2.1 Operations on fuzzy relations

Let $\underset{\sim}{R}$ and $\underset{\sim}{S}$ be fuzzy relations on the Cartesian space $X \times Y$. Then the following operations apply for the membership values for various set operations:

union: $\mu_{\underset{\sim}{R} \cup \underset{\sim}{S}}(x, y) = \max(\mu_{\underset{\sim}{R}}(x, y), \mu_{\underset{\sim}{S}}(x, y)),$ (2.10)

intersection: $\mu_{\underset{\sim}{R} \cap \underset{\sim}{S}}(x, y) = \min(\mu_{\underset{\sim}{R}}(x, y), \mu_{\underset{\sim}{S}}(x, y)),$ (2.11)

complement: $\mu_{\underset{\sim}{\bar{R}}}(x, y) = 1 - \mu_{\underset{\sim}{R}}(x, y),$ (2.12)

containment: $\underset{\sim}{R} \subset \underset{\sim}{S} \Rightarrow \mu_{\underset{\sim}{R}}(x, y) \leq \mu_{\underset{\sim}{S}}(x, y).$ (2.13)

2.2.2   Fuzzy Cartesian product and composition

Because fuzzy relations, in general, are fuzzy sets, we can define the Cartesian product to be a relation between two or more fuzzy sets. Let $\underset{\sim}{A}$ be a fuzzy set in universe X and $\underset{\sim}{B}$ a fuzzy set in universe Y; then the Cartesian product between fuzzy sets $\underset{\sim}{A}$ and $\underset{\sim}{B}$ will result in a fuzzy relation $\underset{\sim}{R}$, which is contained within the full Cartesian product space, or

$$\underset{\sim}{A} \times \underset{\sim}{B} = \underset{\sim}{R} \subset X \times Y, \qquad (2.14)$$

where the fuzzy relation $\underset{\sim}{R}$ has membership function

$$\mu_{\underset{\sim}{R}}(x, y) = \mu_{\underset{\sim}{A} \times \underset{\sim}{B}}(x, y) = \min(\mu_{\underset{\sim}{A}}(x), \mu_{\underset{\sim}{B}}(y)). \qquad (2.15)$$

The Cartesian product defined by $\underset{\sim}{A} \times \underset{\sim}{B} = \underset{\sim}{R}$ in (2.14) is implemented in the same way as the cross-product of two vectors. The Cartesian product is *not* the same operation as the arithmetic product; in the case of two-dimensional relations, the former employs the pairing of elements among sets, whereas the latter uses actual arithmetic products between elements of sets. More can be found on fuzzy arithmetic in Ross (1995). Each of the fuzzy sets could be thought of as a vector of membership values; each value is associated with a particular element in each set. For example, for a fuzzy set (vector) $\underset{\sim}{A}$ that has four elements and hence a column vector of size 4×1, and for a fuzzy set (vector) $\underset{\sim}{B}$ that has five elements and hence a row vector size of 1×5, the resulting fuzzy relation $\underset{\sim}{R}$ will be represented by a matrix of size 4×5; i.e., $\underset{\sim}{R}$ will have four rows and five columns.

A composition is an operator on relations much like an integral is an operator on functions. For example, in probability theory, if we have a joint probability density function (PDF) defined on x and y, we can find the marginal PDF on x by integrating the joint PDF over all y. With this analogy in mind, suppose we have a crisp relation R defined on universes x and y and we want to find a crisp set that is defined only on x. To do this, we can perform a composition on the crisp relation R with another crisp set or relation, say, A, that is defined only on the universe Y.

Fuzzy composition can be defined just as it is for crisp (binary) relations (Ross (1995)). Suppose $\underset{\sim}{R}$ is a fuzzy relation on the Cartesian space $X \times Y$, $\underset{\sim}{S}$ is a fuzzy relation on $Y \times Z$, and $\underset{\sim}{T}$ is a fuzzy relation on $X \times Z$; then fuzzy max–min composition (there are also other forms of the composition; see Ross (1995)) is defined in terms of the membership function-theoretic notation

$$\mu_{\underset{\sim}{T}}(x, z) = \bigvee_{y \in Y} (\mu_{\underset{\sim}{R}}(x, y) \wedge \mu_{\underset{\sim}{S}}(y, z)), \qquad (2.16a)$$

and the fuzzy max-product composition is defined, in terms of the membership function-theoretic notation, as

$$\mu_{\underset{\sim}{T}}(x, z) = \bigvee_{y \in Y} (\mu_{\underset{\sim}{R}}(x, y) \bullet \mu_{\underset{\sim}{S}}(y, z)). \qquad (2.16b)$$

It should be pointed out that neither crisp nor fuzzy compositions have inverses in general; that is,

$$\underline{R} \circ \underline{S} \neq \underline{S} \circ \underline{R}. \tag{2.17}$$

Result (2.17) is general for any matrix operation, fuzzy or otherwise, which must satisfy consistency between the cardinal counts of elements in respective universes. Even for the case of square matrices, the composition inverse represented by (2.17) is not guaranteed.

2.3   Fuzzy and classical logic

2.3.1   Classical logic

In classical predicate logic, a simple proposition P is a linguistic, or declarative, statement contained within a universe of elements, say, X, which can be identified as a collection of elements in X which are strictly true or strictly false. Hence a proposition P is a collection of elements, that is, a set, where the truth values for all elements in the set are either all true or all false. The veracity (truth) of an element in the proposition, P, can be assigned a binary truth value, called $T(P)$, just as an element in a universe is assigned a binary quantity to measure its membership in a particular set. For binary (Boolean) predicate logic, $T(P)$ is assigned a value of 1 (truth) or 0 (false). If U is the universe of all propositions, then T is a mapping of the elements u in these propositions (sets) to the binary quantities (0, 1), or

$$T : u \in U \rightarrow \{0, 1\}. \tag{2.18}$$

All elements u in the universe U that are true for proposition P are called the truth set of P. Those elements u in the universe U that are false for proposition P are called the falsity set of P.

In logic we need to postulate the boundary conditions of truth values just as we do for sets; that is, in function-theoretic terms we need to define the truth value of a universe of discourse. For a universe Y and the null set $\emptyset$, we define the truth values $T(Y) = 1$ and $T(\emptyset) = 0$.

Now let P and Q be two simple propositions on the same universe of discourse that can be combined using the five logical connectives

- disjunction ($\vee$),

- conjunction ($\wedge$),

- negation ($-$),

- implication ($\rightarrow$),

- equivalence ($\leftrightarrow$ or $\Leftrightarrow$)

to form logical expressions involving the two simple propositions. These connectives can be used to form new propositions from simple propositions.

The disjunction connective, the logical "or," is the term used to represent what is commonly referred to as the "inclusive or." The natural language term "or" and the logical "or" are different. The natural language "or" differs from the logical "or" in that the former implies exclusion (denoted in the literature as the *exclusive or*; see Ross (1995) for more details). For example, "soup or salad" on a restaurant menu implies choosing one or the other option, but not both. The "inclusive or" is the one most often employed in logic; the inclusive or ("logical or" as used here) implies that a compound proposition is true if either of the simple propositions is true or if both are true.

The equivalence connective arises from dual implication; that is, for some propositions P and Q, if $P \rightarrow Q$ and $Q \rightarrow P$, then $P \leftrightarrow Q$.

Now define sets A and B from universe X, where these sets might represent linguistic ideas or thoughts. A *propositional calculus* (sometimes called the *algebra of propositions*) will exist for the case where proposition P measures the truth of the statement that an element, x, from the universe X is contained in set A, and where the proposition Q measures the truth of the statement that this element, x, is contained in set B, or more conventionally,

$$P : \text{truth that } x \in A,$$

$$Q : \text{truth that } x \in B,$$

where truth is measured in terms of the truth value; i.e.,

$$\text{if } x \in A, \quad T(P) = 1; \qquad \text{otherwise, } T(P) = 0,$$

$$\text{if } x \in B, \quad T(Q) = 1; \qquad \text{otherwise, } T(Q) = 0,$$

or, using the indicator function to represent truth (1) and false (0), the following notation results:

$$\chi_A(x) = \begin{cases} 1, & x \in A, \\ 0, & x \notin A. \end{cases} \tag{2.19}$$

A notion of *mutual exclusivity* arises in crisp logic; that is, there is no provision for a proposition to be partly true and partly false; it has to be one or the other. For the situation involving two propositions P and Q, where $T(P) \cap T(Q) = \emptyset$, we have that the truth of P always implies the falsity of Q and vice versa; hence P and Q are mutually exclusive propositions.

The five logical connectives defined previously can be used to create compound propositions, where a compound proposition is defined as a logical proposition formed by logically connecting two or more simple propositions. Just as we are interested in the truth of a simple proposition, predicate logic also involves the assessment of the truth of compound propositions. For the case of two simple propositions, the resulting compound propositions are defined below in terms of their binary truth values.

Given a proposition $P : x \in A$, $\bar{P} : x \notin A$, we have for the logical connectives

disjunction:
$$P \vee Q \Rightarrow x \in A \text{ or } B;$$
$$\text{hence } T(P \vee Q) = \max(T(P), T(Q)), \tag{2.20a}$$

conjunction:
$$P \wedge Q \Rightarrow x \in A \text{ and } B;$$
$$\text{hence } T(P \wedge Q) = \min(T(P), T(Q)), \tag{2.20b}$$

negation:
$$\text{if } T(P) = 1, \quad \text{then } T(\bar{P}) = 0;$$
$$\text{if } T(P) = 0, \quad \text{then } T(\bar{P}) = 1, \tag{2.20c}$$

implication:
$$P \rightarrow Q \Rightarrow x \in A, B;$$
$$\text{hence } T(P \rightarrow Q) = T(\bar{P} \cup Q), \tag{2.20d}$$

equivalence:
$$P \leftrightarrow Q \Rightarrow x \in A, B;$$
$$\text{hence } T(P \leftrightarrow Q) = \begin{cases} 1 & \text{for } T(P) = T(Q), \\ 0 & \text{for } T(P) \neq T(Q). \end{cases} \tag{2.20e}$$

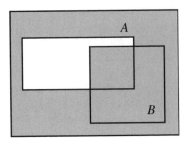

Figure 2.8. *Graphical analogue of the classical implication operation.*

The logical connective "implication," i.e., $P \rightarrow Q$ (P implies Q), presented here also is known as the classical implication to distinguish it from several other forms (see Ross (1995)). In this implication, the proposition P also is referred to as the *hypothesis* or the *antecedent*, and the proposition Q also is referred to as the *conclusion* or the *consequent*. The compound proposition $P \rightarrow Q$ is true in all cases except where a true antecedent P appears with a false consequent.

Hence the classical form of the implication is true for all propositions of P and Q except for those propositions which are in both the truth set of P and the false set of Q; i.e.,

$$T(P \rightarrow Q) = \overline{(T(P) \cap T(\bar{Q}))}. \tag{2.21a}$$

This classical form of the implication operation requires some explanation. For a proposition P defined on set A and a proposition Q defined on set B, the implication "P implies Q" is equivalent to taking the union of elements in the complement of set A with the elements in set B. (This result also can be derived by using DeMorgan's laws shown in (2.8).) That is, the logical implication is analogous to the set-theoretic form

$$P \rightarrow Q \equiv \bar{A} \cup B \text{ is true} \equiv \text{either "not in } A\text{" or "in } B\text{"}$$
$$\text{so that } T(P \rightarrow Q) = T(\bar{P} \vee Q) = \max(T(\bar{P}), T(Q)). \tag{2.21b}$$

This is linguistically equivalent to the statement, "$P \rightarrow Q$ is true" when either "not A" or "B" is true (logical or). Graphically, this implication and the analogous set operation are represented by the Venn diagram in Figure 2.8. The shaded region in Figure 2.8 represents the collection of elements in the universe where the implication is true.

Now with two propositions (P and Q), each being able to take on one of two truth values (true or false, 1 or 0), there will be a total of $2^2 = 4$ propositional situations. These situations are illustrated, along with the appropriate truth values for the propositions P and Q and the various logical connectives between them, in Table 2.1. The values in the last five columns of the table are calculated using the expressions in (2.20). In Table 2.1, T *or* 1 denotes true and F *or* 0 denotes false.

Suppose the implication operation involves two different universes of discourse; P is a proposition described by set A, which is defined on universe X, and Q is a proposition described by set B, which is defined on universe Y. Then the implication $P \rightarrow Q$ can be represented in set-theoretic terms by the relation R, where R is defined by

$$R = (A \times B) \cup (\bar{A} \times Y) \equiv \text{IF } A, \text{ THEN } B;$$
$$\text{if } x \in A, \quad \text{where } x \in X \text{ and } A \subset X, \tag{2.22}$$
$$\text{then } y \in B, \quad \text{where } y \in B \text{ and } B \subset Y.$$

Table 2.1. *Truth table for various compound propositions.*

P	Q	$\bar{P}$	$P \vee Q$	$P \wedge Q$	$P \to Q$	$P \leftrightarrow Q$
$T(1)$	$T(1)$	$F(0)$	$T(1)$	$T(1)$	$T(1)$	$T(1)$
$T(1)$	$F(0)$	$F(0)$	$T(1)$	$F(0)$	$F(0)$	$F(0)$
$F(0)$	$T(1)$	$T(1)$	$T(1)$	$F(0)$	$T(1)$	$F(0)$
$F(0)$	$F(0)$	$T(1)$	$F(0)$	$F(0)$	$T(1)$	$T(1)$

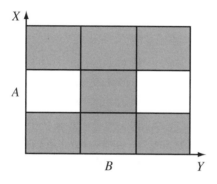

Figure 2.9. *The Cartesian space showing the implication IF A, THEN B.*

Implication (2.22) also is equivalent to the linguistic rule form IF A, THEN B. The graphic shown in Figure 2.9 represents the space of the Cartesian product $X \times Y$, showing typical sets A and B, and superposed on this space is the set-theoretic equivalent of the implication. That is,

$$P \to Q : \text{if } x \in A, \quad \text{then } y \in B \quad \text{or} \quad P \to Q \equiv \bar{A} \cup B.$$

The shaded regions of the compound Venn diagram in Figure 2.9 represent the truth domain of the implication IF A, THEN B ($P \to Q$). Another compound proposition in linguistic rule form is the expression

$$\text{IF } A, \quad \text{THEN } B, \quad \text{ELSE } C.$$

Linguistically, this compound proposition could be expressed as

$$\text{IF } A, \quad \text{THEN } B \quad \text{and} \quad \text{IF } \bar{A}, \quad \text{THEN } C.$$

In predicate logic, this rule has the form

$$(P \to Q) \wedge (\bar{P} \to S),$$
$$\text{where } P : x \in A, \ A \subset X,$$
$$Q : y \in B, \ B \subset Y, \tag{2.23}$$
$$S : y \in C, \ C \subset Y.$$

The set-theoretic equivalent of this compound proposition is given by

$$\text{IF } A, \text{THEN } B, \text{ELSE } C \equiv (A \times B) \cup (\bar{A} \times C) = R = \text{relation on } X \times Y. \tag{2.24}$$

The shaded region in Figure 2.10 illustrates the truth domain for this compound proposition for the particular case where $B \cap C = \emptyset$.

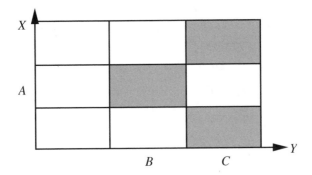

Figure 2.10. *Truth domain for IF A, THEN B, ELSE C.*

Logical inferences

A typical IF–THEN rule is used to determine whether an antecedent (cause or action) infers a consequent (effect or reaction). Suppose we have a rule of the form IF A, THEN B, where A is a set defined on universe X and B is a set defined on universe Y. As discussed before, this rule can be translated into a relation between sets A and B; that is, recalling (2.22), $R = (A \times B) \cup (\bar{A} \times Y)$. Now suppose a new antecedent, say, A', is known. Can we use a logical inference to find a new consequent, say, B', resulting from the new antecedent, that is, in rule form IF A', THEN B'? The answer is, of course, yes, through the use of the composition operation. Since "A implies B" is defined on the Cartesian space $X \times Y$, B' can be found through the set-theoretic formulation (again from (2.22))

$$B' = A' \circ R = A' \circ ((A \times B) \cup (\bar{A} \times Y)), \tag{2.25}$$

where the symbol $\circ$ denotes the composition operation. A logical inference also can be used for the compound rule IF A, THEN B, ELSE C, where this compound rule is equivalent to the relation defined by

$$R = (A \times B) \cup (\bar{A} \times C). \tag{2.26}$$

For this compound rule, if we define another antecedent A' (i.e., another input, different from A), the following possibilities exist, depending on (i) whether A' is fully contained in the original antecedent A, (ii) whether A' is contained only in the complement of A, or (iii) whether A' and A overlap to some extent as described below:

$$\begin{aligned}
&\text{If } A' \subset A, &&\text{then } y = B; \\
&\text{if } A' \subset \bar{A}, &&\text{then } y = C; \\
&\text{if } A' \cap A \neq \emptyset, \ A' \cap \bar{A} \neq \emptyset, &&\text{then } y = B \cup C.
\end{aligned}$$

The rule IF A, THEN B (proposition P is defined on set A in universe X, and proposition Q is defined on set B in universe Y), i.e., $P \to Q \Rightarrow R = (A \times B) \cup (\bar{A} \times Y)$, is then defined in function-theoretic terms as

$$\chi_R(x, y) = \max[(\chi_A(x) \wedge \chi_B(y)), ((1 - \chi_A(x)) \wedge 1)], \tag{2.27}$$

where $\chi(\cdot)$ is the indicator function as defined before.

The compound rule IF A, THEN B, ELSE C also can be defined in terms of a matrix relation as $R = (A \times B) \cup (\bar{A} \times C) \Rightarrow (P \rightarrow Q) \wedge (\bar{P} \rightarrow S)$, given by (2.23) and (2.24), where the membership function is determined as

$$\chi_R(x, y) = \max[(\chi_A(x) \wedge \chi_B(y)), ((1 - \chi_A(x)) \wedge \chi_C(y))]. \tag{2.28}$$

2.3.2 Fuzzy logic

The restriction of classical logic to two-valued logic has created many interesting paradoxes over the ages. For example, does the "liar from Crete" lie when he claims, "All Cretans are liars"? If he is telling the truth, then his statement is false, but if his statement is false, he is not telling the truth. The statement can't be both true and false. The only way for the "liar from Crete" paradox to be solved is if the statement is both true and false simultaneously.

This paradox can be illustrated mathematically using set notation (Kosko (1992)). Let S be the proposition that a Cretan lies and $\bar{S}$ (not S) that he does not. Then since $S \rightarrow \bar{S}$ (S implies not S) and $\bar{S} \rightarrow S$, the two propositions are logically equivalent: $S \leftrightarrow \bar{S}$. Equivalent propositions have the same truth value; hence

$$T(S) = T(\bar{S}) = 1 - T(S),$$

which yields

$$T(S) = \frac{1}{2}.$$

As can be seen, paradoxes reduce to half-truths mathematically (or half-falsities). In classical binary (bivalued) logic, however, such conditions are not allowed; i.e., only

$$T(S) = 1 \text{ or } 0$$

is valid.

There is a more subtle form of paradox that also can be addressed by a multivalued logic. Consider the classical *sorites* (a heap of syllogisms) paradoxes, for example, the case of a liter-full glass of water. Often this example is called the optimist's conclusion: Is the glass half-full or half-empty when the volume is at 500 milliliters? Is the liter-full glass still full if we remove one milliliter of water? Is the glass still full if we remove two, three, four, or 100 milliliters of water? If we continue to answer yes, then eventually we will have removed all the water, and an empty glass will still be characterized as full! At what point did the liter-full glass of water become empty? Perhaps when 500 milliliters full? Unfortunately, no single milliliter of liquid provides for a transition between full and empty. This transition is gradual so that as each milliliter of water is removed, the truth value of the glass being full gradually diminishes from a value of 1 at 1000 milliliters to 0 at 0 milliliters. Hence for many problems, we need a multivalued logic other than the classic binary logic that is so prevalent today.

A fuzzy logic proposition $\underset{\sim}{P}$ is a statement involving some concept without clearly defined boundaries. Linguistic statements that tend to express subjective ideas, and that can be interpreted slightly differently by various individuals, typically involve fuzzy propositions. Most natural language is fuzzy in that it involves vague and imprecise terms. Statements describing a person's height or weight, or assessments of people's preferences about colors or foods, can be used as examples of fuzzy propositions. The truth value assigned to $\underset{\sim}{P}$ can be any value on the interval [0, 1]. The assignment of the truth value to a proposition is actually a mapping from the interval [0, 1] to the universe U of truth values T, as indicated in the equation

$$T = u \in U \rightarrow \{0, 1\}. \tag{2.29}$$

As in classical binary logic, we assign a logical proposition to a set on the universe of discourse. Fuzzy propositions are assigned to fuzzy sets. Suppose that proposition $\underset{\sim}{P}$ is assigned to fuzzy set $\underset{\sim}{A}$; then the truth value of a proposition, denoted $T(\underset{\sim}{P})$, is given by

$$T(\underset{\sim}{P}) = \mu_{\underset{\sim}{A}}(x), \quad \text{where } 0 \le \mu_{\underset{\sim}{A}} \le 1. \tag{2.30}$$

Equation (2.30) indicates that the degree of truth for the proposition $\underset{\sim}{P} : x \in \underset{\sim}{A}$ is equal to the membership grade of x in the fuzzy set $\underset{\sim}{A}$.

The logical connectives of negation, disjunction, conjunction, and implication also are defined for a fuzzy logic. These connectives are given in (2.31a)–(2.31d) for two simple propositions, proposition $\underset{\sim}{P}$ defined on fuzzy set $\underset{\sim}{A}$ and proposition $\underset{\sim}{Q}$ defined on fuzzy set $\underset{\sim}{B}$:

$$\text{negation:} \quad T(\bar{\underset{\sim}{P}}) = 1 - T(\underset{\sim}{P}), \tag{2.31a}$$

$$\text{disjunction:} \quad \begin{array}{l} \underset{\sim}{P} \vee \underset{\sim}{Q} \Rightarrow x \text{ is } \underset{\sim}{A} \text{ or } \underset{\sim}{B}, \\ T(\underset{\sim}{P} \vee \underset{\sim}{Q}) = \max(T(\underset{\sim}{P}), T(\underset{\sim}{Q})), \end{array} \tag{2.31b}$$

$$\text{conjunction:} \quad \begin{array}{l} \underset{\sim}{P} \wedge \underset{\sim}{Q} \Rightarrow x \text{ is } \underset{\sim}{A} \text{ and } \underset{\sim}{B}, \\ T(\underset{\sim}{P} \wedge \underset{\sim}{Q}) = \min(T(\underset{\sim}{P}), T(\underset{\sim}{Q})), \end{array} \tag{2.31c}$$

$$\text{implication:} \quad \begin{array}{l} \underset{\sim}{P} \rightarrow \underset{\sim}{Q} \Rightarrow x \text{ is } \underset{\sim}{A}, \text{ then } x \text{ is } \underset{\sim}{B}, \\ T(\underset{\sim}{P} \rightarrow \underset{\sim}{Q}) = T(\bar{\underset{\sim}{P}} \vee \underset{\sim}{Q}) = \max(T(\bar{\underset{\sim}{P}}), T(\underset{\sim}{Q})). \end{array} \tag{2.31d}$$

As stated before in (2.22), in binary logic the implication connective can be modeled in rule-based form,

$$\underset{\sim}{P} \rightarrow \underset{\sim}{Q} : \text{IF } x \text{ is } \underset{\sim}{A}, \text{THEN } y \text{ is } \underset{\sim}{B},$$

and it is equivalent to the following fuzzy relation $\underset{\sim}{R}$ just as it is in classical logic: $\underset{\sim}{R} = (\underset{\sim}{A} \times \underset{\sim}{B}) \cup (\bar{\underset{\sim}{A}} \times Y)$ (recall (2.22)), whose membership function is expressed by the formula

$$\mu_{\underset{\sim}{R}}(x, y) = \max[(\mu_{\underset{\sim}{A}}(x) \wedge \mu_{\underset{\sim}{B}}(y)), (1 - \mu_{\underset{\sim}{A}}(x))]. \tag{2.32}$$

When the logical conditional implication is of the compound form

$$\text{IF } x \text{ is } \underset{\sim}{A}, \quad \text{THEN } y \text{ is } \underset{\sim}{B}, \quad \text{ELSE } y \text{ is } \underset{\sim}{C},$$

then the equivalent fuzzy relation $\underset{\sim}{R}$ is expressed as

$$\underset{\sim}{R} = (\underset{\sim}{A} \times \underset{\sim}{B}) \cup (\bar{\underset{\sim}{A}} \times \underset{\sim}{C})$$

in a form just as in (2.22), whose membership function is expressed by the formula

$$\mu_{\underset{\sim}{R}}(x, y) = \max[(\mu_{\underset{\sim}{A}}(x) \wedge \mu_{\underset{\sim}{B}}(y)), ((1 - \mu_{\underset{\sim}{A}}(x)) \wedge \mu_{\underset{\sim}{C}}(y))]. \tag{2.33}$$

2.3.3 Approximate reasoning

The ultimate goal of fuzzy logic is to form the theoretical foundation for reasoning about imprecise propositions; such reasoning is referred to as approximate reasoning (Zadeh (1976, 1979)). Approximate reasoning is analogous to predicate logic for reasoning with precise propositions, and hence is an extension of classical propositional calculus that deals with partial truths.

Suppose we have a rule-based format to represent fuzzy information. These rules are expressed in the following conventional antecedent-consequent forms:

- **Rule 1:** IF x is $\underset{\sim}{A}$, THEN y is $\underset{\sim}{B}$,

where $\underset{\sim}{A}$ and $\underset{\sim}{B}$ represent fuzzy propositions (sets). Now suppose we introduce a new antecedent, say, $\underset{\sim}{A}'$, and consider the following rule:

- **Rule 2:** IF x is $\underset{\sim}{A}'$, THEN y is $\underset{\sim}{B}'$.

From information derived from Rule 1, is it possible to derive the consequent in Rule 2, $\underset{\sim}{B}'$, resulting from a new input, $\underset{\sim}{A}'$? The answer is yes, and the procedure is fuzzy composition. The consequent $\underset{\sim}{B}'$ can be found from the composition operation $\underset{\sim}{B}' = \underset{\sim}{A}' \circ \underset{\sim}{R}'$, which is analogous to (2.25) for crisp sets.

2.3.4 Fuzzy systems

In the field of artificial intelligence (machine intelligence) there are various ways to represent knowledge of complex, or *fuzzy*, systems. Perhaps the most common way to represent human knowledge in these systems is to cast the knowledge into natural language expressions of the type

IF premise (antecedent),   THEN conclusion (consequent).

This expression is commonly referred to as the IF–THEN "rule-based" form. It typically expresses an inference such that if a fact (premise, hypothesis, antecedent) is known, then we can infer, or derive, another fact called a conclusion (consequent). This form of knowledge representation, characterized as "shallow knowledge," is quite appropriate in the context of linguistics because it expresses human empirical and heuristic knowledge in our own language of communication. It does not, however, capture the "deeper" forms of knowledge usually associated with intuition, structure, function, and behavior of the objects around us simply because these latter forms of knowledge are not readily reduced to linguistic phrases or representations. The rule-based system is distinguished from classical expert systems in the sense that the rules comprising a rule-based system might derive from sources other than human experts and, in this context, are distinguished from expert systems. This chapter confines itself to the broader category of fuzzy rule-based systems (of which expert systems could be seen as a subset) because of their prevalence and popularity in the literature, because of their preponderant use in engineering practice, and because the rule-based form makes use of linguistic variables as its antecedents and consequents. As illustrated earlier in this chapter, these linguistic variables can be naturally represented by fuzzy sets and logical connectives of these sets.

2.3.4.1 Aggregation of fuzzy rules

Most rule-based systems involve more than one rule. The process of obtaining the overall consequent (conclusion) from the individual consequents contributed by each rule in the rule-base is known as aggregation of rules. In determining an aggregation strategy, the following two simple extreme cases exist (Vadiee (1993)):

(i) *Conjunctive system of rules*: In the case of a system of rules that have to be jointly satisfied, the rules are connected by "and" connectives. In this case the aggregated output (consequent) y is found by the fuzzy intersection of all individual rule consequents y^i, where $i = 1, 2, \ldots, r$, as

$$y = y^1 \text{ and } y^2 \text{ and } \ldots \text{ and } y^r \qquad (2.34)$$

or

$$y = y^1 \cap y^2 \cap \cdots \cap y^r, \qquad (2.35)$$

which is defined by the membership function

$$\mu_y(y) = \min(\mu_{y^1}(y), \mu_{y^2}(y), \ldots, \mu_{y^r}(y)) \quad \text{for } y \in Y. \qquad (2.36)$$

(ii) *Disjunctive system of rules*: For the case of a disjunctive system of rules where the satisfaction of at least one rule is required, the rules are connected by the "or" connectives. In this case, the aggregated output is found by the fuzzy union of all individual rule contributions as

$$y = y^1 \text{ or } y^2 \text{ or } \ldots \text{ or } y^r$$

or

$$y = y^1 \cup y^2 \cup \cdots \cup y^r, \qquad (2.37)$$

which is defined by the membership function

$$\mu_y(y) = \max(\mu_{y^1}(y), \mu_{y^2}(y), \ldots, \mu_{y^r}(y)) \quad \text{for } y \in Y. \qquad (2.38)$$

2.3.4.2 Graphical techniques of inference

Fuzzy relations use mathematical procedures to conduct inferencing of IF–THEN rules (see (2.32) and (2.33)). These procedures can be implemented on a computer for processing speed. Sometimes, however, it is useful to be able to conduct the inference computation manually with a few rules to check computer programs or to verify the inference operations. Conducting the matrix operations illustrated in (2.32) and (2.33) for a few rule sets can quickly become quite onerous. Graphical methods have been proposed that emulate the inference process and that make manual computations involving a few simple rules straightforward (Ross (1995)). To illustrate this graphical method, we consider a simple two-rule system where each rule is comprised of two antecedents and one consequent. This is analogous to a two-input and single-output fuzzy system. The graphical procedures illustrated here can be easily extended and will hold for fuzzy rule-bases (or fuzzy systems) with any number of antecedents (inputs) and consequents (outputs). A fuzzy system with two noninteractive inputs x_1 and x_2 (antecedents) and a single output y (consequent) is described by a collection of "r" linguistic IF–THEN propositions

$$\text{IF } x_1 \text{ is } \underset{\sim}{A}_1^k \quad \text{and} \quad x_2 \text{ is } \underset{\sim}{A}_2^k, \quad \text{THEN } y^k \text{ is } \underset{\sim}{B}^k \quad \text{for } k = 1, 2, \ldots, r, \qquad (2.39)$$

where $\underset{\sim}{A}_1^k$ and $\underset{\sim}{A}_2^k$ are the fuzzy sets representing the kth antecedent pairs, and $\underset{\sim}{B}^k$ are the fuzzy sets representing the kth consequent.

In the following material, we consider a two-input, one-output system. The inputs to the system are represented by fuzzy sets, and we will illustrate a max–min inference method.

The inputs input(i) and input(j) are fuzzy variables described by fuzzy membership functions. The rule-based system is described by (2.39), so for a set of disjunctive rules, where $k = 1, 2, \ldots, r$, the aggregated output using a Mamdani implication (Mamdani and Gaines (1981)) will be given by

$$\mu_{B^k}(y) = \max_k[\min\{\max[\mu_{A_1^k}(x) \wedge \mu(x_1)], \max[\mu_{A_2^k}(x) \wedge \mu(x_2)]\}], \qquad (2.40)$$

where $\mu(x_1)$ and $\mu(x_2)$ are the membership functions for inputs $i(x_1)$ and $j(x_2)$, respectively. Equation (2.40) has a very simple graphical interpretation which is illustrated in Figure 2.11. In this figure the fuzzy inputs are represented by triangular membership functions (input(i) and input(j) in the figure). The intersection of these inputs and the stored membership functions for the antecedents (A_{11}, A_{12} for the first rule and A_{21}, A_{22} for the second rule) results in triangles. The maximum value of each of these intersection triangles results in a membership value, the minimum of which is propagated for each rule (because of the "and" connective between the antecedents of each rule; see (2.39)). Figure 2.11 shows the aggregated consequent resulting from a disjunctive set of rules (the outer envelope of the individual truncated-consequents) and a defuzzified value, Y^*, resulting from a centroidal defuzzification method (Ross (1995)). Defuzzification is the mathematical process whereby a fuzzy membership function is reduced to a single scalar quantity that best summarizes the

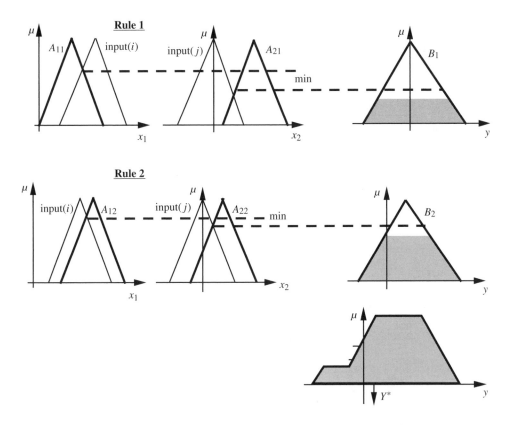

Figure 2.11. *Graphical Mamdani (max–min) implication method with fuzzy inputs.*

function, if a single scalar value is needed. Of the various methods of defuzzification, the centroidal method is intuitive since it represents the center of gravity of the area defined by the membership function.

2.3.5 An example numerical simulation

Most physical processes in the real world are nonlinear. It is our abstraction of the real world that leads us to the use of linear systems in modeling these processes. These linear systems are simple, understandable, and, in many situations, provide acceptable simulations of the actual processes. Unfortunately, only the simplest of linear processes and only a very small fraction of the nonlinear having verifiable solutions can be modeled with linear systems theory. The bulk of the physical processes that we must address are, unfortunately, too complex to reduce to algorithmic form—linear or nonlinear. Most observable processes have only a small amount of information available with which to develop an algorithmic understanding. The vast majority of information that we have on most processes tends to be nonnumeric and nonalgorithmic. Most of the information is fuzzy and linguistic in form.

If a process can be described algorithmically, we are able to describe the solution set for a given input set. If the process is not reducible to algorithmic form, perhaps the input–output features of the system are at least observable or measurable. This section deals with systems that cannot be simulated with conventional crisp or algorithmic approaches but that can be simulated using fuzzy nonlinear simulation methods because of the presence of linguistic information.

We wish to use fuzzy rule-based systems as suitable representations of simple and complex physical systems. For this purpose, a fuzzy rule-based system consists of a set of rules representing a practitioner's understanding of the system's behavior, a set of input data observed going into the system, and a set of output data coming out of the system. The input and output data can be numeric or nonnumeric observations. Figure 2.12 shows a general static physical system that could be a simple mapping from the input space to the output space or could be an industrial control system, a system identification problem, a pattern recognition process, or a decision-making process.

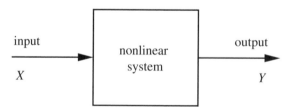

Figure 2.12. *A general static physical system with observed inputs and outputs.*

We illustrated that a fuzzy relation also can represent a logical inference. The fuzzy implication IF $\underline{A}$, THEN $\underline{B}$ is a general form of inference. There are numerous techniques for obtaining a fuzzy relation $\underline{R}$ that will represent this inference in the form of a fuzzy relational equation.

Suppose our knowledge concerning a certain nonlinear process is not algorithmic but rather is in some other more complex form. This more complex form could be data observations of measured inputs and measured outputs. Relations can be developed from this data, which are analogous to a look-up table. Alternatively, the complex form of the knowledge of a nonlinear process also could be described with some linguistic rules of the form IF $\underset{\sim}{A}$, THEN $\underset{\sim}{B}$. For example, suppose that we are monitoring a thermodynamic process involving an input heat of a certain temperature and an output variable, such as pressure. We observe that when we input heat of a "low" temperature, we see "low" pressure in the system; when we input heat of a "moderate" temperature, we see "high" pressure in the system; when we place "high" temperature heat into the thermodynamics of the system, the output pressure reaches an "extremely high" value; and so on. This process is shown in Figure 2.13, where the inputs are now *not* points in the input universe (heat) and the output universe (pressure) but *patches* of the variables in each universe. These patches represent the fuzziness in describing the variables linguistically. Obviously, the mapping (or relation) describing this relationship between heat and pressure is fuzzy. That is, patches from the input space map, or relate, to patches in the output space, and the relations R_1, R_2, and R_3 in Figure 2.13 represent the fuzziness in this mapping. In general, all of the patches, including those representing the relations, overlap because of the ambiguity in their definitions.

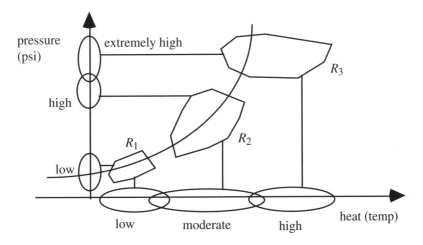

Figure 2.13. *A fuzzy nonlinear relation matching patches in the input space to patches in the output space.*

Each of the patches in the input space shown in Figure 2.13 could represent a fuzzy set, say, $\underset{\sim}{A}$, defined on the input variable, say, x; each of the patches in the output space could be represented by a fuzzy set, say, $\underset{\sim}{B}$, defined on the output variable, say, y; and each of the patches lying on the general nonlinear function path could by represented by a fuzzy relation, say, $\underset{\sim}{R}^k$, where $k = 1, 2, \ldots, r$ represents r possible linguistic relationships between input and output. Suppose we have a situation where a fuzzy input, say, x, results in a series of fuzzy outputs, say, y^k, depending on which fuzzy relation $\underset{\sim}{R}^k$ is used to determine the mapping. Each of these relationships, as listed in Table 2.2, could be described by what is called a *fuzzy relational equation* (or a *fuzzy IF–THEN rule*), where y^k is the output of the system contributed by the kth rule, and whose membership function is given by $\mu_{y^k}(y)$.

2.3.5.1   Fuzzy associative memories (FAMs)

Consider a fuzzy system with n inputs and a single output. Also assume that each input universe of discourse, i.e., $X_1, X_2, \ldots, X_n$, is partitioned into k fuzzy partitions. Based on the canonical fuzzy model given in Table 2.2 for a nonlinear system, the total number of

Table 2.2. *System of fuzzy relational equations (IF–THEN rules).*

$\underline{R}^1$	: $y^1 = x \circ \underline{R}^1$
$\underline{R}^2$	: $y^2 = x \circ \underline{R}^2$
...	...
$\underline{R}^r$	: $y^r = x \circ \underline{R}^r$

possible rules governing this system is given by

$$L = k^n, \tag{2.41a}$$

$$L = (k + 1)^n, \tag{2.41b}$$

where L is the maximum possible number of canonical rules. Equation (2.41b) is to be used if the partition "anything" is to be used; otherwise, (2.41a) determines the number of possible rules. The actual number of rules r necessary to describe a fuzzy system is much less than L, i.e., $r \ll L$. This is due to the interpolative reasoning capability of the fuzzy model and the fact that the fuzzy membership functions for each partition overlap. If each of the n noninteractive inputs are partitioned into a different number of fuzzy partitions, say, X_1 is partitioned into k_1 partitions and X_2 is partitioned into k_2 partitions, and so forth, then the maximum number of rules is given by

$$L = k_1 k_2 k_3 \cdots k_n. \tag{2.42}$$

For a small number of inputs, e.g., $n = 1$, $n = 2$, or $n = 3$, there exists a compact form of representing a fuzzy rule-based system. This form is illustrated for $n = 2$ in Figure 2.14, where there are seven partitions for input A ($A1$ to $A7$), five partitions for input B ($B1$ to $B5$), and four partitions for the output variable, C ($C1$ to $C4$). This compact graphical form is called a fuzzy associative memory (FAM) table (Kosko (1992)). As can be seen from the FAM table, the rule-based system actually represents a general nonlinear mapping from the input space of the fuzzy system to the output space of the fuzzy system. In this mapping, the patches of the input space are related to the patches in the output space. Each rule or, equivalently, each fuzzy relation from input to output, actually represents a fuzzy point of data that characterizes the nonlinear mapping from input to output.

input B \ input A	A1	A2	A3	A4	A5	A6	A7
B1	C1		C4	C4		C3	C3
B2		C1				C2	
B3	C4		C1			C2	C2
B4	C3	C3		C1		C1	C2
B5	C3		C4	C4	C1		C3

Figure 2.14. *FAM table for a two-input, single-output fuzzy rule-based system.*

In the FAM table shown in Figure 2.14, we see that the maximum number of rules for this situation using (2.42) is $L = k_1 k_2 = 7(5) = 35$; but as seen in the figure, the actual number of rules is only $r = 21$ (i.e., the blank spaces in the figure do not contain rules).

2.3.5.2 Example simulation using fuzzy rule-based systems

For the nonlinear function $y = 10 \sin x_1$, we will develop a fuzzy rule-based system using four simple fuzzy rules to approximate the output y. In general, we would not know the form of the nonlinearity, but to illustrate a typical simulation we use a known function. The universe of discourse for the input variable x_1 will be the interval $[-180°, 180°]$ in degrees, and the universe of discourse for the output variable y is the interval $[-10, 10]$.

First, we will partition the input space x_1 into five simple partitions on the interval $[-180°, 180°]$, and we will partition the output space y on the interval $[-10, 10]$ into three membership functions, as shown in Figures 2.15(a) and 2.15(b), respectively.

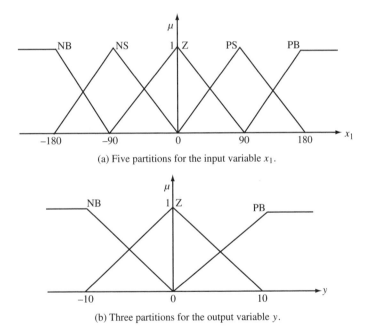

(a) Five partitions for the input variable x_1.

(b) Three partitions for the output variable y.

Figure 2.15. *Fuzzy membership functions for the* (a) *input and* (b) *output spaces.*

In Figures 2.15(a) and 2.15(b), the acronyms NB, NS, Z, PS, and PB refer to the linguistic variables negative big, negative small, zero, positive small, and positive big, respectively.

Second, we develop four simple rules, listed in Table 2.3, that we think emulate the system dynamics (in this case the system is the nonlinear equation $y = 10 \sin x_1$, and we are observing the harmonics of this system) and that use the linguistic variables shown in Figure 2.15. The FAM table for these rules is given in Table 2.4.

The FAM table of Table 2.4 is one-dimensional because there is only one input variable, x_1. As seen in Table 2.4, all rules listed in Table 2.3 are accommodated. The four rules expressed above are not all expressed in canonical form (some have disjunctive antecedents), but if they were transformed into canonical form, they would represent the five rules provided in the FAM table in Table 2.4.

Table 2.3. *Four simple rules for $y = 10 \sin x_1$.*

1	IF x_1 is Z or PB, THEN y is Z.
2	IF x_1 is PS, THEN y is PB.
3	IF x_1 is Z or NB, THEN y is Z.
4	IF x_1 is NS, THEN y is NB.

Table 2.4. *FAM for the four simple rules.*

x_1	NB	NS	Z	PS	PB
y	Z	NB	Z	PB	Z

In developing an approximate solution for the output y, we select a few input points and conduct a graphical inference method similar to that illustrated earlier. We will use the centroid method for defuzzification. Let us choose four crisp singletons as the input:

$$x_1 = \{-135°, -45°, 45°, 135°\}.$$

For input $x_1 = -135°$, rules (3) and (4) are fired, as shown in Figures 2.16(c) and 2.16(d). For input $x_1 = -45°$, rules (1), (3), and (4) are fired. Figures 2.16(a) and 2.16(b) show the graphical inference for input $x_1 = -45°$, which fires rule (1), and for $x_1 = 45°$, which fires rule (2), respectively.

For input $x_1 = -45°$, rules (3) and (4) also are fired, and we get results similar to those shown in Figures 2.16(c) and 2.16(d) after defuzzification:

$$\text{rule (3)} : y = 0,$$
$$\text{rule (4)} : y = -7.$$

For $x_1 = 45°$, rules (1), (2), and (3) are fired (see Figure 2.16(b) for rule (2)), and we get the following results for rules (1) and (3) after defuzzification:

$$\text{rule (1)} : y = 0,$$
$$\text{rule (3)} : y = 0.$$

For $x_1 = 135°$, rules (1) and (2) are fired and, after defuzzification, we get results that are similar to those shown in Figures 2.16(a) and 2.16(b):

$$\text{rule (1)} : y = 0,$$
$$\text{rule (2)} : y = 7.$$

When we combine the results, we get an aggregated result, summarized in Table 2.5 and shown graphically in Figure 2.17. The y values in each column of Table 2.5 are the defuzzified results from various rules firing for each of the inputs x_i. When we aggregate the rules using the union operator (disjunctive rules), this has the effect of taking the maximum value for y in each of the columns in Table 2.5. The plot in Figure 2.17 represents the maximum y for each of the x_i, and it represents a fairly accurate portrayal of the true solution. More rules would result in a closer *fit* to the true sine curve, but in this case we *know* the nonlinearity.

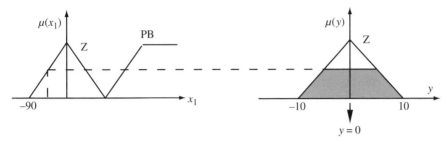

(a) Input $x_1 = -45°$ fires rule (1).

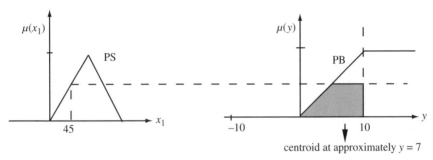

(b) Input $x_1 = 45°$ fires rule (2).

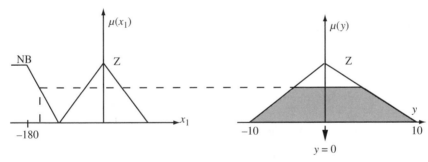

(c) Input $x_1 = -135°$ fires rule (3).

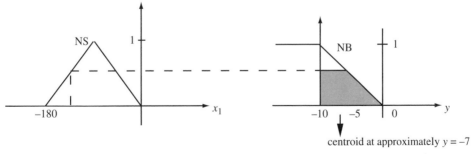

(d) Input $x_1 = -135°$ fires rule (4).

Figure 2.16. *Graphical inference method showing membership propagation and defuzzification.*

Table 2.5. *Defuzzified results for simulation of* $y = 10 \sin x_1$.

x_1	$-135°$	$-45°$	$45°$	$135°$
y	0	0	0	0
	-7	0	0	7
		-7	7	

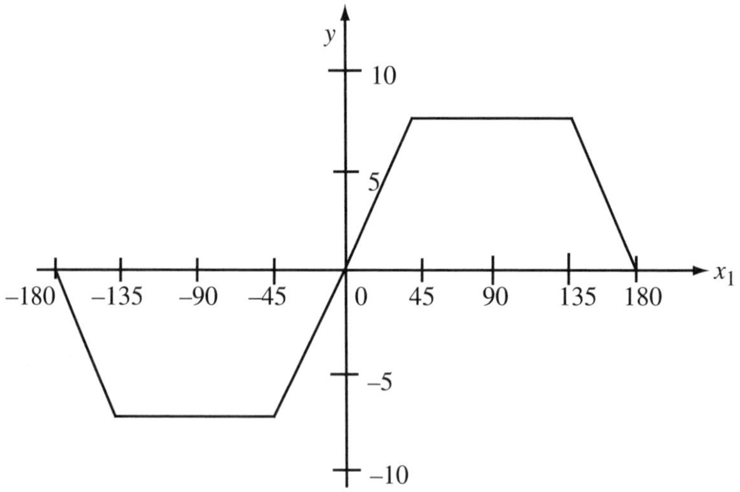

Figure 2.17. *Simulation of nonlinear system* $y = 10 \sin x_1$ *using four-rule fuzzy rule-base.*

2.3.6   Summary

This chapter has presented the basics of fuzzy sets, membership, fuzzy relations, fuzzy logic, fuzzy rule-based systems, and fuzzy simulation. The important area of *fuzzy control* is introduced in Chapter 8 in the applications section. The utility of fuzzy logic transcends all engineering disciplines and all scientific fields as well. Perhaps the greatest current interest in fuzzy logic in the commercial marketplace is the use of fuzzy logic in the development of new "smart" consumer products. Some of these products employ the elements of fuzzy control in their designs.

Beyond the material provided in this chapter lies a new era in fuzzy logic, where this new technology will be combined with the latest digital forms for manifesting automated decision-making. For example, new generations of fuzzy logic controllers are based on the integration of conventional digital and fuzzy controllers (Sugeno (1985)). Digital fuzzy clustering techniques can be used to extract the linguistic IF–THEN rules from numerical data. In general, the trend is toward the compilation and fusion of different forms of knowledge representation for the best possible identification and control of ill-defined complex systems. The two paradigms of artificial neural networks (ANN) and fuzzy logic (FL) try to understand a real world system by starting from the very fundamental sources of knowledge, i.e., patient and careful observations, digital measurements, experience, and intuitive reasoning and judgments rather than starting from a preconceived theory or mathematical model. Advanced cognitive systems characterized by learning (using ANN) and by reasoning (using FL) will use adaptation capabilities to learn, then tune, the membership functions' vertices or supports, or to add or delete rules to optimize the performance and compensate for the effects of any internal or external perturbations to the systems. Finally, principles of genetic algorithms (GA) will be used to find the best string representing an optimal class of digital input or output membership functions in digital form. All these new paradigms (ANN, FL, and GA) fall under the rubric now known as *soft computing*.

References

J. BEZDEK (1993), Editorial: Fuzzy models—what are they, and why? *IEEE Trans. Fuzzy Systems*, 1, pp. 1–5.

B. KOSKO (1992), *Neural Networks and Fuzzy Systems*, Prentice–Hall, Englewood Cliffs, NJ.

E. H. MAMDANI AND R. R. GAINES, EDS. (1981), *Fuzzy Reasoning and Its Applications*, Academic Press, London.

T. ROSS (1995), *Fuzzy Logic with Engineering Applications*, McGraw–Hill, New York.

M. SUGENO, ED. (1985), *Industrial Application of Fuzzy Control*, North-Holland, New York.

N. VADIEE (1993), Fuzzy rule-based expert systems I, in *Fuzzy Logic and Control: Software and Hardware Applications*, M. Jamshidi, N. Vadiee, and T. Ross, eds., Prentice–Hall, Englewood Cliffs, NJ, pp. 51–85.

L. ZADEH (1965), Fuzzy sets, *Inform. and Control*, 8, pp. 338–353.

L. ZADEH (1976), The concept of a linguistic variable and its application to approximate reasoning: Part 3, *Inform. Sci.*, 9, pp. 43–80.

L. ZADEH (1979), A theory of approximate reasoning, in *Machine Intelligence*, J. Hayes, D. Michie, and L. Mikulich, eds., Halstead Press, New York, pp. 149–194.

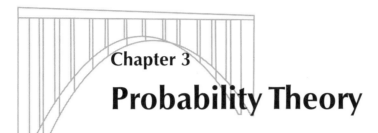

Chapter 3

Probability Theory

Nozer D. Singpurwalla, Jane M. Booker, and Thomas R. Bement

3.1   The calculus of probability

It is convenient to use the following example to describe what is meant by the rules for governing probability. The example of determining the reliability of a system is reasonable because reliability is the probability that the system will perform as required, and because this example also appears in Chapter 11. Probability theory has its basis in (crisp) set theory, but it can be defined in terms of rules or axioms called the *calculus* of probability (Bement et al. (1999)).

The probability of an event, say, X, is a number denoted by $P(X)$, and it is required to satisfy certain rules or axioms. The probability of that same event in light of historical information H at a time τ is a number denoted by $P^\tau(X; H)$.

The event X can pertain to the performance of a certain function surviving to a specified mission time[1] or producing a specified level of output. Then $P^\tau(X; H)$ is known as the *reliability* of the item or unit that is required to function or to survive. Thus reliability is, de facto, the probability of a certain type of an event. (When the item in question is a human subject, the term *survival analysis*, rather than reliability, is commonly used.)

The calculus of probability consists of the following three rules of *convexity*, *addition*, and *multiplication*:

(i) $0 \le P^\tau(X; H) \le 1$ for any event X;

(ii) $P^\tau(X_1 \text{ or } X_2; H) = P^\tau(X_1; H) + P^\tau(X_2; H)$ for any two events X_1 and X_2 that are mutually exclusive—that is, they cannot simultaneously occur; and

(iii) $P^\tau(X_1 \text{ and } X_2; H) = P^\tau(X_1|X_2; H) \cdot P^\tau(X_2; H)$, where $P^\tau(X_1|X_2; H)$ is a quantification via probability of the uncertainty about the occurrence of event X_1, assuming that event X_2 has occurred.

---

[1]The concept of mission time need not be measured in units of time; it could be measured in other metrics such as miles traveled, rounds fired, or cycles completed.

The quantity $P^\tau(X_1|X_2; H)$ is known as the *conditional probability* of X_1, given X_2. It is important to note that conditional probabilities are subjunctive. This is the disposition of X_2 at time τ, for were it to be known, it would become a part of the history H at time τ. The vertical line between X_1 and X_2 represents a supposition or assumption about the occurrence of X_2. Finally, $P^\tau(X_1 \text{ and } X_2; H)$ also can be written as $P^\tau(X_2|X_1; H) \cdot P^\tau(X_1; H)$ because at time τ, both X_1 and X_2 are uncertain events. One can contemplate the uncertainty of X_1 assuming that X_2 is true or, vice versa, contemplating X_2 by assuming that X_1 is true.

These basic rules, along with conditional probability, are rearranged and presented as the 16 axioms of probability in Chapter 5.

The calculus of probability does not interpret probability; that is, it does not tell us what probability means. It is not concerned with issues such as the nature of uncertainty, whose uncertainty, whose history, how large H should be, and, most important, how to determine $P^\tau(X; H)$ and how to make this number operational. The calculus simply provides a set of rules by which the uncertainties about two (or more) events combine or *cohere*. Any set of rules for combining uncertainties that are in violation of the rules given above are said to be *incoherent* with respect to the calculus of probability.

Most researchers agree that the calculus of probability starts with its axioms and do not dwell on the matter of how to assign initial probabilities. Then the matter becomes how to deduce probabilities of events that are compositions of the basic events whose initial probabilities are assumed known. However, there do exist systems for combining and specifying various types of uncertainties that are different from probability. Examples of these are Jeffrey's rule of combination (Jeffrey (1983)), possibility theory and fuzzy logic (Zadeh (1979)), upper and lower probabilities (Smith (1961)), and belief functions (Dempster (1968)).

Questions arise as to why one must subscribe to the calculus of probability, and what is the justification for this calculus. The answers encompass a vast amount of literature, with contributions from gamblers, philosophers, mathematicians, decision theorists, behavioral scientists, and experts in artificial intelligence and knowledge acquisition. The book by Howson and Urbach (1989) provides an overview of some of these. Some brief arguments to justify the probability calculus include the following:

Gambling: Since the time of Cardano in the 1500s, it has been known that, in games of cards, coins, and dice, an avoidance of the rules of probability results in a "sure-loss," also known as a *Dutch book*. That is, heads you lose, tails I win.

Scoring rules: De Finetti (cf. Lindley (1982)) used a *scoring rule* argument to justify the calculus of probability. With this argument, an individual assessing an uncertainty is asked to declare a number that best describes that uncertainty. When the uncertainty reveals itself or is resolved, the individual is rewarded or penalized (i.e., scored) according to how close the declared number was to actuality. De Finetti's argument is that under some very general conditions, an individual faced with a collection of uncertainties must use the calculus of probability to maximize an overall score. The above claim is true for a large class of scoring schemes.

Horse racing: Also known as "horse lotteries," horse racing uses certain numbers called *betting coefficients*. The betting coefficients are the odds for or against a particular horse or horses in a race. It can be shown that to maximize one's winnings the betting coefficients must obey the calculus of probability (Howson and Urbach (1989)).

Behavioristic axioms: The three arguments above involve the issues of gambling or scoring, both of which could be morally or ethically objectionable to some individuals. As an alternative, Ramsey (1934) and Savage (1972) have proposed a system of *behavioristic axioms* to justify the calculus of probability in terms of principles relating to *coherence* and consistency. The Ramsey–Savage argument is mathematical; an excellent exposition of it is given by De Groot (1970). Both Ramsey and Savage expand the use and definition of probability into the behaviorist realm, which represents reality better than games of chance and scoring rules. However, there are two disadvantages to this argument. The first is that the intuitive and natural elements of gambling and scoring are lost, and axiomatic arguments tend to be abstract and therefore less appealing. The second (and perhaps more serious) disadvantage is that the behavioristic axioms of Ramsey and Savage are in reality violated by (most) individuals (cf. Kahneman, Slovic, and Tversky (1986)). A counterpoint to the above disadvantages is that the behavioristic axioms prescribe normative behavior, which tells us how we *should* act, not how we *do* act.

3.2   Popular views of probability

3.2.1   The interpretation of probability

The interpretation, or the meaning, of probability and approaches for assigning initial probabilities has been the subject of much discussion and debate. According to Good (1965), there are approximately 11 ways of interpreting probability, with the following four being most prominent:

- the classical theory,

- the a priori or logical theory,

- the relative frequency theory,

- the personalistic or subjective theory.

Each of these theories subscribes to a particular interpretation of probability, but the calculus in section 3.1 is common to all four. It is important to note that, for all intents and purposes, the calculus of probability is devoid of interpretation, so the meaning of probability is not relevant to the calculus. However, the assignment of initial probabilities (needed to make the calculus operational) depends on one's interpretation of probability. Thus to a user of probability, such as a physicist, an engineer, or a statistician, the interpretation of probability does greatly matter. In what follows, we shall give an overview of the key features of the four theories mentioned above, focusing more on the latter two. For more details consult Fine (1973), Good (1965), Maistrov (1974), Hacking (1974), or Gigerenzer et al. (1989).

3.2.2   The classical theory of probability

The founders of this theory, Cardano, Pascal, Fermat, Huygens, Bernoulli, DeMoiure, Bayes, Laplace, and Poisson, were "determinists" influenced by Newton. They believed there was no such thing as chance. Nature, what Laplace called a "genie," knows it all. Probability was simply a measure of an individual's partial knowledge.

A formal definition of the classical theory of probability is due to Laplace. Loosely speaking, it is the ratio of favorable cases to the number of *equipossible* or equally likely cases. The cases are equipossible if we have no reason to expect the occurrence of one over the other. An equipossible case is also known as

- the principle of indifference,

- the principle of insufficient reason,

- the Bayes postulate.

The classical theory of probability has some merit in games of chance such as coin tossing, dice throwing, or card shuffling. However, there has been some criticism of the theory. First, the principle of indifference seems to be circular, i.e., equipossible implies *equiprobable*, and vice versa. Second, there is a difficulty regarding the dividing up of alternatives. For example, what is the probability of throwing a 5 on the toss of a die? The answer is $\frac{1}{6}$, if the alternatives considered are 1, 2, . . . , 6, but it is $\frac{1}{2}$ if the alternatives considered are a 5 or not a 5. Third, this theory has limited applications. For example, what must we do if the die is loaded? More important, under this theory, how must we interpret a number such as 10^{-6} for the failure rate of a system? Must we think in terms of 10^6 equally likely cases? Clearly, this interpretation of probability poses conceptual difficulties in the context of rare and unobservable events.

Finally, it is important to note that the classical theory of probability involves a personal judgment about the equally likely nature of events. In reality, coins and dice are not perfect and hence not equally likely. Nonetheless, the theory continues to be used, particularly for the teaching of probability.

3.2.3 The a priori theory of probability

The original insights about the a priori theory of probability are due to Keynes, the famous economist. However, there is a version of this theory due to Carnap, who was trained as a physicist but who also worked in logic, syntax, semantics, and formal languages. Others who have been attracted to this theory are Jeffreys (1961), Koopman, Kemney, Good, and Ramsey.

It is difficult to summarize in words the essence of this theory because it involves the notions of logic and syntax. The basic tenet of this theory, however, boils down to the theory that probability is a logically derived entity. When one does not agree with its conclusion, one is wrong because of a violation of the logic. This theory, though much discussed, is not in use because it is difficult to make it operational. Thus we shall not dwell on it further.

3.2.4 The relative frequency theory

The origins of the relative frequency theory of probability can be traced back to Aristotle, although Venn officially announced the idea in 1866. The mathematical development of this theory has been attributed to Von Mises (1957), and its philosophical discourse is due to Reichenbach (1949).

The key ideas of this theory are the following:

- Probability is a measure of an empirical, objective, and physical fact of the external world, independent of human attitudes, opinions, models, and simulations. To von Mises, it was a part of a descriptive physical science; to Reichenbach it was a part of the theoretical structure of physics.

- Probability is never relative to evidence or opinion. Like mass or volume, it is determined by observations on the nature of the real world.

- All probabilities can only be known a posteriori, i.e., only known through observation.

3.2.4.1 The definition of probability: Relative frequency theory

In the relative frequency theory, probability is a property of a *collective*, i.e., scenarios involving events that repeat again and again. Thus it excludes from consideration one-of-a-kind and individual events because such events do not possess a repetitive feature. Games of chance, like coin tossing, and social mass phenomena (like actuarial and insurance problems) are considered collectives.

A collective is a long sequence of observations for which there is sufficient reason to believe that the relative frequency of an observed attribute will tend to a limit if the observations are *indefinitely* continued. This limit is called the probability of the *attribute* within the collective.

The essential requirement in the relative frequency theory is the existence of a collective. In addition, one must establish the existence of limits. Finally, one can determine the probability of encountering a certain attribute in the collective.

3.2.4.2 Source of initial probabilities: Relative frequency theory

To von Mises, the role of probability theory is to derive probabilities from the old (initial probabilities[2]) using the calculus of probability. Both von Mises and Reichenbach agreed that initial probabilities are to be obtained as relative frequencies.

To von Mises, the equally likely values in dice games were a consequence of historical observations based on the word-of-mouth experience of generations. In actuarial applications, where a sizeable amount of data exist, von Mises requires that a collective be identified, that a stable value of a relative frequency be identified over groups and over time, and that the stable value be used as the initial probability. Thus to summarize, the heart of the relative frequency theory of probability lies in the notion of *repetitive events* (actual or conceived), and this is where both the good and bad features of this theory can be fully appreciated.

3.2.4.3 Virtues and criticisms of the relative frequency theory

The most practical virtue of the relative frequency theory of probability is that it applies in cases where the indifference principle fails to hold (as in the situation where the coins and dice are loaded). Its psychological virtue is that it claims to be objective and therefore scientific. Physical scientists are therefore attracted to it because probability is considered to be located in the objects themselves, as a property such as mass or volume, and not in our attitudes. Also, like mass and volume, probability can be discovered using the key tools of science, namely, experimentation, observation, and confirmation by experimental replication.

Criticisms of the theory stem from the fact that, to invoke it, the following need to occur:

- introduction of a random collective;

- definition of a probability that is a random collective;

- specification of a probability that is a property of the collective without being a member of the collective.

Collectives are difficult to construct in real life. For example, tossing a coin an infinite number of times raises the question of how similar the tosses should be. If they are identical, the same outcome must always be observed. If they are dissimilar, how much dissimilarity

---

[2]The specification of initial probabilities is the job of a statistician.

is allowed (if this can be assessed at all) before the tosses can no longer be viewed as a collective? Finally, relative frequency probability is never known, can never be known (limits of sequences is an abstract mathematical notion), and its value can never be confirmed or disputed.

To appreciate implementation of the relative frequency definition of probability, take the example of how to interpret a number such as 10^{-6} (for the failure rate). We must (i) first conceptualize a collective (such as an infinite number of almost identical lunar probes), (ii) focus on an attribute of this collective (say, loss of navigation control), and (iii) be prepared to accept the notion that probability is a property of this collective with respect to the attribute and not to any particular member (a probe) of this collective. The number 10^{-6} reflects the feature of any individual encountering this attribute in the collective. Thus 10^{-6} is really a measure of encounter. This number can be neither verified nor proved nor disproved. It exists only as a mathematical limit.

Whereas collectives can be conceptualized with mass social phenomena (like actuarial tables, IQs of individuals, etc.) and in physics (such as in the movement of gas particles), this conceptualization is often difficult in many other scenarios, such as rare or one-of-a-kind events. Indeed, it was a sociologist, Quetelet, who introduced the idea of a collective. This notion was first embraced by physicists (who may have influenced von Mises) but was then rejected by individuals like Bohr and Schrodinger in light of Heisenberg's "principle of uncertainty," which defined uncertainty and probability without the collective concept and in a way closer to the subjectivist view described next.

Clearly, under the relative frequency view of probability, τ and H play no role so that $P(X; H) = P(X)$. Similarly, expert testimonies, corporate memory, mathematical models, and scientific information do not matter; only hard data on actual events go into assessing the initial probabilities.

3.2.5 The personalistic or subjective theory

The personalistic or subjective theory was first proposed by Ramsey (1934), though Borel may have alluded to it in 1924. The theory was more fully developed by De Finetti (1937, 1974) and by Savage (1972). The key idea of this theory is that there is no such a thing as an objective probability. Probability is a degree of belief of a given person at a given time. The degree of belief must be measured in some sense, and a person's degree of belief must conform to others in a certain way. The person in question is an idealized one, namely, one who behaves normatively.

The intensity of belief is difficult to quantify; thus we must look at some property related to it. Ramsey and De Finetti both favored the behavioristic approach, where the degree of belief is expressed via a willingness to bet. Thus the probability of an event is the amount, say, p, that you are willing to bet on a two-sided bet in exchange for \$1 should the event occur. By a two-sided bet, we mean staking $(1 - p)$ in exchange for \$1 should the event not occur.

The feature of *coherence* that is a part of this theory is the normative feature. It ensures that the degrees of belief do not conflict and the Dutch book is avoided. This is achieved by adhering to the calculus of probability.

Because probability in the personalistic theory is one person's opinion, there is no such thing as an unknown probability, or a correct probability, or an objective probability. A person's probabilities may be elicited by invoking the principle of indifference, or by a system of carefully conducted comparative wagers, or simply by asking. In this theory, any factor that an individual chooses to consider is relevant, and any coherent value is as good as another.

3.2.5.1   Key features of the subjective theory

The subjective theory of probability permits the probability of a simple unique event or the probability of repetitive events. Because there is no notion of an absolute probability, the theory gives no guidance on how to obtain initial probabilities. Personal probabilists claim that objectivity in statistics is a fallacy, because model choice, the judgment of indifference (i.e., the notion of equipossibility), the choice of p-values, significance levels, Type I and Type II errors, etc. are all subjective. The phrase "we know nothing before observing" motivates the relative frequency theory of probability and is viewed by personalistic probabilists as impossible to define.

Finally, under the personalistic theory the number 10^{-6} for the failure rate has an unambiguous interpretation. This means that, to the individual declaring such a number, and based on all of H (to include expert testimony, corporate memory, mathematical modeling, simulation, and hard data, if available) at time τ, the person is indifferent to the following two bets:

- Stake 10^{-6} in exchange for $1 if the event occurs.

- Stake $(1 - 10^{-6})$ in exchange for $1 if the event does not occur.

3.2.5.2   Arguments against the subjective theory

According to some researchers, the hallmark of science is consistency, and scientific inference is based on hard (test or observational) data. Personalistic probabilities are not consistent with this view. Furthermore, declared probabilities may not reflect true belief. Also, it is literally impossible to ensure coherence in real situations that tend to be complex, unlike games of chance. Another argument against this theory stems from the theory that some individuals do not like to bet, especially when considering how the bet is selected, and thus may be reluctant to declare their probabilities in such terms. Also, the theory has no provision for ensuring that individuals with identical background information will declare identical probabilities and that, given an individual's action, it is difficult to separate the individual's probabilities from his/her utilities.

Perhaps the most important argument against the theory of personalistic or subjective probability is that experiments by psychologists have shown that individually declared probabilities are not coherent; i.e., they do not act according to the dictates of the calculus of probability. A counterargument to the above criticism is that the theory of personalistic probability is normative; it prescribes how we *should* act, not how we *do* act.

3.2.6   Choosing an interpretation of probability

Claims of "objectivity" seem to be the persuading argument in choosing the relative frequency interpretation of probability. This position also has been reinforced by the peer review processes of many applied scientific journals. Traditional reliability has been used, for example, in testing the reliability of medical implant devices, aging facilities (power plants), software, the Space Shuttle, the Hubble space telescope, overcrowded transportation systems, and waste transport systems. However, use of the alternatives is motivated by the need to minimize the amount and the time of testing, to make decisions in advance of testing (especially in the automobile industry), to make performance assessments for one-of-a-kind units (such as in aerospace applications), and to incorporate engineering and scientific knowledge into decisions (via simulation and science-based models). To meet such needs, there has been a gradual shift towards the personalistic view (cf. Martz and Waller (1982)).

The impetus for this shift began slowly with the nuclear reactor industry (see, for example, United States Nuclear Regulatory Commission (1975)) and continues with economics, where pressure is being put on defense and environmental protection spending by the U.S. government. It continues because issues surrounding decision making are surfacing in complex and dynamic environments, where obtaining hard data is expensive or prohibitive.

From a philosophical standpoint, the personalistic interpretation of probability does not lead to the logical inconsistencies and difficulties of communication often mentioned by its critics. Furthermore, the personalistic view

- permits statements of uncertainty about one-of-a-kind items and unobservable events;

- allows incorporation of information from all sources deemed appropriate;

- does not demand the availability of a large amount of hard data;

- permits the incorporation of all knowledge available at any given time with the ability to update probabilities (and hence reliabilities) as new knowledge becomes available.

A prime example in which the formal use of all knowledge might have presented a different decision is the *Challenger* Space Shuttle tragedy. Instead of relying completely on the solid rocket boosters' hard data (and a relative frequency view), using a personalistic approach may have revealed the potential problems that led to the disaster.

From a pragmatic point of view, the expansion of our computational capabilities over the last few years has made knowledge and information available in a variety of forms, both qualitative and quantitative. Large-scale simulations of complex, physical systems (such as transportation simulation (Beckman (1997))) are now available, providing gigabytes of information which must be analyzed and condensed for decision making. Taking advantage of all available information, including hard data, is what further motivates our point of view in favor of subjective probability.

3.2.7   The use of expert testimonies in personalistic/subjective probability

Once we adopt a personalistic interpretation of probability, the calculus of probability facilitates a use of informed testimonies, which can be based on judgments, experience, simulations, or mathematical models. We often need to involve expert testimonies in our assessments because hard data on the event of interest, such as failure, may be unavailable or even impossible to obtain. Furthermore, it is comforting to engineers when their technical knowledge and previous experiences can be formally incorporated into the decision process. However, this expertise must be properly elicited and analyzed (cf. Meyer and Booker (2001)). Informed testimonies do not obviate the role of hard data when available. Rather, the personalistic view fuses the import of informed knowledge and hard data, the latter enhancing the former, via the calculus of probability and its extensions; see Lindley and Singpurwalla (1986) and Singpurwalla (1988) for an illustration.

We have stated before that, in many important situations, the computer revolution will provide us with an abundant amount of auxiliary information. However, because of the economics of testing and other political considerations, few, if any, direct hard data will be available. A paradigm for quantifying uncertainty that can function with the above constraints will find a natural place in the eyes of the user. The subjectivist view of probability can provide such a paradigm for quantification of uncertainty and information/data integration for decision making.

The personalistic or subjective probability permits the use of all forms of data, knowledge, and information. Therefore, its usefulness in applications where the required relative frequency data are absent or sparse becomes clear. This view of probability also includes Bayes' theorem (see Chapter 4) and of all the views of probability comes the closest to the interpretation used in fuzzy logic. We therefore feel that this point of view is the most appropriate for addressing the complex decision problems of the 21st century, and we adopt it as our view of probability in the case studies of Part II.

3.3 Concepts for probability theory

3.3.1 Concepts of a random variable and sample space

Having chosen the calculus of probability with the subjective or personalistic view, we find that some basic concepts are necessary for its implementation. The first of these is the notion of *stochastic* behavior. An activity is said to be stochastic if all repetitions or trials of a defined experiment, observation, or computation may not yield identical results. The reasons for these nonidentical results are varied, including simple random chance.

If identical repetitions of an activity produce the exact same result time after time, that activity is said to be *deterministic*. For example, if one inputs the same values for mass (m) and the speed of light (c), then the energy (E) from Einstein's equation is always the same value (deterministic) using the computation $E = mc^2$. However, if one is measuring the masses of objects, these will vary (stochastic) even if the same item is being measured under the same conditions. The variation is due to various sources such as errors in recording, differences in the measuring process, or simply random chance. It should be noted that the speed of light itself is not necessarily a deterministic quantity because of variation due to the medium in which it travels and scientific disputes over the exact value of the quantity. Therefore, if both m and c in the equation have variability due to their stochastic nature, the resultant product E is also a stochastic quantity. We refer to this stochastic nature of quantities through the concept of random variables. In this case, there are random variables for the quantities m, c, and E.

Mathematically, a *random variable* is any real-valued function defined on the sample space of an experiment or activity. A sample space is the set of points in a continuum or a set of discrete values that characterize the possible outcomes of the experiment, trial, computation, or activity. Random variables are denoted by capital letters, e.g., X, which refers to the function. The specific values for the random variable are represented by lowercase letters, e.g., x.

Random variables are defined as discrete or continuous depending on the nature of the possible values they can take on. Random variables can be discrete if the values they can take on are discrete values (such as the faces of a die) whose sample space is the crisp set of discrete values $\{1, 2, 3, 4, 5, 6\}$. A continuous random variable would be the mass of a certain everyday object that we encounter, such as a book. Masses of books are on a continuous scale beginning with zero up to, potentially, infinity (or some defined limit such as the total mass of the universe). However, we rarely encounter masses of books in everyday life as large as kilotons. Likewise, they rarely are less than several grams. These extreme values of books with masses on the order of kilotons or less than a few grams have very low chances of existing. Nonetheless, those values of X are possible but have low probabilities associated with them. Specifying the probabilities for the various values of masses in the everyday sample space leads to the concept of another function—the probability distribution.

3.3.2 Probability distribution functions

Another kind of function specifies the probability or chance associated with each possible value in the sample space of a discrete random variable or with possible ranges of values in the sample space of a continuous random variable. This function is a *probability distribution function* (a PDF or CDF); "PDF" is referred to as *probability density function* in Chapters 1, 2, 4, 7–9, and 14. What defines a PDF is the property that the total summation (for discrete values) or integration (for continuous values) over all the probabilities in the sample space equals 1.0. A CDF is the cumulative or integral of the PDF.

In the example of the die, the PDF would be a table or plot of the values in the sample space (1–6) versus the probability or chance of realizing those values, denoted by p. In this case, each probability is equally likely at $\frac{1}{6}$. The summation of all probabilities is equal to 1.0:

$$\sum_{i=1}^{6} P(X = x_i) = 1.0.$$

The equally likely structure of the probabilities defines a discrete uniform distribution. Other examples of discrete PDFs include the binomial distribution, the Poisson distribution, and the hypergeometric distribution.

Continuous PDFs, $f(x)$, map each value of X into a probability density value. The probability is calculated by integrating $f(x)$ over a range of values in the sample space. There is zero probability associated with each individual value of X in the sample space. Here the integral over the entire sample space (which may include limits of $-\infty$ and $+\infty$) must equal 1.0:

$$\int_{-\infty}^{\infty} f(x)dx = 1.0.$$

Commonly used continuous PDFs include the normal (or Gaussian), beta, gamma, lognormal, and exponential distributions.

The cumulative (probability) distribution function (CDF) is denoted by $F(x_t)$, where

$$F(x_t) = P(X \leq x_t) = \alpha$$

for all x in the sample space and all probabilities (α) ranging from 0 to 1. $F(x)$ is increasing in x and has the value of 1.0 at the maximum value of x (which may be ∞).

For discrete random variables,

$$F(x_t) = P(X \leq x_t) = \sum_{i=1}^{t} P(X = x_i),$$

and for continuous random variables, the summation is replaced by an integral:

$$F(x_t) = \int_{-\infty}^{x_t} f(x)dx$$

(Hastings and Peacock (1975)).

Distribution functions (PDFs and CDFs) can be formulated empirically from data without belonging to a particular form, such as the Gaussian, a.k.a. the normal:

$$f(x) = \frac{\exp\left\{\frac{-(x-\mu)^2}{(2\sigma^2)}\right\}}{\sigma\sqrt{2\pi}}, \quad -\infty \le x \le \infty.$$

Common forms include the beta, gamma, lognormal, uniform, exponential, and Weibull distributions. Each has a set of parameters that determine its unique shape and scaling appearance.

Regardless of form, probability distributions have characteristics based on moments, which can be calculated from their formulas or estimated from data. The most common of these is the first moment, the mean, often referred to as the average. For continuous distributions, the mean μ is

$$\mu = \int_{-\infty}^{\infty} x \cdot f(x)dx.$$

The next most common distribution characteristic is the variance σ^2, which is based on the second moment. For continuous distributions, the variance is

$$\sigma^2 = \int_{-\infty}^{\infty} (x - \mu)^2 \cdot f(x)dx.$$

Similar forms of distributions belong to families of distribution types. For example, the exponential, normal, and gamma distributions all belong to the very broad exponential family.

As discussed in Chapter 4, likelihoods are based on the *likelihood principle* (section 4.2.2) and are not PDFs. They are mentioned here because they are distributions only in the sense of being functions of the likelihood of every value in the sample space. It is not required that they sum or integrate to 1.0.

3.3.2.1   Joint and marginal distributions

If X_1 and X_2 are two random variables, each has its own PDF; however, together they can form a joint PDF $f(x_1, x_2)$, where the point (x_1, x_2) is in two-dimensional Euclidean space. For any number n of random variables, their *joint distribution* is $f(x_1, x_2, x_3, \ldots, x_n)$ with the properties

$$f(x_1, x_2, x_3, \ldots, x_n) \ge 0, \quad -\infty \le x \le \infty, \quad i = 1, 2, \ldots, n,$$

and

$$\int_{-\infty}^{\infty} f(x_1, x_2, x_3, \ldots, x_n)dx_1 dx_2 \cdots dx_n = 1.$$

For a given joint distribution function, the PDF for any individual variable X_i can be found by integrating the joint PDF over the other variables. This so-called *marginal distribution* is found by

$$f(x_i) = \int_{-\infty}^{\infty} \int_{-\infty}^{\infty} \cdots \int_{-\infty}^{\infty} f(x_1, x_2, x_3, \ldots, x_n)dx_1 dx_2 \cdots dx_{i-1} dx_{i+1} \cdots dx_n.$$

3.3.2.2   Conditional distribution

A conditional PDF is defined in terms of two random variables X_1 and X_2, where X_1 is conditioned on (or given) X_2:

$$f(x_1|x_2) = \frac{f_{12}(x_1, x_2)}{f_2(x_2)},$$

where $f(x_2)$ is the PDF of x_2 and $f_{12}(x_1, x_2)$ is the joint PDF of both random variables.

3.3.3   Conditional probability and dependence

The equation for probability distributions in section 3.3.2.2 is derived from its probability counterpart for two events X_1 and X_2 in a sample space. The conditional probability of event X_1 given event X_2 has occurred is defined by

$$P(X_1|X_2) = \frac{P(X_1 \cap X_2)}{P(X_2)}$$

provided that $P(X_1) > 0$. X_1 is said to be *independent* of X_2 if

$$P(X_1|X_2) = P(X_1)$$

or

$$P(X_1 \cap X_2) = P(X_1) \cdot P(X_2).$$

Two random variables are said to be independent if their joint PDF factors into the product of the individual PDFs:

$$f(x_1, x_2) = f_1(x_1) \cdot f_2(x_2).$$

A sample $x_1, x_2, x_3, \ldots, x_n$ taken from the same distribution function $f(x)$ is a random sample if the joint PDF $f(x_1, x_2, x_3, \ldots, x_n)$ of the observations factors as

$$f(x_1, x_2, x_3, \ldots, x_n) = f(x_1) \cdot f(x_2) \cdot f(x_3) \cdots f(x_n).$$

The corresponding random variables are said to be independent and identically distributed.

3.3.4   Comparing distributions

When characterizing uncertainties using PDFs, the assessor may need to compare different distributions. Distribution characteristics, such as parameters or moments, are customarily tested one at a time using traditional statistical inference tests. While these are useful, they do not address the problem of how to compare entire distributions rather than comparing only their characteristic properties (e.g., their means). Specifically, what is needed are metrics that compare the *information* contained in the two or more uncertainty distributions. This measure or metric should also be on an invariant scale that enables one to say, for example, that with this metric a score of 0.01 implies that the distributions contain similar information, while a score of 2.0 implies that the distributions contain very different information. The metric should have properties that make it useful independent of the underlying distribution to whatever extent possible, invariant to singular transformations, symmetric in operation, and independent of the number of equally spaced partitions of the generated distributions.

Jeffreys's invariant J (Jeffreys (1946, 1998)) is such a metric meeting these requirements for the very broad family of distribution functions called the exponential family (of which the normal distribution is a member). While the metric J still is under development for use in other distribution families and for empirical distributions, we demonstrate its potential use as a metric for comparing distributions generated by different fuzzy combination methods.

For two distributions $f_1(x)$ and $f_2(x)$, J is defined as

$$ J = \int [f_1(x) - f_2(x)] \ln \left[\frac{f_1(x)}{f_2(x)} \right] dx. $$

The popular Kullback–Leibler (1951) KL information measure is the first term of the integral above:

$$ KL = \int f_1(x) \ln \left[\frac{f_1(x)}{f_2(x)} \right] dx. $$

Because of the lack of symmetry in KL, the use of J is recommended for comparison of two functions $f_1(x)$ and $f_2(x)$.

Both KL and J are measures of information content of PDFs—a concept that stems from the entropy of the distribution:

$$ -\int f(x) \ln[f(x)] dx. $$

Other methods for comparing PDFs include relative distributions (Handcock and Morris (1999)) and using quantiles (Mason, Gunst, and Hess (1989)).

3.3.5   Representing data, information, and uncertainties as distributions

Distribution functions are used to represent data, information, and uncertainties. As noted above, these do not have to follow a particular distribution form (e.g., normal) or family.

Distributions can be empirical (or data-based) as a histogram from a set of test or observational data. These distributions represent the nonrepeatability in the data set. Sources include random variation, measurement or observational imprecision, and any number of known or unknown effects (e.g., treatments, conditions, or manufacturing differences). Data from outputs of computer codes or models are subject to uncertainties in those models/calculations. Changes in inputs can propagate through the code/model and affect the output. All of these sources of uncertainty can be represented by PDFs. Often overlooked, but most important in this list, is uncertainty from lack of knowledge. While often difficult to quantify, this uncertainty also can be represented by probability distributions (see Chapters 12 and 17 of Meyer and Booker (2001)).

Distributions can be drawn by an expert according to any form, depending on that expert's knowledge of how the probability or its density is distributed. This is the same sort of exercise used in asking experts to provide (i.e., draw) membership functions (as seen in Chapter 2).

Eliciting uncertainties from experts in the form of distribution functions requires that the experts be comfortable with subjectivist or personalistic probability and understand PDFs and CDFs. This is not always the case. Therefore, it is the jobof the assessor to transform the experts' information into distributions and then meticulously explain to the experts what that transformation means so as to not misrepresent the original information. If expert knowledge is in the form of IF–THEN rules (as described in Chapter 2), then that uncertainty may be best captured using membership functions. As noted in Chapter 4 and described in Chapter 6, membership functions can serve as likelihoods to be translated into probability distributions through the use of Bayes' theorem.

3.4   Information, data, and knowledge

Information has a particular definition in probability theory originating from physics (Jaynes (1957)). For our purposes, information has a more popular meaning: it is any fact, observation, or opinion that provides value to the existing collective of what is understood about a system, phenomenon, or decision problem. Knowledge (what is known), judgment (how knowledge is interpreted by experts), and data (numerical and qualitative measurements or observations) are all sources of information.

Data is defined in the popular science/technology sense as follows:

- *hard* data—from experiments, measurements, or observations;

- *soft* data—expert judgment (defined in section 6.1 of Chapter 6), estimates of quantities, assumptions, boundary conditions, utilities, heuristics, and rules.

Data, knowledge, and information can exist in the following forms:

- *quantitative*—such as numbers, ranks, ratings, or any form of numerical value;

- *qualitative*—descriptions, categories, or any form of words.

Categorical data/information can be qualitative or quantitative. For example, qualitative categories can imply a quantitative-like ordering and yet have qualitative words such as low, medium, and high.

All information (data plus knowledge) have uncertainties attached to them. Some of these uncertainties may be estimable from known sources, while others are not so easily characterized. Often it is the job of the assessor, in concert with experts, to estimate these uncertainties given the current state of knowledge (which in some cases may be very poor).

Information should be gathered in such a way as to preserve its original content and ensure the best quality possible. For data, this requires statistically sound sampling methods—selecting units out of the population (the entire sample space) in such as way as to ensure all attributes of the population are represented. To provide such representative selection, some aspect of random selection is involved. Attributes of the population that contribute to uncertainties should either be controlled (sampling done in such a way as to fix or control these uncertainties) or measured for estimation (if control is not possible, then recording certain values of attributes provides a means of estimating their uncertainty). For example, in a typical sample survey such attributes as age, income, family size, and gender may affect the respondents' answers. The survey is then either designed to ask about these attributes or implemented to control them (such as polling certain individuals with specified ages, genders, etc.).

In expert elicitation, many attributes can affect the information being gathered and are potential sources of uncertainty. Examples include

- how experts solve problems;

- the assumptions, cues, and heuristics experts use;

- how the questions are asked;

- experts' personal agendas and motivations.

Techniques for minimizing the biases that can result from these and other examples can be found in Meyer and Booker (2001).

Expert judgment is not sampling in the traditional statistical sense. While uncertainties exist in the information elicited, these do not arise from drawing samples of expert judgment from the population of experts. These uncertainties are inherent in the information itself. Nonetheless, the assessor can characterize uncertainties by utilizing the probability-based methods from this chapter and/or the fuzzy-based methods from Chapter 2. As will be seen in some of the application chapters in Part II of this book, often a combination of probability- and fuzzy-based methods is required to adequately represent all uncertainties for a given state of knowledge.

References

R. J. BECKMAN (1997), *Transportation Analysis Simulation System: The Dallas–Ft. Worth Case Study*, TRANSIMS Release 1.0, Report LA-UR-97-4502, Los Alamos National Laboratory, Los Alamos, NM.

T. R. BEMENT, J. M. BOOKER, N. D. SINGPURWALLA, AND S. A. KELLER-MCNULTY (1999), *Testing the Untestable: Reliability in the 21st Century*, Report LA-UR-00-1766 and Report GWU/IRRA Serial TR99/6, Los Alamos National Laboratory and George Washington University, Los Alamos, NM and Washington, DC.

B. DE FINETTI (1937), Foresight: Its logical flaws, its subjective sources, in *Studies in Subjective Probability*, H. E. Kyburg and H. E. Smokler, eds., John Wiley, New York, 1964, pp. 93–158.

B. DE FINETTI (1974), *Theory of Probability, Vol.* I, John Wiley, New York.

M. DE GROOT (1970), *Optimal Statistical Decisions*, McGraw–Hill, New York.

A. P. DEMPSTER (1968), A generalization of Bayesian inference (with discussion), *J. Roy. Statist. Soc. Ser.* B, 30, pp. 205–247.

T. FINE (1973), *Theories of Probability*, Academic Press, New York.

G. GIGERENZER, Z. SWIJTINK, T. PORTER, L. DASTON, J. BEALTY, AND L. KRUGER, (1989), *The Empire of Chance*, Cambridge University Press, Cambridge, UK.

I. J. GOOD (1965), *The Estimation of Probabilities*, MIT Press, Cambridge, MA.

N. A. J. HASTINGS AND J. B. PEACOCK (1975), *Statistical Distribution: A Handbook for Students and Practitioners*, John Wiley, New York.

I. HACKING (1974), *The Emergence of Probability*, Cambridge University Press, London, New York.

M. S. HANDCOCK AND M. MORRIS (1999), *Relative Distribution Methods in the Social Sciences*, Springer-Verlag, New York.

C. HOWSON AND P. URBACH (1989), *Scientific Reasoning: The Bayesian Approach*, Open Court, La Salle, IL.

E. T. JAYNES (1957), Information theory and statistical mechanics, *Phys. Rev.*, 106, pp. 620–630.

R. C. JEFFREY (1983), *The Logic of Decision*, 2nd ed., University of Chicago Press, Chicago.

H. JEFFREYS (1946), An invariant form for the prior probability in estimation problems, *Proc. Roy. Soc. London Ser.* A, 186, pp. 453–461.

H. JEFFREYS (1961), *Theory of Probability*, Oxford University Press, New York.

D. KAHNEMAN, P. SLOVIC, AND A. TVERSKY (1986), *Judgement Under Uncertainty: Heuristics and Biases*, Cambridge University Press, Cambridge, UK.

S. KULLBACK AND R. LEIBLER (1951), On information and sufficiency, *Ann. Math. Statist.*, 22, pp. 79–86.

D. V. LINDLEY (1982), Scoring rules and the inevitability of probability, *Internat. Statist. Rev.*, 50, pp. 1–26.

D. V. LINDLEY AND N. D. SINGPURWALLA (1986), Reliability (and fault tree) analysis using expert opinion, *J. Amer. Statist. Assoc.*, 81, pp. 87–90.

R. L. MASON, R. F. GUNST, AND R. L. HESS (1989), *Statistical Design and Analysis of Experiments*, John Wiley, New York.

L. E. MAISTROV (1974), *Probability Theory: A Historical Sketch*, Academic Press, New York (translated from the Russian and edited by Samuel Kotz).

H. F. MARTZ AND R. A. WALLER (1982), *Bayesian Reliability Analysis*, John Wiley, New York.

M. A. MEYER AND J. M. BOOKER (2001), *Eliciting and Analyzing Expert Judgment: A Practical Guide*, ASA–SIAM Series on Statistics and Applied Probability, SIAM, Philadelphia, ASA, Alexandria, VA.

F. P. RAMSEY (1934), Truth and probability, in *The Logical Foundations of Mathematics and Other Essays*, Routledge and Kegan Paul, London; reprinted in *Studies in Subjective Probability*, H. E. Kyburg and H. E. Smokler, eds., John Wiley, New York, 1964, pp. 61–92.

H. REICHENBACH (1949), *The Theory of Probability*, University of California Press, Berkeley, CA.

L. J. SAVAGE (1972), *The Foundations of Statistics*, 2nd Ed., Dover, New York.

N. D. SINGPURWALLA (1988), An interactive PC-based procedure for reliability assessment incorporating expert opinion and survival data, *J. Amer. Statist. Assoc.*, 83, pp. 43–51.

C. A. B. SMITH (1961), Consistency in statistical inference and decision (with discussion), *J. Roy. Statist. Soc. Ser.* B, 23, pp. 1–37.

R. Von Mises (1957), *Probability, Statistics, and Truth*, 2nd ed., George Allen and Unwin, London; reprinted by Dover, New York, 1981.

United States Nuclear Regulatory Commission (1975), *The Reactor Safety Study: An Assessment of the Accident Risks in U.S. Commercial Nuclear Power Plants*, Report WASH-1400 (NUREG-75/014), Washington, DC.

L. A. Zadeh (1979), *Possibility Theory and Soft Data Analysis*, Memo UCB/ERL M79/66, University of California at Berkeley, Berkeley, CA.

Chapter 4

Bayesian Methods

Kimberly F. Sellers and Jane M. Booker

The use of Bayesian methods as both an information combination scheme and an updating tool has become widespread, combining or updating prior information with existing information about events. Bayesian methods stem from the application of Bayes' theorem in probability. As will be noted in this chapter, not only do these methods provide ways of handling various kinds of uncertainties, but they also can serve as a link between subjective-based probability theory and fuzzy logic.

In the Bayesian paradigm, uncertainty is quantified in terms of a personal or subjective probability following the axioms of probability theory (see Chapter 3, section 3.2.5). The probability of an event X is denoted $P(X; H)$, where H represents the assessor's information, often called the prior information. *Prior* refers to the knowledge that exists prior to the acquisition of information about event X. Uncertainties combine via rules of probability that stem from the axiomatic behavioristic interpretation of probability (see Chapter 3, section 3.1). The fundamental Bayesian philosophy is that prior information, H, is valuable and can be mathematically combined with information about X and, with such combination, uncertainties can be reduced.

Understanding the uses of Bayesian methods begins with the historical development of the theory.

4.1   Introduction

In 1763, the Reverend Thomas Bayes of England made a significant contribution to the field of probability when he published a paper describing a relationship among probabilities of events (X and Y) in terms of conditional probability. This relationship,

$$P(X \text{ and } Y; H) = P(X; H) \cdot P(Y|X; H),$$

stems from a basic probability law about the probability of two events occurring (Iversen (1984)). The interpretation of this relationship is as follows: In light of information H, the probability of events X and Y occurring is equal to the probability of event X occurring multiplied by the probability of event Y occurring if event X has occurred. For a simple example, suppose that a fair six-faced die is rolled. Let $X = \{$an even number is rolled$\}$ and $Y = \{$the roll produces a 6$\}$. Then $P(X; H) = \frac{1}{2}$ and $P(Y|X; H) = \frac{1}{3}$. Therefore,

$$P(X \text{ and } Y) = P(X) \cdot P(Y|X) = \left(\frac{1}{2}\right)\left(\frac{1}{3}\right) = \left(\frac{1}{6}\right).$$

Another expression for $P(X \text{ and } Y; H)$ from probability laws is

$$P(X \text{ and } Y; H) = P(Y; H) \cdot P(X|Y; H).$$

That is, the probability that both events X and Y occur equals the probability that Y occurs multiplied by the probability of X given Y occurring.

By equating the two expressions for $P(X \text{ and } Y; H)$ and rearranging the associated terms, we get a form of Bayes' theorem that states

$$P(X|Y; H) = \frac{P(Y|X; H) \cdot P(X; H)}{P(Y; H)}.$$

Bayes' theorem expresses the probability that event X occurs if we have observed Y in terms of the probability of Y if given that X occurred.

History indicates that Laplace may have independently established another form of Bayes' theorem by considering k events, $X_1, X_2, \ldots, X_k$. Then the probability of Y, $P(Y; H)$ can be rewritten as

$$P(Y|X_1; H) \cdot P(X_1; H) + P(Y|X_2; H) \cdot P(X_2; H) + \cdots + P(Y|X_k; H) \cdot P(X_k; H).$$

This is the law of total probability (i.e., law of the extension of conversation): For two events X and Y,

$$P(Y = y_i; H) = \sum_{\text{all } x_j} P(Y = y_i | X = x_j; H) P(X = x_j; H),$$

which is abbreviated as

$$P(Y; H) = \sum_{\text{all } x} P(Y|X; H) P(X; H).$$

Proof of the law of total probability for two events, x and y. By the definition of conditional probability,

$$\sum_{\text{all } x_j} P(Y = y_i | X = x_j; H) P(X = x_j; H)$$

$$= \sum_{\text{all } x_j} \left[\frac{P(Y = y_i \text{ and } X = x_j; H)}{P(X = x_j; H)} \right] P(X = x_j; H)$$

$$= \sum_{\text{all } x_j} P(Y = y_i \text{ and } X = x_j; H)$$

$$= P\left[\bigcup_{\text{all } x_j} (Y = y_i \text{ and } X = x_j) \right]$$

since the events $(X = x_i)$ are mutually exclusive. Finally,

$$P\left[\bigcup_{\text{all } x_j} (Y = y_i \text{ and } X = x_j)\right] = P(Y = y_i; H);$$

thus the equation is satisfied.   □

The law of total probability allows the assessor to consider the probability of an event Y, $P(Y; H)$ by summing over all possible conditional probabilities $P(Y|X; H)$.

Thus for one event X_i, Bayes' theorem becomes

$P(X_i|Y; H)$

$$= \frac{P(Y|X_i; H) \bullet P(X_i; H)}{P(Y|X_1; H) \bullet P(X_1; H) + P(Y|X_2; H) \bullet P(X_2; H) + \cdots + P(Y|X_k; H) \bullet P(X_k; H)}$$

or, equivalently,

$$P(X|Y; H) = \frac{P(Y|X; H)P(X; H)}{\sum_x P(Y|X; H)P(X; H)}.$$

This is the discrete form of Bayes' theorem (which is also referred to as the law of inverse probability): For two events X and Y,

$$P(X = x_i|Y = y_j; H) = \frac{P(Y = y_j|X = x_i; H)P(X = x_i; H)}{\sum_{\text{all } x_i} P(Y = y_j|X = x_i; H)P(X = x_i; H)},$$

which is abbreviated as

$$P(X|Y; H) = \frac{P(Y|X; H)P(X; H)}{\sum_{\text{all } x} P(Y|X; H)P(X; H)} \propto P(Y|X; H)P(X; H).$$

Proof of the discrete form of Bayes' theorem. By the multiplication rule,

$$\frac{P(Y = y_j|X = x_i; H)P(X = x_i; H)}{\sum_{\text{all } x_i} P(Y = y_j|X = x_i; H)P(X = x_i; H)} = \frac{P(X = x_i \text{ and } Y = y_j; H)}{\sum_{\text{all } x_i} P(X = x_i \text{ and } Y = y_j; H)}$$

$$= \frac{P(X = x_i \text{ and } Y = y_j; H)}{P(Y = y_j; H)}$$

by the additivity rule because $(X = x_i \text{ and } Y = y_i)$ are mutually exclusive for all i. Finally,

$$\frac{P(X = x_i \text{ and } Y = y_j; H)}{P(Y = y_j; H)} = P(X = x_i|Y = y_j; H)$$

by the definition of conditional probability.   □

The implications of Bayes' theorem are significant in that it does the following:

- It shows the proportional relationship between the conditional probability $P(X|Y; H)$ and the product of probabilities $P(X; H)$ and $P(Y|X; H)$.

- It tells the assessor how to relate the two uncertainties about X: one prior to knowing Y, the other posterior to knowing Y.

- It tells the assessor how to change the opinion about X were Y to be known; this is also called "the mathematics of changing your mind."

- It gives the assessor a vehicle for incorporating additional information, i.e., expanding it.

- It prescribes a procedure for the assessor, i.e., how to bet on X should Y be observed or known. That is, it prescribes the assessor's behavior before actually observing Y.

4.2   Probability theory of Bayesian methods

In order to understand the foundation of Bayesian analysis, the interpretations of probability and conditional probability must first be discussed using the context of events X and Y and prior information H:

- How do we interpret $P(X; H)$?

- How do we interpret $P(X|Y; H)$?

- What is the probability of an event that is known to have occurred?

Consider the event X with $P(X; H) = p$ as the assessor's subjective probability as defined in Chapter 3, section 3.2.5. By definition, p implies that the assessor is prepared to bet an amount p in exchange for 1 if a challenger accepts the bet and the outcome is in favor of the assessor. With this in mind, suppose that the assessor is very sure (but not certain) about the outcome of X. Then p should be small (i.e., $p = \varepsilon$, where $\varepsilon > 0$ is a small number) because this bet will allow the assessor to make the largest profit. However, with ε small, the assessor is very likely to lose $(1 - \varepsilon)$. To avoid this type of inequity, the challenger will require that the assessor accept a two-sided bet, where the sides are specified:

(a) Stake ε in exchange for 1 if the outcome is in favor of the assessor.

(b) Stake $(1 - \varepsilon)$ in exchange for 1 if the outcome is in favor of the challenger.

Under (a), the assessor is most likely to gain $(1 - \varepsilon)$, but under (b), the assessor is most likely to lose $(1 - \varepsilon)$. This is because the challenger gets to specify (a) or (b). Therefore, to avoid the possibility of a loss, the assessor's probability should be $(1 - \varepsilon)$ so that the two-sided bet is as follows:

(a') Stake $(1 - \varepsilon)$ in exchange for 1 if the outcome is in favor of the assessor.

(b') Stake ε in exchange for 1 if the outcome is in favor of the challenger.

Under this scheme, the assessor is likely to gain ε if (a') is chosen and likely to lose ε if (b') is chosen.

From the basic interpretation of probabilities as a bet, $P(X|Y; H)$ says that the gamble involving X is contingent on Y occurring. In other words, the interpretation of $P(X|Y; H)$ is identical to that for $P(X; H)$ but under the requirement that Y occurs. Therefore, if Y does not occur, then the bet is off.

Finally, if the assessor is absolutely sure (i.e., certain) of the outcome, then by adhering to the methodology discussed to establish $P(X; H)$, $\varepsilon = 0$. However, this now implies that (b') vanishes. No challenger would be willing to accept such a bet that they would surely lose, so the idea of such a probability cannot be discussed; i.e., there is no such entity as the probability of a sure event.

4.2.1 The incorporation of actual information (expansion of H)

Suppose that $Y = y$ has been observed. What is the assessor's uncertainty about X; i.e., what is $P(X; y, H)$?

There are two possible approaches to answer such a question. The first option is to try to reassess X in light of H and y. However, there are difficulties associated with this approach: it is a cumbersome process, and the assessor cannot ensure that $P(X; H)$ and $P(X; y, H)$ are coherent (i.e., consistent) with each other.

The second option, attributed to Bernoulli, is to first prescribe our uncertainty about X if we were to observe Y (but assuming that $Y = y$ has not yet occurred). Using Bayes' rule, we obtain

$$P(X|Y; H) \propto P(Y|X; H) \cdot P(X; H).$$

However, since the assessor has actually observed $Y = y$, the left-hand side should be written as $P(X; y, H)$. Therefore,

$$P(X; y, H) \propto P(Y = y|X = x; H) \cdot P(X = x; H)$$

and the assessor now has a problem because $P(Y = y|X = x; H)$ cannot be interpreted as a probability. Thus $P(Y = y|X = x; H)$ is called the *likelihood* that $X = x$ in light of $Y = y$, and H and is denoted $L(X = x; y, H)$. Furthermore, when viewed as a function of x, $L(X = x; y, H)$ is called a *likelihood function* of X for a fixed value of y. For example, the likelihood of a test resulting in a particular failure rate would be expressed in terms of $L(X = x; y, H)$.

As a result, this formulation of Bayes' theorem,

$$P(X = x; y, H) \propto L(X = x; y, H) \cdot P(X = x; H),$$

provides a convenient method for combining different sources of information. It is a systematic scheme to update probabilities in light of data. $P(X = x; H)$ is called the *prior probability distribution* of X, i.e., the source for information that exists "prior" to test data in the form of expert judgment and other historical information. By definition, the *prior distribution* $P(X = x; H)$ represents the possible values and associated probabilities for the quantity of interest, X. For example, X might represent the average failure rate of a particular manufactured item (expressed in the earlier example as failures per 1000 items). The likelihood function, $L(X = x; y, H)$, is formed from the test data, and $P(X = x; y, H)$ is the posterior distribution in light of y (the data) and H produced from the prior information and the data.

Note that the likelihood function is not a probability and therefore does not need to obey the laws of probability. It simply connects the two probabilities: the prior distribution $P(X; H)$ and the posterior distribution $P(X; y, H)$. The likelihood function is a purely subjective function that enables the assessor to assign relative weights to different values of $X = x$. As a result, the likelihood provides the assessor with a mechanism for learning from the data.

An advantage of using Bayes' theorem to combine distribution functions of different information sources is that the spread (uncertainty) in the posterior distribution is reduced when the information in the prior and likelihood distributions are consistent with each other. In other words, the combined information from the prior distribution and the data has less uncertainty because the prior distribution and data are two different information sources that support each other.

4.2.2   The likelihood principle

Implicit in the use of Bayes' theorem is the likelihood principle, whose development stems from R. A. Fisher:

> "The likelihood principle: All the information about X obtainable from an experiment is contained in the likelihood function for X given y. Two likelihood functions for y (from the same or different experiments) contain the same information about y if they are proportional to one another."

This principle states that all evidence obtained from an experiment or observation about some unknown quantity X is contained in the likelihood function of X for the given data y (Berger and Wolpert (1988)). Because Bayesian inference is based on the posterior distribution, the likelihood principle (and hence the likelihood) becomes extremely important, almost to the exclusion of the assessor's choice of prior distribution. "Consequently the whole of the information contained in the observation that is relevant to the posterior probabilities of different hypotheses is summed up in the values that they give to the likelihood" (Jeffreys (1961)). The likelihood contains all of the sample information even when the prior is unknown. It plays an important role in the distribution function formulation of Bayes' theorem, where the test or observational data (continuous or discrete) forms the likelihood.

4.2.3   Distribution function formulation of Bayes' theorem

Bayes' theorem has been provided for the discrete form for two random variables X and Y. For continuous X and Y, the probability statements are replaced by probability density functions (PDFs), and the likelihood is replaced by a likelihood function. If Y is a continuous random variable whose PDF depends on the variable X, then the conditional PDF of Y given X is $f(y|x)$. If the prior PDF of X is $g(x)$, then for every y such that $f(y) > 0$ exists, the posterior PDF of X given $Y = y$ is

$$g(x|y; H) = \frac{f(y|x; H) \cdot g(x; H)}{\int f(y|x; H) \cdot g(x; H)},$$

where the denominator integral is a normalizing factor so that $g(x|y; H)$, the posterior distribution, integrates to 1 (as a proper PDF).

Alternatively, utilizing the likelihood notation, we get

$$g(x|y; H) \propto L(x|y; H)g(x; H)$$

so that the posterior is proportional to the likelihood function multiplied by the prior distribution.

Bayes' theorem can be interpreted as a weighting mechanism. The theorem mathematically weights the likelihood and prior distributions, combining them to form the posterior. If these two distributions overlap to a large extent, this mathematical combination produces a desirable result: the variance of the posterior distribution is smaller than that produced by a simple weighted combination, $w_1 \cdot f_{\text{prior}} + w_2 \cdot f_{\text{likelihood}}$, for example. The reduction in the variance results from the added information of combining two distributions that contain similar information (overlap).

If the prior and likelihood distributions are widely separated, then the posterior will fall into the gap between the two functions. This is an undesirable outcome because the resulting combination falls in a region unsupported by either the prior or the likelihood. The analyst may want to reconsider using Bayesian combination in this case and seek to resolve the differences between the prior and likelihood, or may want to use some other combination method, such as a simple weighting scheme; for example, consider $w_1 \cdot f_{\text{prior}} + w_2 \cdot f_{\text{likelihood}}$.

4.3   Issues with Bayes' theory

Recent use of Bayes' theorem has exploded, as evidenced by citations in the literature (Malakoff (1999)). However, with increasing use comes increasing misuse.

4.3.1   Criticisms and interpretations of Bayes' theorem

The difficulty in using Bayes' theorem is the determination of what information should be labeled as prior and what should be labeled within the likelihood function. Because the likelihood does not need to satisfy the axioms of probability, it is often misunderstood or misspecified. According to the theory, the likelihood should be the data or information collected from an experiment, or observation, or new information gained since collection of the old (prior) information.

The most prominent argument brought against Bayesian-based methods by those with the more frequentist view is for the use of the subjectivist or degree of belief probability theory required to interpret the meaning of the prior and posterior probabilities. These issues were mentioned in Chapter 3, section 3.2.5 and are the subject of much debate in the statistical community.

Another related argument against Bayesian methods revolves around the role and determination of the prior distribution. According to the theorem and its formulations in section 4.2 above, the prior represents the uncertainty (as a probability distribution) in the parameter(s) of the likelihood function. How the assessor adequately represents what is known prior to collecting data so that it is consistent with this definition is often difficult to determine. Prior information usually is in the form of historical information and/or expert judgment and may not be readily translatable into a probability distribution function for the parameter(s) found in the likelihood. Such a representation calls for a parametric interpretation of the prior and likelihood. This interpretation actually violates the basic ground rules of subjectivist probability. To overcome this violation, the assessor should really be using a parameter-free approach to Bayes' theory. However, this approach is only theoretical and implementing it in practice is not feasible until more research is done (Singpurwalla (1999)).

On the other hand, the frequentist interpretation of probability violates some basic principles of that theory. Frequentists cannot accommodate any existing historical information, and they rely solely on experimental or observational data that may be too sparse for formulating conclusions. For example, frequentists commonly use the method of maximum likelihood; however, from a philosophical standpoint, this has no merit.

Consider the likelihood function $L(X = x; y, H)$, where y and H are fixed and known. The maximum likelihood estimator of X, x^* is the value of x where $L(X = x; y, H)$ attains a maximum value. Statistically speaking, it is the value of x that is most supported by the data y. Frequentists consider the method of maximum likelihood for a variety of reasons. In general, the maximum likelihood estimator has good large sample properties such as consistency. Also, it generally has strong sampling properties, i.e., unbiasedness and minimum variance. However, to prove these properties violates the likelihood principle because the data have already been observed, and thus other possible values for y cannot be considered.

Empirical Bayesian methods offer a partial solution to approaching the theoretical implementation of parameter-free Bayes, and it offers a solution to the difficulties of forming a prior. Here prior distributions are empirically (or information-) based distributions. For overviews on this topic, see Martz (2000) and Martz, Parker, and Rasmuson (1999).

Critics of prior determination methods have good cause for judgment when avid Bayesians insist on the use of *noninformative* priors. Here the prior distributions are constructed when there is no prior information available and solely for the sake of using Bayes' theorem. Noninformative priors often are called vague, flat, or diffuse priors. The result is that the likelihood dominates the posterior. However, the use of such priors violates the philosophical foundations of Bayesian analysis, which is to take advantage and utilize all available information. If no information is available for constructing a prior, then Bayes' theorem is not recommended.

A counterexample would be the use of the uniform prior. Here, perhaps the only prior information known is that the parameter lies within some specified range (the range chosen for the uniform distribution), and the assessor is willing to specify that the parameter can lie with equal probability within that range. Here the assessor has prior information: equal probability and a chosen range captured through the choice of the uniform prior. This is not an example of specifying a noninformative prior.

4.3.2   Uses and interpretations

There are many uses for and interpretations of Bayes' theorem extolling its virtues. A few are summarized below (Martz (1998)):

- Bayes' theorem indicates how point estimates (and their associated uncertainties) are updated (combined) in light of additional pertinent information or data (such as relevant information from computer models).

- Bayes' theorem is a statistical method for combining different kinds of data and/or information about some quantity of interest (such as reliability of a system).

- Bayes' theorem describes how uncertainties in data regarding a quantity of interest (such as a performance measure of a system) are modified in light of other available information about the quantity of interest.

- Bayes' theorem operationally describes how to combine statistical data with other sources of pertinent information, such as computer models and/or expert opinion, regarding a quantity of interest.

- Bayes' theorem provides a mechanism for inverting conditional probability distributions of data in light of additional prior information and/or data.

- Bayes' theorem provides a theoretically based mechanism for transforming fuzzy membership functions (which serve as the likelihood) into probability space (as a posterior, using an appropriate prior probability distribution). Section 4.6 below discusses this particular use of Bayes' theorem in relation to the topics of this book.

4.4   Bayesian updating: Implementation of Bayes' theorem in practical applications

One of the most practical uses of Bayes' theorem (mentioned in the first bullet above) is updating. In many dynamic applications (e.g., an aging system, the design phases of a development system), new data or information becomes available for a variety of reasons (new tests, new design features, etc.), and it is desirable to update an existing probability distribution in light of this new data/information. In such cases, the existing distribution becomes the prior and the new information becomes the likelihood, with the resulting posterior providing the updated version.

Before the days of modern computers, calculating Bayes' theorem was computationally cumbersome. For those times, it was fortunate that certain choices of PDFs for the prior and likelihood produced easily obtained posterior distributions. For example, a beta prior with a binomial likelihood produces a beta posterior whose parameters are simple functions of the prior beta and binomial parameters. Examples of these so-called *conjugate priors* follow. With modern computational methods, these analytical shortcuts are not as necessary as they were in the past; however, the interesting application-oriented interpretations of the parameters for some of these conjugate priors retain their usefulness today.

4.4.1   Binomial/beta example

Suppose that we prototype a system, building 20 units, and subject these to a stress test. All 20 units pass the test. The estimate of success/failure rates from test data alone are

$$n_1 = 20 \text{ tests} \quad \text{with } x_1 = 20 \text{ successes.}$$

Using just this information, the success rate is $\frac{20}{20} = 1$, and the failure rate is $\frac{0}{20} = 0$. This simple estimate, based on only 20 units, does not reflect the uncertainty in the reliability for the system and does not account for any previously existing information about the units before the test.

Using a Bayesian approach, the probability of a success is denoted as p. There exists some prior knowledge, relevant data, specifications, etc. that can be used to formulate another distribution for p—the prior distribution, $g(p)$. In this example, that prior is a beta(x_0, n_0) distribution

$$g(p) = \frac{\Gamma(n_0)}{\Gamma(x_0)\Gamma(n_0 - x_0)} x^{x_0-1}(1 - x)^{n_0-x_0-1}.$$

The beta distribution is often chosen as a prior for a probability because it ranges from 0 to 1 and can take on many shapes (uniform, J shape, U shape, and Gaussian-like) by adjusting its two parameters n_0 and x_0.

The assessor knows something about this system before the test. Prior information is in the form of an estimate of the failure rate from the test data done on a similar system that is considered relevant for this new system:

$$n_0 = 48 \text{ tests on a similar system,}$$

$$x_0 = 47 \text{ successes.}$$

The new test data form the likelihood $L(p; x)$ and represent the number of successes x_1 in n_1 trials conforming to the binomial distribution with parameter p. The beta distribution

$g(p)$ is a conjugate prior when combined with the binomial likelihood $L(p; x)$ using Bayes' theorem. As noted above, *conjugate prior distribution* is a distribution such that the prior and posterior come from the same family of distributions. Thus the resulting or posterior distribution $g(p|x)$ is also a beta distribution. For this example, the beta posterior distribution has parameters $(x_0 + x_1, n_0 + n_1)$ as follows:

$$g(p|x) = \frac{\Gamma(n_0 + n_1)}{\Gamma(x_0 + x_1)\Gamma(n_0 + n_1 - x_0 - x_1)} p^{(x_0+x_1)-1}(1 - p)^{(n_0+n_1-x_0-x_1)-1}$$

or

$$g(p|x) = \frac{\Gamma(68)}{\Gamma(67)\Gamma(1)} p^{66}(1 - p)^0 = 67p^{66},$$

$$\frac{x_0 + x_1}{n_0 + n_1} = \frac{67}{68} \approx 0.985,$$

or in terms of a failure rate, the beta posterior has a mean failure rate of approximately $1 - 0.985 = 0.015$. The variance of the beta posterior distribution is

$$\frac{(x_0 + x_1)[(n_0 + n_1) - (x_0 + x_1)]}{(n_0 + n_1)^2(n_0 + n_1 + 1)} \approx 0.00021.$$

The engineering reliability community gravitates toward the binomial/beta conjugate prior because many of the failures are binomial in nature and the parameters of the prior and posterior can have a reliability-based interpretation. As noted above, $n_0 =$ number of tests and $x_0 =$ number of successes for the prior parameter interpretation. Similarly, $n_0 + n_1 =$ number of *pseudo*tests and $x_0 + x_1 =$ number of *pseudo*successes for the posterior parameter interpretation, provided these values are greater than 1.

4.4.2 Exponential/gamma example

Another popular conjugate prior example in reliability is the gamma distribution

$$g(\lambda; \alpha, \theta) = \frac{\theta^\alpha}{\Gamma(\alpha)} \lambda^{\alpha-1} e^{-\theta\lambda}.$$

It is the conjugate prior distribution for λ (which can represent a failure rate) in the exponential distribution

$$f(t; \lambda) = \lambda e^{-\lambda t}, \quad t, \lambda > 0,$$

(Martz and Waller (1982)).

The parameter α can be interpreted as the *pseudo*number of failures corresponding to prior information, and parameter θ interpreted as the *pseudo*total time on test. If test data are obtained for time t on a test, resulting in s failures, the prior distribution can be updated and the posterior distribution for λ is a gamma distribution, $g(\lambda; \alpha + s, \theta + t)$, where $t = \sum_i t_i$ and t_i is the time on test for the ith test unit.

4.4.3 Normal/normal example

Because some physical or random phenomena are Gaussian (or normally)distributed, consider data from a random sample of size n with a normal (μ, σ^2) distribution, where the mean μ has a prior distribution that is also normally distributed with mean τ and variance v^2. Then

$$f(\bar{x}|\mu; \sigma^2) = \frac{1}{\sqrt{2\pi\left(\frac{\sigma^2}{n}\right)}} \exp\left[-\frac{(\bar{x}-\mu)^2}{2\left(\frac{\sigma^2}{n}\right)}\right]$$

is the PDF for the data mean $\bar{x}$, and

$$g(\mu|\tau; v^2) = \frac{1}{\sqrt{2\pi}\,v} \exp\left[-\frac{(\mu-\tau)^2}{2v^2}\right]$$

is the prior distribution for μ. Thus the posterior distribution is

$$g(\mu|\bar{x}; \sigma^2, v^2) \doteq N(T, V^2),$$

where $T = \frac{\sigma^2\tau + nv^2\bar{x}}{\sigma^2 + nv^2}$ is the posterior mean and $V^2 = \frac{\sigma^2 v^2}{\sigma^2 + nv^2}$ is the posterior variance.

For example, suppose that the failure times of 22 bushings of a power generator can be described by an $N(\mu, \sigma^2 = 9)$ distribution. Furthermore, suppose that an $N(\tau = 15, v^2 = 4)$ prior distribution is assigned to μ. Then the posterior distribution with $\bar{x} = 14.01$ years is normal with mean

$$T = \frac{(9)(15) + (22)(4)(14.01)}{9 + (22)(4)} \approx 14.10 \text{ years}$$

and variance

$$V^2 = \frac{(9)(4)}{9 + 22(4)} \approx 0.371.$$

It should be noted that these formulas arise from assuming a squared-error loss function (Martz and Waller (1982)). The Bayes point estimate of μ is then 14.10 years.

4.5 Bayesian networks

Analyses of today's complex systems must accommodate more than two variables X and Y. Directed graphs or causal diagrams indicating the relationships between the various components (called Bayesian *networks*) could represent such complex systems. These systems could contain different levels, which can, in turn, influence or affect other levels. The basic levels, or parents levels, provide the conditions that influence the parts of the system at the children levels. As such, these parents levels form the priors to be used in Bayes' theorem.

Bayes' network also is used when new evidence (data or information from knowledge) enters into the system at any level, affecting any part of the system (e.g., a component or node). Incorporating that new evidence as a likelihood can be done using Bayes' theorem, where the existing state of knowledge serves as the prior distribution.

A Bayesian network can be thought of as a graphical model (see Figure 4.1) that encodes the joint probability distribution for a large set of interrelated variables (Heckerman, (1996)). The network structure contains a set of conditional independence relations about these variables and a set of probabilities associated with each variable. The network consists of the following features (Jensen (1996)):

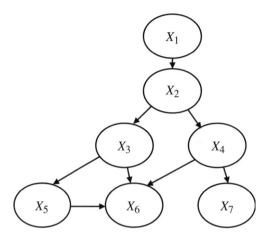

Figure 4.1. *Example of a Bayesian network.*

- a set of variables $U = \{X_1, \ldots, X_m\}$;

- a set of directed edges (relationships) between the variables;

- each variable has a finite set of mutually exclusive states;

- the variables with the directed edges form a directed acyclic (no feedback loops) graph;

- each *child* variable has *parent(s)* with conditional probabilities of the form $P(\text{child}|\text{parents})$.

The joint probability distribution $P(U)$ is the product of all the conditional probabilities,

$$P(U) = \prod_i P(X_i | \text{pa}(X_i)),$$

where $\text{pa}(X_i)$ is the parent set of X_i.

Networks are useful for demonstrating concepts such as conditional independence, as illustrated in Figure 4.2. Here the relationships between the random variables are denoted by X_1 and X_3, where X_1 is independent of X_3 given (conditional upon) X_2 if $P(X_1|X_2) = P(X_1|X_2, X_3)$. In other words, if X_2 is known, then no knowledge of X_3 will affect the probability of X_1.

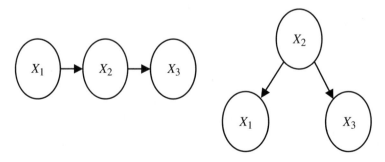

Figure 4.2. *Two examples of conditional independence of X_1 and X_3.*

Because of the conditional Bayesian reasoning and the updating capability for new evidence entering into the network, these networks are useful in constructing expert systems (Cowell et al. (1999)).

4.6   Relationship with fuzzy sets

Bayes' theorem and the associated Bayesian methods can be used as a link between probability theory and fuzzy set theory. Specifically, probability measures relate to fuzzy sets through the use of membership functions. The associated membership function of a fuzzy set can be interpreted as a likelihood function of X for a fixed set S. This relationship is true because the process of defining a membership function is subjective in nature, reflecting the assessor's opinion regarding the level of membership in the set S. Furthermore, because membership functions are merely nonnegative, they can serve as likelihoods. Then for a given probability-based prior distribution, the membership functions, as likelihoods, can be combined with that prior through Bayes' theorem and produce a probability-based posterior.

This methodology can be applied in a number of areas, including reliability theory and biostatistics. In reliability theory, we can gain more information regarding system reliability as opposed to learning whether or not the system simply works or fails; we can measure its degradation level within certain categories of membership. Chapter 11 provides a case study in reliability (which is a probability-based concept) demonstrating the use of membership functions to characterize uncertainty and how these are mapped into probability space for use in calculating reliability. In biostatistics, we can use probability measures of fuzzy sets to gain more information on test results that are deemed inconclusive.

In both these applications, the knowledge that exists in experts' experience may be better captured using the rule-based structure of membership functions. However, other data or information on the system may be more easily represented using PDFs. This bridge between fuzzy logic and probability is necessary to combine all the data/information (probability-based or membership function–based) to perform a complete analysis of the system.

References

J. O. BERGER AND R. L. WOLPERT (1988), *The Likelihood Principle*, 2nd ed., Lecture Notes Monograph Series 6, Institute of Mathematical Statistics, Hayward, CA.

R. G. COWELL, A. P. DAWID, S. L. LAURITZEN, AND D. J. SPIEGELHALTER (1999), *Probabilistic Networks and Expert Systems*, Springer-Verlag, New York.

D. HECKERMAN (1996), *A Tutorial on Learning with Bayesian Networks*, Technical Report MSR-TR-95-06, Advance Technology Division, Microsoft Corporation, Redmond, WA; available online from ftp://ftp.research.microsoft.com/pub/dtg/david/tutorial.ps.

G. R. IVERSEN (1984), *Bayesian Statistical Inference*, Quantitative Applications in the Social Sciences Series 43, Sage Publications, Beverly Hills, CA.

H. JEFFREYS (1961), *Theory of Probability*, Clarendon, Oxford, UK.

F. V. JENSEN (1996), *An Introduction to Bayesian Networks*, Springer-Verlag, New York.

D. MALAKOFF (1999), Bayes offers a "new" way to make sense of numbers, *Science*, 286, pp. 1460–1464.

H. F. MARTZ (1998), private communication based on Martz and Waller (1982).

H. F. MARTZ (2000), An introduction to Bayes, hierarchical Bayes, and empirical Bayes statistical methods in health physics, in *Applications of Probability and Statistics in Health Physics*, T. Borak, ed., Medical Physics Publishing, Madison, WI, pp. 55–84.

H. F. MARTZ AND R. A. WALLER (1982), *Bayesian Reliability Analysis*, Wiley, New York.

H. F. MARTZ, R. L. PARKER, AND D. M. RASMUSON (1999), Estimation of trends in the scram rate at nuclear power plants, *Technometrics*, 41, pp. 352–364.

N. D. SINGPURWALLA (1999), private communication based on T. R. Bement, J. M. Booker, N. D. Singpurwalla, and S. A. Keller-McNulty, *Testing the Untestable: Reliability in the 21st Century*, Report LA-UR-00-1766 and Report GWU/IRRA Serial TR99/6, Los Alamos National Laboratory and George Washington University, Los Alamos, NM and Washington, DC.

Chapter 5

Considerations for Using Fuzzy Set Theory and Probability Theory

Timothy J. Ross, Kimberly F. Sellers, and Jane M. Booker

Most of the applications chapters in this book deal with the quantification of various forms of uncertainty, both numeric and nonnumeric. Uncertainty in numerical quantities can be random in nature, where probability theory is very useful, or it can be the result of bias or an unknown error, in which case fuzzy set theory, evidence theory, or possibility theory might prove useful. Probability theory also has been used almost exclusively to deal with the form of uncertainty due to *chance* (randomness), sometimes called *variability*, and with uncertainties arising from eliciting and analyzing expert information. Three other prevalent forms of uncertainty are those arising from *ambiguity*, *vagueness*, and *imprecision*. While vagueness and ambiguity can arise from linguistic uncertainty, they also can be associated with some numerical quantities, such as "approximately 5." Imprecision is generally associated with numerical quantities, although applications may exist where this type of uncertainty is nonnumeric (i.e., qualitative), for example, "the missile was close to the target." How do variability, ambiguity, vagueness, and imprecision differ as forms of uncertainty?

In the sections that follow, we detail the most popular methods used to address these various forms of uncertainty whether they are quantitative or qualitative. While fuzzy set theory and probability theory have been used for all these forms of uncertainty, this chapter will extend this scope somewhat by commenting on the use of possibility theory in the characterization of ambiguity.

5.1 Vagueness, imprecision, and chance: Fuzziness versus probability

Often vagueness and imprecision are used synonymously, but they can also differ in the following sense. Vagueness can be used to describe certain kinds of uncertainty associated with linguistic information or intuitive information. Examples of vague information are that the data quality is "good" or that the transparency of an optical element is "acceptable."

Moreover, in terms of semantics, even the terms *vague* and *fuzzy* cannot be generally considered synonyms, as explained by Zadeh (1995). For example, "I shall return sometime" is vague and "I shall return in a few minutes" is fuzzy. The former is not known to be associated with any unit of time (seconds, hours, days), whereas the latter is associated with an uncertainty that is at least known to be on the order of minutes. However, notwithstanding this distinction, the terms *vague* and *fuzzy* used herein will represent the same kind of uncertainty; i.e., they will be used interchangeably in our discussions of this general form of uncertainty. As explained by Zadeh (1995), "Usually a vague proposition is fuzzy, but the converse is not generally true."

5.1.1   A historical perspective on vagueness

Black (1939) defines a *vague proposition* as a proposition where the possible states (of the proposition) are not clearly defined with regard to inclusion. For example, consider the proposition that a person is young. Since the term "young" has different interpretations to different individuals, we cannot decisively determine the age(s) at which an individual is young versus the age(s) at which an individual is not considered to be young. Thus the proposition is vaguely defined. Classical (binary) logic does not hold under these circumstances; therefore, we must establish a different method of interpretation.

Black introduced a consistency profile as a graphical representation of the membership of a proposition in a set of states where 1 denotes absolute membership and 0 denotes absolute complementary membership. These profiles are established to represent a precise (crisp) or vague proposition: a precisely defined proposition has a consistency profile that is a step function (Figure 5.1), while a vague proposition has a consistency profile that gradually ranges between the two extremes (Figure 5.2).

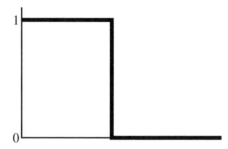

Figure 5.1. *Precise consistency profile.*

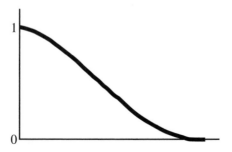

Figure 5.2. *Vague consistency profile.*

5.1.2   Imprecision

Imprecision can be associated with quantitative, or countable, data as well as noncountable data. An example of the latter might be, "the length of a flaw in an integrated circuit is long." An example of countable imprecision would be to report the length of such a flaw as 30 microns. If we use a microscope and measure the length of the flaw 100 times we will likely come up with 100 different values; the differences in the numbers will no doubt be on the order of the precision of the resolution limitations of the microscope and other measurement factors. Measurements using a light microscope (which has accuracy on the order of microns) will be less precise than those developed from an electron microscope (which has resolution on the order of nanometers). If we plot the flaw lengths on some sort of probit paper and develop a Gaussian distribution to describe the length of this flaw, we could state the imprecision in probabilistic terms. In this case, the length of the flaw is uncertain to some degree of precision that is quantified in the language of probability. Since we are not able to make this measurement an infinite number of times, there is also uncertainty in the statistics describing the flaw length. Hence, imprecision can be used to quantify random variability in the quantitative uncertainty, and it also can be used to describe a lack of knowledge for descriptive entities (e.g., acceptable transparency, good data quality). Vagueness is usually related to nonmeasurable issues.

Many practitioners suggest that the uncertainty arising from vagueness makes use of fuzzy set theory, since this theory was developed to handle the vagueness inherent in natural language. Klir and Folger (1988) have shown that the relationships among fuzzy set theory, probability theory, evidence theory, and possibility theory stem from a common framework of fuzzy measures in that they all have been used to characterize and model various forms of uncertainty. That they all are related mathematically (as fuzzy measures) is an especially crucial advantage in their use in quantifying the uncertainty spectrum. As more information about a problem becomes available, the mathematical description of uncertainty easily can transform from one theory to the next in the characterization of the uncertainty as the uncertainty diminishes or, alternatively, as the information granularity increases and becomes specific. There are various forms of fuzzy measures such as belief, plausibility, possibility, and probability. Yager (1993) has discussed the relationship between a possibility distribution and a fuzzy set. Ongoing research continues to provide theoretically based methods for translating uncertainties established in one of these measures into uncertainties in another without violating the fundamental axioms of either.

5.2   Chance versus vagueness

Suppose that you are a basketball recruiter and are looking for a "very tall" player for the center position on a men's team. One of your information sources tells you that a hot prospect in Oregon has a 95% chance[1] of being over 7 feet tall. Another source tells you that a good player in Louisiana has a high membership in the set "very tall" people. The problem with the information from the first source is that it is a quantity constrained by the law of the excluded-middle. There is a 5% chance that the Oregon player is not over 7 feet tall; this person could conceivably be of extremely short stature. The second source of information would, in this case, contain less uncertainty for the recruiter because if the player turned out to be less than 7 feet tall, then there is a high likelihood that this player would still be quite tall.

---

[1]Here the use of the term *chance* refers to the relative frequency interpretation of probability (Chapter 3, section 3.4). For example, there will be 40% rotten apples in the refrigerator. If it were known how many total apples there are and the term "rotten" were defined, then rottenness would be measured unambiguously.

As discussed in Chapter 1 for a similar example comparing two glasses of liquids, what philosophical distinction can be made regarding this form of information? Suppose that we are allowed to measure the basketball players' heights. The prior probability of 0.95 becomes a posterior probability of 1.0 or 0.0; i.e., the player either is or is not over 7 feet tall. However, the membership value of 0.95 that measures the extent to which the player's height is over 7 feet remains 0.95 after measuring. This example and the one from Chapter 1 illustrate very clearly the difference in the information content between chance and vague events.

This brings us to the clearest distinction between fuzziness and randomness: *Fuzziness describes the vagueness of an event, whereas chance describes the uncertainty in the occurrence of the event.* The event will or will not occur; but is the description of the event clear and specific enough to measure its occurrence or nonoccurrence?

5.3   Many-valued logic

Binary logic pertains to propositions that are either true or false. However, Lukasiewicz (1930) recognized the existence of propositions that are simultaneously true and false at various levels. For example, a degraded yet functioning system cannot be classified as being in state 1 (perfect functioning), nor can it be classified as being in state 0 (total failure). For such propositions, their membership is neither 1 nor 0. Therefore, Lukasiewicz assigned a third number, $\frac{1}{2}$, as its associated truth value (i.e., membership value) to account for any deviation from states 0 or 1. For example, a degraded component (system) would be considered in state $\frac{1}{2}$. This concept defines a *three-valued logic*.

Lukasiewicz (1930) defined the truth table corresponding to the implication $X \to Y$ for two propositions X and Y ("if X, then Y") and negation X' (propositions complementary to X); see Tables 5.1 and 5.2.

Table 5.1. *Three-valued truth table for $X \to Y$.*

$X \to Y$		Values of proposition Y		
		0	$\frac{1}{2}$	1
Values of proposition X	0	1	1	1
	$\frac{1}{2}$	$\frac{1}{2}$	1	1
	1	0	$\frac{1}{2}$	1

Table 5.2. *Three-valued truth table for X'.*

Values of proposition X	Values of proposition X'
0	1
$\frac{1}{2}$	$\frac{1}{2}$
1	0

From the logical equivalencies for conjunction ("AND," denoted $\wedge$) and disjunction ("OR," denoted $\vee$), we have

$$X \wedge Y = (X \to Y) \to Y \quad \text{and} \quad X \vee Y = (X' \wedge Y')'.$$

Truth tables can be found in Malinowski (1993); also see Tables 5.3 and 5.4. As a result,

Table 5.3. *Three-valued truth table for $X \wedge Y$.*

$X \wedge Y$		Values of proposition Y		
		0	$\frac{1}{2}$	1
Values of proposition X	0	0	0	0
	$\frac{1}{2}$	0	$\frac{1}{2}$	$\frac{1}{2}$
	1	0	$\frac{1}{2}$	1

Table 5.4. *Three-valued truth table for $X \vee Y$.*

$X \vee Y$		Values of proposition Y		
		0	$\frac{1}{2}$	1
Values of proposition X	0	0	$\frac{1}{2}$	1
	$\frac{1}{2}$	$\frac{1}{2}$	$\frac{1}{2}$	1
	1	1	1	1

$$X \wedge Y = \min(X, Y)$$

and

$$X \vee Y = \max(X, Y).$$

These ideas can be generalized to define an n-valued logic in that propositions are true and false simultaneously but with varying degrees of truth (or falsity). Furthermore, extensions of the n-valued logic to the continuum lead to the notions of vagueness and fuzzy logic.

5.4 Axiomatic structure of probability and fuzzy logics

To attempt to bring some relevance to the historical confusion between the logics associated with probability theory and fuzzy set theory, a paper by Gaines (1978) does an eloquent job of addressing this issue. Historically, probability and fuzzy sets have been presented as distinct theoretical foundations for reasoning and decision-making in situations involving uncertainty. Yet when one examines the underlying axioms of both probability and fuzzy set theories, the two theories differ only by one axiom in a total of 16 axioms needed for a complete foundation! The material that follows is a brief summary of Gaines' paper which established a common basis for both forms of logic of uncertainty in which *a basic uncertainty logic* is defined in terms of truth valuation on a lattice of propositions. Addition of the axiom of the excluded-middle to the basic logic defines *standard probability logic*. Alternatively, addition of a requirement for strong truth functionality yields a *fuzzy logic*.

In this discussion, fuzzy logic is taken to be a many-valued extension of Boolean logic based on fuzzy set theory in which truth values are a continuous function between the endpoints of the interval [0, 1]. This is seen to be an extension of the idea proposed by Lukasiewicz (1930) discussed in the previous section. The normal logical operations of both probability theory and fuzzy logic are defined in terms of arithmetic operations on values $x \in [0, 1]$; the values are regarded as degrees of membership to truth. The logic operations and associated arithmetic operations are those of conjunction, disjunction, and negation. Gaines (1978) points out that the use of the max and min operators in fuzzy logic is not sufficient to distinguish it from probability theory—both operators arise naturally in the calculation of the conjunction and disjunction of probabilistic events. Our association of addition and multiplication as natural operations upon probabilities comes from our frequent interest in statistically independent events, not from the logic of probability itself.

In developing a basic uncertainty logic (one that embraces both a fuzzy set theory and a probability theory), we begin first by defining a lattice consisting of a universe of discourse X, a maximal element T, a minimal element F, a conjunction $\wedge$, and a disjunction $\vee$. This lattice will be denoted $L(X, T, F, \wedge, \vee)$. For the axioms (or postulates) to follow, the lower-case letters x, y, and z denote specific elements of the universe X within the lattice. The following 15 axioms completely specify a basic uncertainty logic. The basic uncertainty logic begins with the lattice L satisfying the following axioms:

1. idempotency: for all $x \in L$, $x \vee x = x \wedge x = x$;

2. commutativity: for all $x, y \in L$, $x \vee y = y \vee x$; $x \wedge y = y \wedge x$;

3. associativity: for all $x, y, z \in L$, $x \vee (y \vee z) = (x \vee y) \vee z$;

4. absorption: for all $x, y \in L$, $x \vee (x \wedge y) = x$ and $x \wedge (x \vee y) = x$;

5. definition of maximal and minimal elements: for all $x \in L$, $x \vee T = T$, $x \wedge T = x$.

The usual order relation may also be defined as follows:

6. $x, y \in L$, $x \leq y$ if there exists a $z \in L$ such that $y = x \vee z$.

Now suppose that every element of the lattice L is assigned a truth value (for various applications this truth value would be called a probability, degree of belief, etc.) in the interval $[0, 1]$ by a continuous order-preserving function $p : L \to [0, 1]$ with constraints

7. $p(F) = 0$, $p(T) = 1$,

8. for all $x, y \in L$, $x \leq y$; then $p(x) \leq p(y)$,

and an additivity axiom

9. for all $x, y \in L$, $p(x \wedge y) + p(x \vee y) = p(x) + p(y)$.

We note that for p to exist, we must have the postulate

$$p(x \wedge y) \leq \min[p(x), p(y)] \leq \max[p(x), p(y)] \leq p(x \vee y).$$

Now a logical equivalence (or congruence) is defined by

10. for all $x, y \in L$, $x \leftrightarrow y$ if $p(x \wedge y) = p(x \vee y)$.

The general structure so far provided by the first 10 axioms is common to virtually all forms of logic. To finalize Gaines' basic uncertainty logic, we now need to define implication and negation, for it is largely the definition of these latter two operations that distinguish various multivalued logics (Gaines (1978)). We also note that postulate 9 still holds when the outer inequalities become equalities—a further illustration that the additivity of probability-like valuations is completely compatible with and closely related to the minimum and maximum operations of fuzzy logic.

To define implication and negation, we make use of a metric on the lattice L that measures *distance* between the truth values of two different propositions. This is based on the notion that logically equivalent propositions should have a zero distance between them. Thus we define a distance measure $d(x, y)$ satisfying

$$d(x, x) = 0,$$
$$0 \leq d(x, y) \leq 1,$$

and

$$d(x, y) + d(y, z) = d(x, z)$$

such that

11. for all $x, y \in L$, $d(x, y) = p(x \vee y) - p(x \wedge y)$.

Therefore, a measure of equivalence between two elements can be 1 minus the distance between them:

12. for all $x, y \in L$, $p(x \leftrightarrow y) = 1 - d(x, y) = 1 - p(x \vee y) + p(x \wedge y)$.

Hence if $d = 0$, the two elements x and y are equivalent.

To measure the strength of an implication, we measure a distance between x and $x \wedge y$:

13. for all $x, y \in L$, $p(x \rightarrow y) = p(x \leftrightarrow x \wedge y) = 1 - d(x, x \wedge y) = 1 - p(x) + p(x \wedge y) = 1 + p(y) - p(x \vee y) = 1 - d(y, x \vee y)$.

Negation (the complement of x is denoted x') now can be defined in terms of equivalence and implication:

14. for all $x \in L$, $p(x') = p(x \leftrightarrow F) = 1 - p(x) = 1 - d(x, F)$.

We note that, by combining axioms 9 and 14, if element y is replaced by element x', we get

$$p(x \vee x') + p(x \wedge x') = p(x) + p(x') = 1.$$

Finally, we add a postulate of distributivity:

15. for all $x, y, z \in L$, $x \wedge (y \vee z) = (x \wedge y) \vee (x \wedge z)$,

which will prove useful for the two specializations of this basic uncertainty logic described in the following axioms.

Axioms 1–15 provide for a basic distributive uncertainty logic. The addition of the following special 16th axiom (denoted 16.1), known in the literature as the law of the excluded-middle (see Chapter 3),

16.1. for all $x \in L$, $p(x \vee x') = 1$,

leads to Rescher's standard probability logic (Rescher (1969)). Alternatively, if we add another special 16th axiom (denoted 16.2) to the basic axioms 1–15, we get a special form of fuzzy logic,

16.2. for all $x, y \in L$, $\{p(x \rightarrow y) = 1 \vee p(y \rightarrow x) = 1\}$, known in the literature as the Lukasiewicz infinite-valued logic (Rescher (1969)).

Axiom 16.2 is called the *strong truth functionality* (*strict implication*) in the literature. It should be pointed out that a weaker form of axiom 16.2,

$$\text{for all } x, y \in L, \quad p((x \to y) \vee (y \to x)) = 1,$$

is embraced by both probability and fuzzy logic (Gaines (1978)).

In summary, the preceding material has established a formal relationship between probability and fuzzy logic and has illuminated axiomatically that their common features are more substantial than their differences. It should be noted that while axiom 9 is common to both probability and fuzzy logic, it is rejected in the Dempster–Shafer theory of evidence (Gaines (1978)), and it often presents difficulties in human reasoning. For probability logic, the law of the excluded-middle (or its dual, the law of contradiction, i.e., $p(x \wedge x') = 0$) *must* apply; for fuzzy logic, it *may* or *may not* apply.

5.4.1   Relationship between vagueness and membership functions

As stated in Chapter 2, Zadeh (1965) introduced membership functions as a means to represent an assessor's belief of containment in a fuzzy set. This ideology stems from ideas such as those addressed in the historical paper of Black (1939). A consistency profile can be viewed as a graphical parallel to Zadeh's membership function associated with fuzzy sets. For this section, we assume $\mu_A(x)$ to be a normalized membership function; i.e., for all x, $0 \le \mu_A(x) \le 1$. By definition, a membership function describes a belief of containment in a fixed set A, where 1 represents absolute membership in the set and 0 represents absolute complementary membership. If $\mu_A(x) = 0$ or 1 for all x, then A is a precise (crisp) set. Otherwise, A is a fuzzy set.

Example I

Let $A_1 = \{x \in (1, 2, \ldots, 10) | x \le 4\}$. Given the definition $x \le 4$, it is clear that for any value of x, we can establish absolute membership or no membership in the set A_1. Therefore, A_1 is a precise set because $\mu_{A_1}(x)$ can take only values 0 or 1. The membership function is represented in Table 5.5.

Table 5.5. *Membership table for precise set A_1.*

x	1	2	3	4	5	6	7	8	9	10
$\mu_{A_1}(x)$	1	1	1	1	0	0	0	0	0	0

Example II

In contrast to Example I, consider $A_2 = \{x \in (1, 2, \ldots, 10) | x \text{ is small}\}$. Given the term "small," we cannot easily determine the level of containment in the set A_2. Then a possible membership function $\mu_{A_2}(x)$ is represented in Table 5.6. A_2 is a fuzzy set because the membership function allows for a range of values between 0 and 1 to describe the level of membership.

Zadeh (1968) defines operations of membership functions for two fuzzy sets A and B and for all x as follows:

Table 5.6. *Membership table for precise set A_2.*

x	1	2	3	4	5	6	7	8	9	10
$\mu_{A_2}(x)$	1	1	0.8	0.55	0.3	0.1	0	0	0	0

1. $\mu_{A \cup B}(x) = \max(\mu_A(x), \mu_B(x))$,

2. $\mu_{A \cap B}(x) = \min(\mu_A(x), \mu_B(x))$,

3. $\mu_{A^c}(x) = 1 - \mu_A(x)$,

4. $A \subseteq B \Leftrightarrow \mu_A(x) \leq \mu_B(x)$,

5. $A \equiv B \Leftrightarrow \mu_A(x) = \mu_B(x)$.

5.4.2 Relationship between fuzzy set theory and Bayesian analysis

For membership functions corresponding to precise sets, we can construct probability measures without any loss of coherence because the precise sets guarantee that all subsets of the σ-algebra will be clearly defined. However, this is not the case for fuzzy sets. Due to the loss in precision of fuzzy sets, we must reexamine membership functions in an effort to relate fuzzy set theory to Bayesian analysis (see Chapter 4).

In many respects, probability functions and membership functions follow the same axioms, in particular, $p(x), \mu_A(x) \geq 0$. However, while $\sum_x p(x) = 1$ for probability distributions, $\sum_x \mu_A(x)$ may not necessarily equal 1. Thus $\mu_A(x)$ cannot be interpreted as a probability because it does not satisfy this important property. However, we can interpret $\mu_A(x)$ as a likelihood of x for a fixed set A. Defining a membership function is philosophically consistent with the process of defining a likelihood function in that it is a subjective exercise reflecting the assessor's opinion regarding the level of membership in the set A. Furthermore, A represents an observation while the possible values of x represent the hypotheses. This interpretation of the membership function as a likelihood function may seem unconventional. However, this proposed interpretation is consistent with the foundational notion of a likelihood function as viewed from a more philosophical point of view and, as noted in Chapter 4, permits combinations of uncertainties from probability and membership functions.

5.5 Early works comparing fuzzy set theory and probability theory

Loginov (1966) made the earliest attempt to relate fuzzy set theory and probability theory by interpreting the membership function as a conditional probability. Let A be a fuzzy set with membership function $\mu_A(x)$. Suppose that an experiment X is to be performed and that Ω represents the set of all possible outcomes of X. Let $x \in \Omega$ be one such outcome of X. According to Loginov (1966),

$$\mu_A(x) = P(x \in A | \text{outcome of } X \text{ is } x);$$

i.e., $\mu_A(x)$ is the conditional probability that x is classified in A (should x be the outcome of X).

Loginov, being a member of the Russian school of probability, views probability in the frequentist sense. Thus in order to proceed with this interpretation, he conceptualizes an ensemble of membership function specifiers, each of whom has to vote on $x \in A$ or $x \in A^C$. Zadeh (1995) dismisses Loginov's interpretation on the following grounds:

1. Requiring each voter to classify any observed $x \in A$ or $x \in A^C$, where A is a fuzzy set, is unnatural. Recall that fuzzy sets were introduced to reject the law of the excluded-middle.

2. Membership functions are often specified by one individual based on subjective considerations. The consensus model (i.e., a majority vote) is unrealistic.

The second attempt at making fuzzy set and probability theories work in concert was due to Zadeh (1968). His construction proceeds as follows:

- Let (Ω, F, P) be a probability space with $x \in \Omega$.

- Let $C \in F$ be a crisp set. Then it is true that if $I_C(x)$ is the indicator function of set C, then

$$P(C) = \int_\Omega I_C(x) dP.$$

Motivated by the above, Zadeh defines (or declares) that the P-measure of a fuzzy subset $A \in \Omega$ (called a fuzzy event) with membership function $\mu_A(x)$ is

$$P(A) = \int_\Omega \mu_A(x) dP = E[\mu_A(x)],$$

the expectation taken with respect to the initial probability measure P. Having defined $P(A)$ as above, he shows that the following rules of probability hold for two fuzzy sets A and B:

1. $A \subseteq B \Rightarrow P(A) \leq P(B)$,

2. $P(A \cup B) = P(A) + P(B) - P(A \cap B)$,

3. $P(A + B) = P(A) + P(B) - P(AB)$; note that AB is the product and not the intersection of A and B.

Extensions of the above to the cases of finite and countable additivity follow by induction.

Finally, A and B are declared independent if $P(AB) = P(A) \cdot P(B)$; again, AB refers to the product as opposed to the intersection of the sets A and B. Furthermore,

$$P(A|B) = \frac{P(AB)}{P(B)} \quad \text{if } P(B) > 0$$

is the conditional probability of A were B to occur. This contrasts the definition of a conditional probability as provided in Chapter 4 in that AB replaces $A \cap B$. Thus if A and B are independent (denoted $A \perp B$) in the sense mentioned before, then $P(A|B) = P(A)$. The motivation for introducing the notions of sums and products of fuzzy sets is now clear. However, the interpretation of these two operations remains unclear. As a result, the following concerns arise:

1. In $P(A \cup B) = P(A) + P(B) - P(A \cap B)$, the assessment of $P(A \cap B)$ is never established. Instead, $P(A+B) = P(A) + P(B) - P(AB)$ is addressed with $P(AB) = P(A) \cdot P(B)$ if $A \perp B$ or with $P(AB) = P(A|B) \cdot P(B)$.

2. How can $P(A)$ or, for that matter, $P(A|B)$, be made operational?

3. Given that we are assessing probabilities of fuzzy events, there is no law of probability that leads to $P(A)$. Since $\mu_A(x)$ is not a probability, the law of total probability does not apply when declaring $P(A) = \int_\Omega \mu_A(x) dP$.

5.6 Treatment of uncertainty and imprecision: Treating membership functions as likelihoods

We address this matter by conceptualizing a scenario involving a purely subjective probabilist, say, D, who is interested in assessing the probability of a fuzzy set A. That is, D needs to pin down $P_D(A) = P_D(x \in A)$, where x is the unknown outcome of an experiment X. The subscript D denotes that the probability is subjective with respect to the assessor D. Now D is confronted with two uncertainties: uncertainty about the outcome $X = x$ and uncertainty about classifying the membership of any x in A. As a subjectivist, D views imprecision as simply another uncertainty and, to D, all uncertainties can only be quantified by probability. Thus in keeping with this dictum, D specifies two probabilities:

- D's prior probability that an x will be the outcome of X (denoted $P_D(x)$);

- D's prior probability that an outcome x belongs to A (denoted $P_D(x \in A)$).

Although the assignment $P_D(x)$ is standard and noncontroversial, the assessment of $P_D(x \in A)$ may raise an issue pertaining to the classification of x. Specifically, whereas $P_D(x \in A)$ is D's probability that x is classified in A, the question now arises of who is doing the classification and on what basis the classification is being done.

The point of view we adopt is analogous to the "genie," introduced by Laplace. The genie knows it all and rarely tells it all. In our case, nature can classify any x in any set A or A^C with precision so that, to nature, all sets are crisp, and so that fuzzy sets are only a consequence of our uncertainty about the boundaries of sharp sets. Thus $P_D(x \in A)$ reflects D's uncertainty (or partial knowledge) of the boundaries of a crisp set. Nature will never reveal these boundaries, so D's uncertainty of classification is impossible to ever resolve; i.e., the genie is able to classify any x with precision, but D is unsure of this classification. However, D has partial knowledge of the genie's actions, and this is encapsulated in D's probability, $P_D(x \in A)$. We note that although D's uncertainty about how nature classifies x will never be resolved, this does not make a subjective specification of $P_D(x \in A)$ vacuous. Obtaining probability measures of fuzzy sets is a relevant exercise because probability may be required for decision making under uncertainty. Therefore, as long as probability measures of fuzzy sets are harnessed within a broader framework of decision-making, they are necessary and useful. This viewpoint is also supported in practice in that the biggest impact of fuzzy set theory has been in control theory.

After specifying $P_D(x)$ and $P_D(x \in A)$ for all $x \in \Omega$, D can use the law of total probability to obtain

$$P_D(A) = P_D(X \in A) = \sum_x P_D(X \in A | X = x) P_D(x) = \sum_x P_D(x \in A) P_D(x),$$

which is the expected value of D's classification probability with respect to D's prior probability of X. Note that the probability measure for fuzzy set A does not consider the membership function. The given $P_D(A)$ has been based on D's inputs alone. To incorporate the role of the membership function in the assessment of a probability measure of A, we elicit information from our expert Z, whose expertise lies in specifying membership functions $\mu_A(x)$ for all $x \in \Omega$ and any set A.

Now suppose that after assessing a $P_D(A)$, D consults Z and obtains the function $\mu_A(x)$ for all $x \in \Omega$. D must now update $P_D(A)$ in light of Z's expert testimony. Thus D now needs to assess $P_D(A; \mu_A(x))$. In order to appropriately represent the expansion of knowledge, D will first consider this probability as if $\mu_A(x)$ were unknown, therefore interpreting $P_D(A; \mu_A(x))$ and appealing to the calculus of probability (see Chapter 3, section 3.1) to obtain

$$P_D(X \in A | \mu_A(x)) = \sum_x P_D(X \in A | X = x, \mu_A(x)) \cdot P_D(X = x | \mu_A(x)).$$

In writing out the above, D treats $\mu_A(x)$ as a random quantity. The supposition here is that when D is contemplating probabilities, Z's response is unknown. If D assumes that the process by which X is generated is independent of the manner by which Z specifies $\mu_A(x)$, then

$$P_D(X \in A | \mu_A(x)) = \sum_x P_D(X \in A | X = x, \mu_A(x)) \cdot P_D(x)$$

$$= \sum_x P_D(x \in A | \mu_A(x)) \cdot P_D(x),$$

where, according to Bayes' theorem (see Chapter 4, section 4.2.1),

$$P_D(x \in A | \mu_A(x)) \propto P_D(\mu_A(x) | x \in A) \cdot P_D(x \in A).$$

In actuality, D knows $\mu_A(x)$ to be Z's expert testimony; therefore, using the likelihood,

$$P_D(x \in A; \mu_A(x)) \propto L_D(x \in A; \mu_A(x)) \cdot P_D(x \in A),$$

where $L_D(x \in A; \mu_A(x))$ is D's likelihood that Z specifies $\mu_A(x)$ when nature classifies $x \in A$. The specification of $L_D(x \in A; \mu_A(x))$ is subjective with respect to the assessor D. If D wishes to adopt Z's judgments without any tampering, then $L_D(x \in A; \mu_A(x)) = \mu_A(x)$ for all $x \in \Omega$. Otherwise, if D feels that Z has certain biases, then D will suitably modify $\mu_A(x)$ to $\mu_A^*(x)$ (where $*$ denotes D's modification). Assuming, however, that D chooses not to tamper with $\mu_A(x)$, then

$$P_D(X \in A; \mu_A(x)) \propto \sum_x \mu_A(x) P_D(x \in A) P_D(x)$$

is D's unnormalized probability measure of the fuzzy set A, a measure that encapsulates an (untampered) membership function $\mu_A(x)$. As the above two relationships suggest, the membership function can serve as the likelihood in Bayes' theorem, providing a translation mechanism from fuzzy logic to probability.

5.7   Ambiguity versus chance: Possibility versus probability

A fuzzy measure describes the *ambiguity* in the assignment of an element "x" from a universe of discourse X to one or more crisp subsets on the power set of the universe. For example, if $X = \{a, b, c\}$, then the power set is the collection of all possible subsets in X and is given by

PowerSet$(X) = \{\emptyset, a, b, c, (a \cup b), (a \cup c), (b \cup c), (a \cup b \cup c)\}$. Figure 5.3 illustrates this idea. In the figure, the universe of discourse is comprised of a collection of sets and subsets, or the power set. In a fuzzy measure, what we are trying to describe is the *ambiguity* in assigning this point x to any of the crisp sets on the power set. This is not a random notion; it is a case of ambiguity. The crisp subsets that make up the power set have no uncertainty about their boundaries as fuzzy sets do. The uncertainty in the case of a fuzzy measure lies in the ambiguity in making the assignment. This uncertainty is usually associated with evidence to establish an assignment. The evidence can be completely lacking—the case of total ignorance—or can be complete and specific—the case of a probability assignment. Hence the difference between a fuzzy measure and a fuzzy set on a universe of elements is that in the former the ambiguity is in the assignment of an element to one or more crisp sets, and in the latter the vagueness is in the prescription of the boundaries of a set. We see a distinct difference between the two kinds of uncertainty, ambiguity and vagueness. Hence what follows is a mathematical description of a theory very useful in addressing ambiguity— possibility theory. This section is given for completeness in describing mathematical models useful in addressing various forms of uncertainty and not as a matter of extending the scope of this book beyond a comparison of the utilities of fuzzy set theory and probability theory.

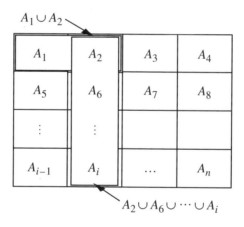

Figure 5.3. *The power set of a universe X.*

To continue, a set function $v(A) \in [0, 1]$ for $A \subseteq X$ is a fuzzy measure if $A \subseteq B \rightarrow v(A) \leq v(B)$. A triangular norm is a conjunctive operator $n : [0, 1]^2 \rightarrow [0, 1]$, which is associative, commutative, and monotonic, with identity 1. Typical examples are the multiplication $(x * y)$ and minimum $(\min(x, y))$ operators. Their dual conorms are disjunctive, have identity 0, and include bounded sum $\min(x + y, 1)$ and maximum $\max(x, y)$. Norms and conorms are used to operate on fuzzy measures in various ways. They also represent the different degrees of dependence or independence among events in a probability space and thus all possible relations among marginal and joint probability distributions (Schweizer (1991)).

A probability measure P is a fuzzy measure, where $P(A \cup B) = P(A) + P(B) - P(A \cap B)$. The other major class of fuzzy measures consists of possibility measures $\prod$, where $\prod(A \cup B) = \max(\prod(A), \prod(B))$. Where a probability measure is characterized by its probability distribution $p(x) = P(\{x\})$ for $x \in X$ such that $P(A) = \sum_{x \in A} p(x)$, a possibility measure is characterized by its possibility distribution $\pi(x) = \prod(\{x\})$ such that $\prod(A) = \max_{x \in A} \pi(x)$. Also, while both probability and possibility measures are generally normal, with $P(X) = \prod(X) = 1$, for a probability measure, this entails additivity

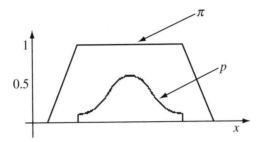

Figure 5.4. *Illustration of possibility function and PDF plotted on same space.*

$\sum_{x \in X} p(x) = 1$, while for possibility measures this is $\max_{x \in X} \pi(x) = 1$. Finally, both p and π can be fuzzy sets.

Possibility theory, with its maximal normalization which is weaker than probability, is sometimes portrayed as equivalent to fuzzy set theory, and thus possibility measures are presumed to be determined by the same methods of expert elicitation. However, our interpretation and that of others (Joslyn (1995)) is that possibility measures and fuzzy sets are representations with distinct formal and semantic domains (which would be reflected in the elicitation process). However, the combination of possibility measures, as measures on sets, and nonadditive fuzzy sets, as functions on points, can have significant advantages over the strict use of probability measures in studies such as system reliability (Ross (2000)).

Joslyn (1994a, 1994b) developed the semantic and methodological bases for possibility theory on the basis of random sets and intervals and applied it to qualitative modeling. This approach is motivated by two important concepts of Gaines and Kohout (1976), who distinguish between events A that are "traditionally possible" in that $P(A) = \varepsilon > 0$ for some small ε and thus *must* occur in the limit of infinite time, and those that are *not impossible* but still *need not ever occur*. This latter case can be identified as "properly possible."

Building from this, we can draw on Zadeh's (1978) measure of compatibility between a probability and possibility distribution as

$$g(p, \pi) = \sum_{x \in X} p(x)\pi(x).$$

Joslyn identified the condition of strong compatibility when $g(p, \pi) = 1$, which, when combined with the respective normalization requirements, requires that possibility is unity wherever probability is positive, and vice versa.

Thus by this view, traditionally possible events, including all events that are actually observed, require total mathematical possibility $\prod(A) = 1$ and positive probability $P(A) > 0$. However, properly possible events are those such that $\prod(A) \in (0, 1)$ yet such that $P(A) = 0$. These are precisely the rare events that are so important, for example, in reliability studies of high-consequence systems (Cooper and Ross (1997)). Figure 5.4 shows an example in which the cutoff tails of a Gaussian distribution represent the rare and thus properly possible events.

In a hybrid approach as we are describing, where a subjective model using a possibility distribution will be married to a probabilistic model derived from frequency data, $g(p, \pi)$ can provide a measure of the fit between the two models. In particular, $g(p, \pi) = 0$ indicates a complete inconsistency between the subjective and objective (probabilistic) models.

Now, consider a random set defined as a probability distribution m on the sets $A \subseteq$

$\mathcal{P}(X)$ (the power set of X) so that $\sum_{A \subseteq \mathcal{P}(X)} m(A) = 1$. Given m, define the plausibility

$$Pl(A) = \sum_{B \cap A \neq \emptyset} m(B)$$

and the belief

$$Bel(A) = \sum_{B \subseteq A} m(B).$$

Pl and Bel are fuzzy measures. If m is specific so that $m(A) > 0 \rightarrow |A| = 1$, then Bel = Pl is recovered as a probability measure. If, on the other hand, m is consonant so that for all A and B such that $m(A), m(B) > 0$ either $A \subseteq B$ or $B \subseteq A$, then Pl is a possibility measure. Strict consonance can be relaxed somewhat to consistency so that $\bigcap_{A:m(A)>0} A \neq \emptyset$ and possibility measures are recovered.

Thus probability and possibility measures are distinct special cases of fuzzy measures in random sets (Joslyn (1996)). In general, nonspecific (cardinality-greater-than-1) evidence yields general plausibilities and beliefs—and possibilities in the case of consonant evidence—but as the evidence becomes more specific, a strictly probabilistic representation is recovered. More generally, Goodman, Mahler, and Nguyen (1997) demonstrated that random sets provide a natural framework in which to combine these fuzzy measures with fuzzy sets by considering the fuzzy sets as "one-point traces" of random sets. This model thereby naturally yields fuzzy measures of the type more appropriate for nonspecific data, but it does so in the context of an overall *probability* model on a higher-order space. This allows both the construction of possibility measures from empirical sources and their combination with fuzzy sets.

In particular, when X is on the real line, a random interval A results. Figure 5.5 shows a consistent random interval as an ensemble of imprecise interval observations resulting in a possibilistic histogram (Joslyn (1997)).

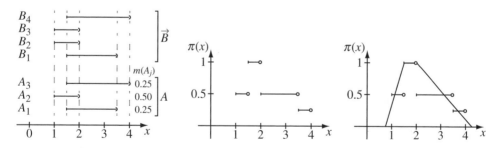

Figure 5.5. *Left top: A collection of four interval observations. Left bottom: Their random interval representation. Center: The raw possibilistic histogram. Right: A trapezoidal approximation.*

Note how this representation uses Zadeh's principle: the source intervals are nonspecific, indicating the occurrence of an event anywhere within them. Where they all agree is on the interval [1.5, 2], which is where $\pi = 1$ and thus where we would concentrate the mass of any probability distributions to be used. Nevertheless, the possibilistic histogram also captures the variance information in the tails (Joslyn (1994b)). As a final note, it is important to point out that if the information content in the uncertainty being addressed *is consonant*, then the possibility distribution has the *same* properties as a fuzzy membership function. If this is the case, then the possibility distribution can take the place of a fuzzy membership function that might be used in the analysis of some problem.

5.8   Conclusions

This chapter has provided a description of the differences between a probability theory and a fuzzy set theory. The basic difference between the two ideas is subtle: Probability theory is based on inherently binary notions of membership and the strict separation and allocation of information to a set and its complement, while fuzzy set theory provides for a gradual transition for processing the information from one set to its complement. Where probability theory has unique membership functions and adheres to a property of coherence, fuzzy set theory has an infinite number of choices for membership functions and is not based axiomatically on coherence. Basically, probability theory adheres to the axiom of the excluded-middle, and fuzzy set theory is not constrained by such an axiom.

Chapter 1 discusses some reasons why adherence to the excluded-middle axiom might be inappropriate in a model. Moreover, for emphasis, we repeat a few sentences here: "A question of whether a fuzzy logic is appropriate is, after all, a question of balancing the model with the nature of the uncertainty contained within it. Problems without an underlying physical model; problems involving a complicated weave of technical, social, political, and economic factors and decisions; and problems with incomplete, ill-defined, and inconsistent information, where conditional probabilities cannot be supplied or rationally formulated, perhaps are candidates for fuzzy logic applications. It has become apparent that the engineering profession is overwhelmed by these sorts of problems." Many of these involve the formal elicitation and use of expert judgment. As noted in Chapter 6, the reasons and steps listed here equally apply to determining whether to address probability or fuzzy set theory.

This chapter concludes with an interesting discussion of ambiguity versus chance, where a possibility theory is introduced. While possibility theory and fuzzy set theory have some common axiomatic similarities, they differ in one important idea: the former measures the ambiguity in assignments to predefined crisp sets, while the latter measures vagueness in the definition of a set itself. While the appropriateness and utility of possibility theory are not central themes in this book, this brief discussion does serve to highlight the fact that there are other theories available to treat various forms of uncertainty of the nonrandom kind— possibility theory being one of those. Possibility can be thought of as being complementary to the *degree of surprise* that a person experiences IF an event occurs, as opposed to *whether* it occurs as a matter of chance. An essential difference between possibility and probability is that the possibility of the union of mutually exclusive events is equal to the maximum of the possibilities of the individual event, whereas for a probability this union is the sum of the mutually exclusive individual probabilities (additivity). Hence the possibility of an event and its complement add to a number greater or equal to unity, whereas the probabilities of an event and its complement must add to 1 (law of excluded-middle). This chapter has shown, however, that possibilities become probabilities as the information granularity becomes more specific, and this is a very nice feature in modeling various forms of uncertainty in the same problem.

To address the question of which theory should be used on what problem, we need to look at the information content of the problem. Is the information countable, measurable, and plentiful, or is the information sparse, vague, ambiguous, or nonnumeric? Once these questions are answered the choice of an approach—or the use of a hybrid approach—might readily become apparent.

References

J. BEZDEK (1993), Editorial: Fuzzy models—What are they, and why?, *IEEE Trans. Fuzzy Systems*, 1, pp. 1–5.

N. BLACK (1939), Vagueness: An exercise in logical analysis, *Philos. Sci.*, 4, pp. 427–455.

D. BLOCKLEY (1983), Comments on "Model uncertainty in structural reliability" by Ove Ditlevsen, *J. Struct. Safety*, 1, pp. 233–235.

A. COOPER AND T. ROSS (1997), Report, Sandia National Laboratory, Albuquerque, NM.

B. GAINES (1978), Fuzzy and probability uncertainty logics, *Inform. and Control*, 38, pp. 154–169.

B. R. GAINES AND L. J. KOHOUT (1976), The logic of automata, *Internat. J. Gen. Systems*, 2, pp. 191–208.

I. R. GOODMAN, R. P. S. MAHLER, AND H. T. NGUYEN (1997), *Mathematics of Data Fusion*, Kluwer Academic Publishers, Dordrecht, The Netherlands.

C. JOSLYN (1994A), *Possibilistic Processes for Complex Systems Modeling*, Ph.D. thesis, Department of Systems Science, SUNY Binghamton, Binghamton, NY.

C. JOSLYN (1994B), A possibilistic approach to qualitative model-based diagnosis, *Telematics and Informatics*, 11, pp. 365–384.

C. JOSLYN (1995), In support of an independent possibility theory, in *Foundations and Applications of Possibility Theory*, G. de Cooman, D. Ruan, and E. E. Kerre, eds., World Scientific, Singapore, pp. 152–164.

C. JOSLYN (1996), Aggregation and completion of random sets with distributional fuzzy measures, *Internat. J. Uncertainty, Fuzziness, and Knowledge-Based Systems*, 4, pp. 307–329.

C. JOSLYN (1997), Measurement of possibilistic histograms from interval data, *Internat. J. Gen. Systems*, 26, pp. 9–33.

G. KLIR AND T. FOLGER (1988), *Fuzzy Sets, Uncertainty, and Information*, Prentice–Hall, Englewood Cliffs, NJ.

V. J. LOGINOV (1966), Probability treatment of Zadeh membership functions and their use in pattern recognition, *Engrg. Cybernet.*, pp. 68–69.

J. LUKASIEWICZ (1930), Philosophische Bemerkungen zu mehrwertigen Systemen des Aussagenkalküls, *C. R. Séances Soc. Sci. Lett. Varsovie Cl.* III, 23, pp. 51–77; English translation: Philosophical remarks on many-valued systems of propositional logic, in *Polish Logic* 1920–1939, X. McCall, ed., Clarendon, Oxford, UK, 1967, pp. 40–65.

G. MALINOWSKI (1993), *Many-Valued Logics*, Clarendon, Oxford, UK.

N. RESCHER (1969), *Many-Valued Logic*, McGraw–Hill, New York.

T. ROSS (2000), *A Method for Developing Empirical Possibility Distributions in Structural Dynamics Models*, NAFIPS Annual Conference, Atlanta, GA, IEEE, Piscataway, NJ.

B. SCHWEIZER (1991), Thirty years of copulas, in *Advances in Probability Distributions with Given Marginals*, G. Dall'aglio, ed., Kluwer Academic Publishers, Dordrecht, The Netherlands, pp. 13–50.

T. WALLSTEN AND D. BUDESCU (1983), Encoding subjective probabilities: A psychological and psychometric review, *Management Sci.*, 29, pp. 151–173.

R. YAGER (1993), Aggregating fuzzy sets represented by belief structures, *J. Intell. Fuzzy Systems*, 1, pp. 215–224.

L. ZADEH (1965), Fuzzy sets, *Inform. and Control*, pp. 338–353.

L. ZADEH (1968), Probability measures of fuzzy events, *J. Math. Anal. Appl.*, 23, pp. 421–427.

L. ZADEH (1978), Fuzzy sets as a basis for a theory of possibility, *Fuzzy Sets Systems*, 1, pp. 3–28.

L. ZADEH (1995), Discussion: Probability theory and fuzzy logic are complementary rather than competitive, *Technometrics*, 37, pp. 271–276.

Chapter 6

Guidelines for Eliciting Expert Judgment as Probabilities or Fuzzy Logic

Mary A. Meyer, Kenneth B. Butterfield, William S. Murray, Ronald E. Smith, and Jane M. Booker

6.1 Introduction

We recommend *formal elicitation of expert judgment* as a method for obtaining probabilities or fuzzy rules from individuals. Formal elicitation of expert judgment draws from the fields of cognitive psychology, decision analysis, statistics, sociology, cultural anthropology, and knowledge acquisition. It entails the use of specific procedures to identify the experts, define the technical problems, and elicit and document experts' judgment.[1] Expert judgment may be expressed as *probabilities* (either point estimates, such as 90%, or as probability distribution functions) or *fuzzy terms* (for example, *low*, *medium*, *high*). The experts' sources of information, considerations, and assumptions are documented as part of their judgment. Formal elicitation can counter common biases arising from human cognition and behavior. The benefits of formal elicitation are added rigor, defensibility of the judgments, and increased ability to update the judgments as new information becomes available. We provide the reader with guidelines and examples of formal elicitation in the following areas:

- determining whether expert judgment can be feasibly elicited;

- determining whether expert judgment can be better elicited in a probabilistic or fuzzy framework;

---

[1]Expert judgment also can be elicited informally, even tacitly. The experts are asked for their best guess, which is then applied to the analysis with no other questions asked. Hence this informal approach has been called "ask and use" (French, McKay, and Meyer (1999)). However, the trend seems to be toward formalizing expert judgment procedures in a number of areas, such as expert testimony in the legal system, medicine, environmental analysis, and nuclear safety analysis (Stanbro and Budlong-Sylvestor (2000)). Ellen Hisdal articulated some of these ideas in the context of logical statements being represented as fuzzy sets; see Hisdal (1988a, 1988b) in Chapter 1.

- formulating technical questions;

- structuring interview situations for one expert, multiple experts, or teams of experts;

- eliciting and documenting the expert judgment;

- representing expert judgment for the experts' review and refinement;

- facilitating the comparison of multiple experts' judgments.

Expert judgment is an expert's informed opinion, based on knowledge and experience, given in response to a technical problem (Ortiz et al. (1991)). Expert judgment can be viewed as a representation—a snapshot of the expert's state of knowledge at the time of his or her response to the technical problem (Keeney and von Winderfeldt (1989)). Expert judgment is used to

- predict future events;

- provide estimates on new, rare, complex, or poorly understood phenomena;

- integrate or interpret existing information;

- learn an expert's problem-solving process or a group's decision-making processes;

- determine what is currently known, how well it is known, or what is worth learning in a field (Meyer and Booker (2001)).

Expert judgment may be expressed in quantitative or qualitative form for fuzzy and probabilistic applications. Examples of judgments given in quantitative form include probabilities, uncertainty estimates, and membership functions, and are often given in reference to other quantities of interest, such as performance, cost, and time. Examples of qualitative form include

- the experts' natural language statements of physical phenomena of interest (for example, "the system performs well under these conditions");

- "if–then" rules (for example, "if the temperature is high, then the system performs poorly");

- textual descriptions of experts' assumptions in reaching an answer;

- reasons for selecting or eliminating certain data or information from consideration.

Whatever its form, expert judgment should include a description of the experts' thinking—their reasoning, algorithms, and assumptions—and the information they considered in arriving at a response. Ideally, expert judgment provides a complete record allowing decision makers, other experts, or even novices to track the experts' problem-solving processes.

6.2   Method

6.2.1   Illustration

We illustrate the phases and steps of expert elicitation with four probability and fuzzy examples. Backgrounds on these four examples are given below.

6.2.1.1   Probability example: Predicting automotive reliability

This example of probabilistic elicitation is an automotive application whose goal is to characterize the reliability of new products during their developmental programs (Kerscher et al. (2000)). Characterizing the reliability in the early developmental phases poses problems because traditional reliability methods require test data to characterize the reliability. Test data typically are not available while the product is in the prototype stage nor later in the development stage. During these early stages, however, another source of reliability information is available—the knowledge of the product experts. We used the award-winning process PREDICT (Meyer, Booker, Bement, and Kerscher (1999)) to elicit initial reliability judgments from the product experts. As the product is developed, information from other sources, such as test data, the supplier, and customer, is folded into the reliability characterization using a Bayesian updating approach.

This automotive application has involved lengthy and formal elicitations of teams of experts from the automotive industry's national and international sites over several years for several different systems. It also has involved working closely with automotive personnel to develop a core group of people with expertise in the elicitation process.

6.2.1.2   Probability example: Comparing expert and trainee performance predictions

The focus of this example of probabilistic elicitation is the performance of an aging defense technology. The goal of this study was to elicit both expert and trainee predictions on how the technology would perform given its potential condition. Its performance was defined metrically, based on its condition and its closeness to the original design specifications (i.e., whether it met the design specs or was potentially divergent from them). In this study, the experts themselves were interested in whether their predictions would differ from those of the trainees and if so, how, particularly because at some point the trainees would become the reigning experts. While the experts had mentored the trainees, the trainees lacked the field training of the experts and the experts expected this to lead to differences in their predictions.

Performance data was sparse or nonexistent, especially for potential conditions outside the original design specifications. Thus the participants' estimates ranged from being based on limited test data, calculations, and simulations for conditions approximating the design specs, to being based entirely on their subjective judgment for conditions greatly diverging from the specs. During short interviews of about an hour, the participants gave their judgments as subjective probability estimates with uncertainty ranges. The participants also described their sources of information, their assumptions leading to their estimates, their years of experience, and the names of their mentors. The number of participants was small, about seven, because this was the number of knowledgeable persons.

6.2.1.3   Fuzzy example: Identifying radioisotopes

This fuzzy elicitation example involves creating an instrument to correctly identify radioisotopes from their gamma-ray spectrum. Gamma-ray spectra are detected indirectly by the ionization they produce in materials. Measurements of the ionization are recorded as a pulse-height distribution. Because gamma-ray spectra can be measured only indirectly, experts must try to identify imprecise features of the pulse-height distribution and match these to precise features of radioisotope spectra. (Please note that we will be using *spectra* and *pulse-height distribution* interchangeably throughout this chapter.)

Identifying radioisotopes is useful to customs agents or law enforcement officers who must deal with suspicious packages. For example, customs agents must verify that packages contain the radioisotopes they are purported to contain and not some other radioisotope that is being shipped illegally. Radioisotopes such as Technetium 99 and Iodine 131 are routinely shipped legally to medical institutions.

In contrast to the automotive application, the gamma-ray application involves only informal elicitations of a few experts. These experts are largely eliciting their own fuzzy rules, using texts on fuzzy logic to understand how rules should be formulated, and texts on spectroscopy, as well as their own experience, to create the content of the rules. They select the best rules by testing them against a large database of radioisotope spectra.

6.2.1.4   Fuzzy example: Comparing experts on performance predictions

The focus of this fuzzy example is similar to the expert–trainee study described above, that is, the reliability of an aging defense technology. However, in this example, two experts were asked to supply their fuzzy rules for predicting how hypothesized conditions of the technology could affect its performance. The goal of this pilot project was to quantify performance, largely based on expert judgment, in the absence of test and other data. In addition, this project addressed the following question posed by project sponsors: How should they interpret and handle cases where the only information available was subjective expert judgment and in which the experts differed?

The fuzzy elicitations, in contrast to those of the expert–trainee study, were intensive, taking about 20 hours per expert over the course of two years. The two experts were interviewed separately and then were brought together in a structured interview situation to review their sources of information, fuzzy rules, assumptions, and uncertainty ranges. They were allowed to amend their judgments—their fuzzy rules, assumptions, and uncertainty ranges—as they so wished. Their judgments were summarized and displayed side by side to facilitate comparison.

6.2.2   Summary table of phases and steps

A summary of the elicitation phases and steps is given in Table 6.1. Some phases and steps are performed differently for fuzzy and probabilistic elicitations; these are prefaced with an asterisk. For example, item 4, "Eliciting and documenting the expert judgment," involves eliciting the fuzzy rules in the former and obtaining probability responses in the latter. Other phases or steps vary according to the situation, such as how the experts are selected. Note that the same fuzzy examples are not used throughout the table.

6.2.3   Phases and steps for expert elicitation

Phase 1: Determining whether expert judgment can be feasibly elicited

To answer the question of whether expert judgment can be feasibly elicited, consider the following:

- The domain. *Recommendation*: Most domains of science and engineering are amenable to eliciting expert judgment. If, in addition, there are articles on expert judgment, or detailed instructions passed on by word of mouth, it bodes well for eliciting judgment.

Table 6.1. *Elicitation phases, steps, and examples.*

Phases, steps	Probability example: Auto reliability	Fuzzy example: Radioisotopes
1. Determining whether expert judgment can be feasibly elicited.	Feasibility indicated by prior (informal) use of experts' judgment.	Feasibility indicated by prior use of expert judgment.
2. Determining whether expert judgment can be better elicited in a probabilistic or fuzzy framework.	Experts thought in terms of numeric likelihoods; the mathematical foundations of subjectivist probabilities were a plus.	Incoming information was imprecise; one advisor expert preferred fuzzy for the quick creation of a robust expert system.
3. Designing the elicitation. 1. Identify the advisor expert(s).	One self-identified advisor expert identified additional advisors at the national and international levels.	One advisor expert volunteered himself and identified another advisor.
2. Construct representations of the way that the experts measure or forecast the phenomena of interest.	Representations included reliability block diagrams, reliability success trees, and failure modes.	Representations focused on features evident in plots of gamma-ray spectrum and of the second derivative of the spectra.
*3. Draft the questions. For fuzzy, this involves identifying the variables; identifying the inputs and outputs to the system; and disaggregating the inputs and outputs into distinct linguistic variables.	What is your expected, number of incidents per thousand vehicles to fail to meet specifications? Best-case number? Worst-case number?	What are your fuzzy rules concerning a peak and these linguistic variables: *low*, *medium*, and *high energy* and *very very good*, *very good*, *good*, *somewhat good*, or *somewhat somewhat good*?
*4. Plan the interview situation. 5. Select the experts.	Team interviews because the experts worked in teams. The advisor selected the auto products for reliability characterization, which determined the selection of teams, already composed of experts.	Separate interviews followed by structured joint interviews. The advisor identified the two locally available and recognized experts.
6. Motivate participation by the experts.	The advisor carefully drafted the formal request for participation by cover memo and followed up with telephone calls.	The motivation of participation by the advisor was very informal because this was an in-house effort and there were only two experts.
7. Pilot test the questions and interview situation.	Extensive pilot tests of the sets of questions and the cover letter (for motivating participation) were performed via teleconference calls.	Pilot tests of the questions were conducted on the advisor expert and led to refinements in how the fuzzy rules were elicited.
*4. Eliciting and documenting the expert judgment.	Advisor and those he designated lead the team interviews, elicited and recorded the subjective probability estimates, assumptions, and failure modes.	The researchers elicited and documented the experts' fuzzy rules, membership functions, the information, and assumptions the experts considered.
*5. Representing the expert judgment for the experts' review and refinement.	Teams' performance estimates were represented as probability distributions. Teams reviewed the probability distributions and updated their estimates as new information became available.	The researchers and experts refined the fuzzy rules and membership functions. The experts refined their fuzzy rules, in structured joint interviews. The experts' reviews led to labels and caveats being placed on their expert judgment.
6. Facilitating the comparison of multiple experts' judgments.	Comparisons were done between proposed designs and options for testing, instead of between experts' judgments.	We compared experts' fuzzy rules, assumptions, qualifications, and the difference to the bottom line in using one expert's judgment over another.

Recommendation: Domains that may *not* be amenable to the elicitation techniques described in this chapter are those in which the experts must quickly respond to control a physical process and are unable to explain their responses, even in retrospect. Jet pilots performing flight simulations would be an example (Shiraz and Sammut (1998)).

- The capabilities of the individual experts. Some individuals are less able than others to articulate their thinking. Generally, individuals can describe their thinking if descriptions are elicited while the problem is at the forefront of their thinking rather than in retrospect. *Recommendation*: In our experience, about 5% of the experts have great difficulty in "thinking aloud," regardless of the elicitation (Meyer and Booker (2001)).

Probability example.   In the automotive application, it was determined that expert judgment could be elicited. Not only was there recognition of "engineering experience"—another term for expert judgment—as a valuable resource but also it was tacitly used in team discussions of the reliability of new automotive products.

Fuzzy example.   In the radioisotope example, gamma-ray spectroscopists had historically given their judgments. For example, an expert might say "that looks like a Bismuth spectrum because there are three well-shaped peaks with about the correct energies."

Phase 2: Determining whether the experts' judgments can be better elicited in a probabilistic or a fuzzy framework

Consider (1) whether the expert is accustomed to and able to think in terms of probabilities, and (2) to what degree the knowledge being elicited is imprecise:

- Ask the experts how they represent the technical problem and look at the solutions or results of their work. *Recommendation*: If quantitative representations are absent and the experts describe results in qualitative linguistic terms, the experts may prefer the fuzzy approach. If the results are represented as probabilities, percentiles, confidence intervals, or points on a plot, the experts may be accustomed to thinking in probabilities. While many scientists, engineers, and mathematicians are accustomed to formulating their thinking in quantitative terms, and even probabilities, they still are prone to the usual biases, such as inconsistency, broadly defined here to mean that their point estimates of mutually exclusive events do not sum to 1.0.

- Also consider the preference of the experts in determining whether to elicit in a fuzzy or probabilistic form. If the experts have a strong preference for either one, it is generally best to use their preference. Ask them their reasons for the preference. If their reasons are based on misconceptions about fuzzy sets or probabilities (for example, contradict some of the recommendations in this section), resolve the misconceptions and ask the experts to reconsider their preference.

Additional considerations are the following:

- Requirement for an expert system: If the application requires a system to run without input from users or experts (for example, a control system such as that described in Parkinson et al. (1998)), the fuzzy approach may be preferable.

- Changeability of the representation: If the way that the experts identify, measure, or forecast the phenomena of interest is likely to change greatly through time, a fuzzy framework may be more flexible.

- Requirement for probability distributions: Techniques are available for deriving probability distributions from judgments elicited in fuzzy form (Booker et al. (2000), Parkinson et al. (1999), Smith et al. (1997, 1998)), so this requirement does *not* by itself necessitate a probability elicitation.

To check your selection of probability or fuzzy elicitation, ask the experts if they would be able to respond in the form you have selected.

Probability example.   In the automotive application, the auto engineers thought in terms of numeric likelihoods (probabilities) of systems succeeding or failing. Also, the researchers had used a subjectivist probability method on a similar reliability application and were asked by the automotive sponsor to tailor it to this application. Additionally, the statistical rigor and defensibility of the subjectivist approach appealed to the sponsor.

Fuzzy example.   In the application whose goal was to create an instrument to identify radioisotopes from their gamma-ray pulse-height distributions, the experts had already considered both probability and fuzzy techniques. They had selected fuzzy and had begun to self-elicit their rules in this framework before our first meeting. Their reasons for preferring fuzzy were valid: they were creating an expert system for pattern recognition, and the inputs to the expert system were likely to be imprecise.

Phase 3: Designing the elicitation

Involve the experts in the steps described below to ensure that the steps reflect the experts' way of thinking about the technical problems.

Step 1: Identify the advisor experts (also known as "champions")

Look for one or two individuals who are knowledgeable about their domain and their culture, who can provide "entree" into their culture, explain its workings, provide guidance on the below-mentioned aspects of the elicitation, motivate wider participation by the experts, and who are willing to act as advisor experts. Often the advisor experts will be the same individuals who initially contacted you.

We identified the advisor experts, almost after the fact, when they began to push the elicitation forward in their work groups or companies. Once the advisors are identified, it is helpful to ask them privately what they personally would like to gain from participating in this elicitation and how they will judge its success or failure.

Probability example.   In the automotive application, the individual who was to become the advisor expert volunteered himself when we described the role of the advisor; he worked along with us (over the telephone) to conduct the steps below and to involve additional advisors from the company's national and international sites. This advisor defined success in terms of his company adopting the process of characterizing reliability as its new way of doing business and applying it to new products. This advisor defined success in terms of (1) developing the approach to the point where the reality of higher reliability products could be demonstrated in the field and (2) of his company's applying the approach to all of his company's new product development programs.

Fuzzy example.   In the identifying radioisotopes application, the first expert who contacted us volunteered to act as an advisor expert. He then involved the principal investigator of this project as another advisor expert. These advisor experts wished our elicitation to lead to some additional rules that would help them distinguish valid peak shapes within a particularly confusing energy region.

Step 2: Construct representations of the way that experts measure or forecast the phenomena of interest

The experts' organization already may have an officially accepted representation of the phenomena of interest, or the experts may have a tacit understanding of what the phenomena are. If the representations do not already exist, they can be created through interaction with the experts. (Note that this step may be done in parallel with step 3.) While these representations are not absolutely necessary for conducting elicitations, they are highly desirable if the goal of the work is to form a common basis of understanding or effect a change in the way of doing business (for example, making decisions). Additionally, the representations provide a mechanism for incorporating all available information and a framework for displaying the results of the expert judgment.

Explain the need for the representation to the advisor, define it, and ask the advisor if such a thing exists. If the advisor is unsure about what you are requesting, or whether such a representation exists, ask (1) for examples of the information that the experts have on the problem; (2) what information they receive, in what form, and from whom; and (3) what they show as evidence of their expertise. *Recommendation*: In our experience, discussions about representing the phenomena of interest have led to detailed explanations of the problem by the advisor, often taking days. We recommend allowing time for these explanations because they will lead to better representations of the phenomena.

Probability example.   In an automotive application, the representations reflected the goal of the project—characterizing the reliability of new automotive products during their developmental programs (Kerscher et al. (2000)). The representations already existed in part and were further elaborated on. The representations had a particularly important role in this application: they provided the common language—the "roadmap" for doing business—between the auto employees and us and the mechanism for incorporating new information as it became available.

The representation focused on the automotive products whose reliability was being modeled. Because the reliability of a product depends on the reliability of its parts—component, subsystem, and system—this step involved representing these parts and their logical relationships in models. These models took the form of reliability logic flow diagrams, typically, reliability block diagrams or reliability success trees. For example, Figure 6.1 (Kerscher et al. (2000)) shows a simple generic subsystem D composed of components A, B, and C. If components A, B, and C are all in series, the reliability of subsystem D will be the product of the reliabilities of the components.

Fuzzy example.   In the application for identifying radioisotopes, the key representations were plots of features of gamma-ray spectra.

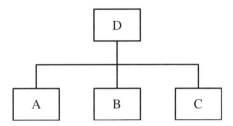

Figure 6.1. *Reliability success tree diagram.*

To learn about the representations in this application, we asked an advisor expert to describe the information the experts have available and how they give their expert judgment. The information available to the experts, the gamma spectroscopists, consisted of pulse-height distributions of the observed gamma rays, the detector response functions, and libraries of photo peak energies associated with specific radioisotopes. The pulse-height distribution is used to identify the observed gamma-ray peaks. The observed peaks are compared to the detector response function to determine if the observed peak is either consistent with the detector response or due to statistics or noise. The experts identify all the features in a pulse-height distribution; that is, features other than peaks are present, such as Compton edges, which are due to the detector response. These additional features are identified as "not a peak" and are eliminated as peaks using fuzzy membership. The library is the knowledge base upon which a particular pulse-height distribution is categorized. For example, an expert might say "that looks like a Bismuth pulse-height distribution because of the observed peaks, and the extra features look like the Compton edges associated with the observed peaks."

In addition, an advisor expert was asked to roughly list the steps that he anticipated the expert system would perform. Theses steps (1) identify "peak shape" from that which is "not a peak"; (2) compare peak energy to library energy using fuzzy membership; (3) tally the peak matches for all isotopes in the library; and (4) determine the best match to identify the isotope.

Step 3: Draft the questions

As a starting point for drafting the questions, ask the advisor "What are the phenomena (variables) of interest; how do you assess these; what metrics or natural language terms do you use?" The endpoint of this step differs for fuzzy and probability elicitations, as illustrated in Table 6.1: for a probability elicitation, it leads to the creation of technical questions that the experts will later answer; for a fuzzy elicitation, it provides the linguistic variables that the experts will use as building blocks in constructing their fuzzy rules. (For details on linguistic variables and fuzzy rules, see Ross (1995)). *Recommendation*: For a fuzzy elicitation, it helps to ask the advisor expert to identify the variables of interest, identify the inputs and outputs to the system, and disaggregate these inputs and outputs into linguistic variables. Note that variables that are to be handled as fuzzy or crisp may emerge at this point. For instance, if the linguistic variables do not have a fuzzy continuum of values but one set value, these may be treated as crisp. For example, one of the early determinations made by the fuzzy expert system for identifying radioisotopes involves a crisp value. The expert system essentially asks "Are there enough counts in a specific energy region to say whether there is a feature, such as peak shape?" The crisp value is defined to be three standard deviations of net counts above the background as determined by adjacent energy regions. *Recommendation*:

In drafting the questions for probability elicitation, include consistency checks (for example, check that the mutually exclusive events sum to 1.0).

Probability example.   In the automotive application, the advisor described how the design and process (manufacturing) engineers thought about product reliability or performance. The design engineers thought in the metric of incidents per thousand vehicles failing to meet specifications. The process engineers thought in the metric of parts per million. The advisor further described how the experts thought in terms of what caused the product to fail, or its "failure modes."

We drafted separate questions for the design and process engineers. For example, the design engineers were asked to estimate the number of incidents per thousand vehicles in which the specifications would not be met. The design engineers also were asked to provide a subjective range on this estimate. We elicited the range by asking them for reasonable worst-case and best-case numbers of incidents. We also asked the engineers to provide textual descriptions of the potential failure modes for the product and to estimate their likelihood (Kerscher et al. (1998, 1999, 2000), Meyer, Booker, Bement, and Kerscher (1999)).

Once we developed the basic questions for those working in design engineering, we modified the questions for those working in process engineering. Later, we further modified the questions for those working in software engineering, that is, on automotive parts run by software.

Fuzzy example.   To draft the questions to elicit the fuzzy rules for this radioisotope application, we asked the advisor expert to do the following:

- Identify the variables of interest: The advisor named peak energy, peak width, peak shape, and peak area as the variables of interest and added that the detector response function had several variables—energy calibration, efficiency, and resolution—that could affect these variables.

- Identify the inputs and outputs to the system: The advisor considered the information coming into the detector and its output (the detector response function) and provided the following list of inputs to the expert system (see Table 6.2).

Table 6.2. *Inputs to the expert system.*

Library data	Detector response	Observation
photon energy	calibration	peak energy
	resolution	peak width
	Compton scattering	peak shape
intensity	efficiency	peak area

The advisor listed the following (Tables 6.3 and 6.4) as outputs that he would like the expert system to provide for expert and novice users, respectively. For expert users, the last row lists the possible isotopes—Barium 133, Xenon 133, and Iodine 131—as determined by a fuzzy step, followed by a curve-fitting step. The fuzzy step, labeled "Fuzzy Energy Attribute ~Match" involves matching the shape and energy between the observed gamma-ray spectra and the library of the photon peak energies. In this case, Barium 133 is the most likely radioisotope because it has the best (highest) match with Barium's peak energy.

Table 6.3. *Sample outputs for expert users.*

	Barium 133	Xenon 133	Iodine 131
Peak Peak **Energy ~Shape**	**Fuzzy Energy** **Attribute ~Match**	**Fuzzy Energy** **Attribute ~Match**	**Fuzzy Energy** **Attribute ~Match**
82.8 1.00	53.2 UNLIKELY 0.00	81.0 MUST_HAVE 1.00	80.2 MIGHT_HAVE 1.00
157.7 0.56	81.0 WILL_HAVE 1.00		284.3 MUST_HAVE 0.87
279.0 1.00	160.6 UNLIKELY 0.56		364.5 MUST_HAVE 0.97
306.9 0.96	223.2 UNLIKELY 0.00		637.0 MIGHT_HAVE 0.00
359.3 1.00	276.4 MIGHT_HAVE 1.00		722.9 MIGHT_HAVE 0.00
389.3 1.00	302.0 MUST_HAVE 0.96		
689.7 0.26	356.0 MUST_HAVE 1.00		
	383.9 MIGHT_HAVE 0.93		
PEAK = 5.78	ISOTOPE NUMBER = 5.44	ISOTOPE NUMBER = 1.00	ISOTOPE NUMBER = 2.84
	Ba133 RESIDUAL = 0.34	Xe133 RESIDUAL = 4.78	I131 RESIDUAL = 2.94
	Ba133 MATCH = 0.94	Xe133 MATCH = 0.17	I131 MATCH = 0.49

Table 6.4. *Sample output for novice users.*

Best match
Barium 133

For novice users, the expert system will only display the best match between the observed gamma-ray spectra and the library of photon peak energies, in this case Barium 133.

Disaggregate these inputs and outputs into distinct linguistic variables. The linguistic variables for the inputs of the observed peak energy could be *low*, *medium*, and *high*, and the output linguistic variables for the fuzzy peak shape membership could be *very very good*, *very good*, *good*, *somewhat good*, and *somewhat somewhat good*.

Step 4: Plan the interview situation

Consult the advisors on the interview situation that will fit their culture, their way of thinking, and their business practices. Advisors will probably suggest whatever situation has worked well in the past. While it may seem obvious to tailor the interview to the experts' culture, we have observed many researchers who attempt the reverse. They try to fit the experts to a particular kind of interview that they, the researchers, prefer. This approach often results in the experts' not participating or the credibility of the project being undermined.

Describe the main ways in which experts can be interviewed, and ask the advisor whether or not

- the experts are to arrive at consensus, such as in a team meeting (for example, such as when their judgments are later to be combined by statistical means);

- a problem is likely to arise with experts unconsciously or unwillingly adjusting their own judgments to match others' judgments (Meyer and Booker (2001)); while this bias usually is not a problem in our applications, it can occur, usually in interview situations in which the participants are not of equal status (for example, managers and their employees, military officers and their staff, mentors and those they have mentored);

- the experts' names are to be associated with individual judgments or whether individual judgments are to be anonymous;

- the expert judgment is to be provided on paper (for example, in response to written sets of questions), during face-to-face meetings, or by electronic means; or

- the researchers or the experts themselves are to document the expert judgment.

By this step, often you will still have questions about which interview situation is best. We recommend that you

- ask the advisors which interview situation they recommend;

- consider using a combination of interview situations (for example, initially interview the experts separately, then bring them and the records of their elicitation together and allow them to amend their judgments); or

- pilot-test the interview situation and let the results answer any remaining questions.

Recommendation: It is easier to conduct detailed and lengthy interviews with one expert at a time. For this reason, experts are typically interviewed separately, at least initially, for fuzzy elicitations. Also, often there is only one expert locally available, anyway. *Recommendation*: It's generally best to interview the experts in the same situation in which they work (for example, if the experts usually work individually, interview them separately; if they work as teams to arrive at judgments, interview them as teams). *Recommendation*: If it is likely that the experts will unconsciously or unwittingly adjust their judgments to those of other experts, it is best to elicit their judgments separately (Meyer and Booker (2001)). *Recommendation*: The experts should document their own judgments if they will be updating their judgments through time (for example, as in the automotive application). Otherwise, the researchers should do it because they tend to be more thorough and because it relieves the experts of the burden.

Probabilistic example. On the automotive application, the advisor explained that the experts worked in teams to design and manufacture products, that team members typically met face-to-face, and that the experts reached some consensus in their product planning. The advisor was not unduly concerned about the possibility of group think (where team members unconsciously acquiesce to a dominant member's decision). However, to minimize the chances of this bias occurring and to maximize the diversity of the judgments, we decided to have the team members individually self-elicit their own judgments. The plan was to have team members individually complete worksheets asking for their expected, worst-case and best-case estimates of incidents per thousand vehicles. Also on the worksheets, the team members were asked to list the failure modes and their associated likelihood of occurrence. They were to bring these worksheets to the face-to-face meetings of their team. Each team was to be led by the advisor, or the advisor's designee, who would record the team's judgments on flip charts for the team's review.

Fuzzy example. In the expert comparison application, we knew that the two experts had been consulted formally and informally about what a particular real or potential "condition might do to the technology's performance." We planned to interview the experts separately because they were often consulted individually. Also, there were two indicators that our interviews of the two experts should, at least initially, be separate: first, fuzzy rules would be elicited, which meant the interviews would be intensive; second, one of the experts had earlier mentored the other, which could have meant that the newer expert was prone to group think bias.

We planned to interview both experts separately for their fuzzy rules and membership functions and also to record their assumptions and sources of information. We would then bring the experts and their judgments together in structured interviews where the experts could view each other's responses and amend their own as they wished. We also planned to monitor the structured joint sessions for signs of the bias, such as the newer expert deferring to the more experiences expert. (For further information on this bias, see Meyer and Booker (2001, pp. 134–135)).

Step 5: Select the experts

The advisor selects the experts to be elicited or advises on the selection strategy (for example, whether other experts are to be selected on the basis of publications, experience, the organization or work group to which they belong, and/or their availability). This step varies more according to the circumstances of the elicitation, rather than to whether the elicitation is fuzzy or probabilistic. When few experts are available or the application is in-house, the process of selecting the experts can be informal. For example, in the study comparing experts and trainees in their performance predictions, the advisor expert recommended selecting individuals with a range of years of experience and who had mentors. The advisor then provided their names, information on their years of experience, and who had mentored them.

Probabilistic example. The advisor selected the auto products for reliability characterization, which in turn determined the teams that would be interviewed. There were teams for each component in a subsystem or system. The teams typically were composed of four to eight experts who saw their part of the auto product from its concept through its development cycle.

Fuzzy example. In the expert comparison application, the advisor identified the locally available and recognized experts. There were only two experts, one of whom was the advisor.

Step 6: Motivate participation by the experts

Ask the advisor whether problems will arise in getting the experts to participate, and if so, how best to motivate participation. For example, in the expert–trainee comparison, the advisor thought experts would be more likely to participate if they knew elicitations would take only an hour. The advisor talked to the individuals he had mentally selected and encouraged them to participate. Given that these individuals were incredibly busy, we probably could not have obtained participation by other more formal means. In essence, the advisor, through his standing in this culture, motivated the participation of about seven individuals.

If problems are anticipated, or if the experts have to be motivated formally (as is often the case if they are employed in industry), ask the advisor

- how the request should be delivered: verbally (in person or via telephone), by hard copy memo or electronic communication, or by some combination of these;

- from whom the communication should come and which letterhead should be used;

- the order in which the communication will be routed to possible participants;

- the timing of the communication (for example, before or after a meeting describing this endeavor).

Show the advisor the checklist of things (Meyer and Booker (2001, pp. 90–93)) that individuals typically want to know in deciding whether they will participate: how they were selected, who is sponsoring the effort, how long it will take, what tasks they will perform, and the anticipated product of the effort and their access to it. This information is usually provided to the experts in a cover letter or e-mail requesting their participation.

Probability example. In the automotive application, we were outsiders, unfamiliar with the culture, and thus relied on what the advisor thought would motivate participation. The advisor carefully drafted the request for participation using the checklist mentioned above. The request for participation was a cover memo followed by a series of telephone calls. The advisor also apprised the participants of the progress of the project, in particular, how well their initial reliability judgments predicted the later test data. Receiving this information helped motivate a large number of experts to participate over several years.

Fuzzy example. In the expert comparison situation, this step was informal because the work was being done in-house and only two experts were involved, one of whom was the advisor expert. The advisor expert was responsible for motivating his and the other expert's involvement in this effort.

Step 7: Pilot-test the questions and the interview situation

The pilot test provides the last check on the elicitation design before it is conducted. Aspects of the elicitation that need pilot testing are the experts' understanding of the technical question, the response mode—fuzzy or probabilities—and any directions, such as how to complete the set of questions. If the expert judgment serves as input to another process, such as decision-making, the decision makers should be included in the pilot tests to ensure that the judgments are on the desired phenomena, at the right level, and in the needed form.

Pilot tests are conducted on the advisor expert and on any other experts or users that the advisor recommends. Pilot tests involve having the selected individuals answer the draft questions, with one major addition. The pilot testers are to "think aloud" as they go through the elicitation to allow the researchers to pinpoint problems in the elicitation, such as where the metrics caused confusion (Meyer and Booker (2001, pp. 155–156)). *Recommendation*: Pilot testing is generally a good idea if you will be interviewing more than a few experts or intensively interviewing more than one expert through time.

Probability example. On the automotive application, pilot tests were necessary because about forty design engineers would be interviewed, followed by process and software engineers. We conducted, during conference calls, extensive pilot tests of the sets of questions (worksheets) and the cover letter for the design engineers. The advisor called other advisor experts, introduced us and the idea of pilot testing to them, and listened in on the pilots, as a means of learning how to conduct these. Later, the advisor conducted the pilot tests of worksheets himself when these needed to be tailored to the process and software engineers.

Fuzzy example.   On the expert comparison application, pilot tests were performed because the fuzzy elicitations were expected to be intensive. (The elicitations later averaged 20 hours per expert.) Pilot tests of the questions were conducted on the advisor expert and led to refinements in how the fuzzy rules were elicited (for example, the linguistic variables describing conditions and performance).

Phase 4: Eliciting and documenting the expert judgment

This stage basically involves administering and documenting the questions as designed and pilot-tested in the earlier stages. Note that this phase differs for the fuzzy and probabilistic elicitations. For fuzzy elicitations, this phase involves eliciting the fuzzy rules, while for probabilistic elicitations, it involves obtaining probability responses to the technical questions. *Recommendation*: Try to keep interviews to about an hour in length so as not to tire the experts. If it appears that that the interview will run over the allotted time, ask the experts if they wish to continue or to schedule another interview.

Probabilistic example.   In the automotive application, the advisor and those he designated lead the team interviews and elicit and record the subjective probability estimates, assumptions, and failure modes.

Fuzzy example.   In the expert comparison application, the researchers elicited and documented the experts' fuzzy rules, membership functions, information sources, and assumptions.

Phase 5: Representing the expert judgment for the experts' review and refinement

The representations mentioned in Step 2 (Phase 3) can form a framework for documenting the expert judgment. Review the experts' judgments with the experts and refine these judgments. In probabilistic applications, the refinements are likely to be made to the probability responses and to caveats concerning their use. In fuzzy applications, the refinements tend to focus on the fuzzy rules, membership functions, and caveats. *Recommendation*: If a membership function is very narrow, this may be the time to redefine it as crisp. *Recommendation*: If it is not possible to do the reviews in person, send the transcriptions of the experts' judgments to each expert for review. We recommend setting a date by which they are to respond, explaining that you will assume they accept the judgment as is unless you hear otherwise.

Probabilistic example.   To probabilistically represent the expert judgment in the automotive application, we transformed the teams' initial performance estimates into probability distributions using Monte Carlo computer simulations. The teams' estimates referred to their part of the automotive product, such as a component. The probability distributions for each part of the automotive product were combined according to how the parts fit together, as represented in Step 2 (for example, see Figure 6.1). This combining of distributions provided an overall reliability characterization for the auto product at a particular point in time and a means for determining if the target reliability was reached. These reliability characterizations were given to the advisor, who passed them to the participating experts, for review.

The advisor kept the experts apprised of how well their initial reliability judgments predicted the later test data on the product.

The teams refined their judgments through time as test data became available. Equations such as Bayes' theorem were used to mathematically update the teams' performance estimates (Meyer, Booker, Bement, and Kerscher (1999)).

Fuzzy example.   In the expert comparison application, representation and refinement included working with the experts separately to refine their fuzzy rules and membership functions. As part of the review and refinement, we plotted the results of applying the fuzzy rules, showing what the predicted performance would be for a given condition or combination of conditions. We reviewed these plots with the experts separately to check that their rules delivered results consistent with their expectations. For instance, one membership function was found to be broad and the experts wanted it to have a crisp edge with no overlap, so it was redone accordingly.

Then we brought the two experts together in a structured interview situation to share their sources of information and judgments. In some cases, the newer expert had not seen some of the simulations of the other expert and modified his fuzzy rules in light of these. We recorded these changes and then, in a last joint meeting, had the experts review the written records again. Then later, as articles were written about this project (Meyer, Booker, Bement, and Kerscher (1999)), the experts again reviewed documentation.

As a result of these reviews, we refined our presentation of the results. The experts had expressed concern that the plotted results, particularly for the more extreme hypothetical conditions, might be misinterpreted as coming from frequentist statistics; that is, from statistically sampled experimental data, which as mentioned earlier did not exist. To address this concern, the labels of *subjective* or *expert judgment* were applied to all results and a caveat was added cautioning individuals on the origin and recommended use of this information.

Phase 6: Facilitating the comparison of multiple experts' judgments

If there are judgments from multiple experts, decision makers or application sponsors may want to evaluate for themselves whether the experts provided significantly different judgments, and if so, the basis for these differences.

We recommend side-by-side comparisons, in which the user can compare different experts' judgments (fuzzy rules, membership functions, probabilities, subjectively estimated uncertainties, assumptions, and sources of information) and if appropriate, the expert's qualifications (years of experience and why they were selected for participation). If possible, provide information on whether the use of one expert's judgment over another makes a difference to the bottom line. In addition, it is often helpful to have statements in the experts' own words as to whether they thought the judgments differed and if so, why.

Probabilistic example.   In the automotive application, judgments were not compared because teams were the unit of elicitation, and there were not duplicate teams whose judgments could be compared. Instead, comparisons were done between proposed designs, such as if a component were made of aluminum as opposed to plastic, to see what the resulting reliabilities would be. These comparisons were called *what ifs* and were also done to anticipate which components should be prototyped and tested.

Fuzzy example.   In the expert comparison application, we presented a side-by-side display of the experts' judgments because the comparison of expert judgments had been of particular interest to our sponsors. The experts' rules were shown side-by-side in simple tables to allow viewers to visually compare them. In addition, a summary table was given of the experts' descriptions in their own words of whether they were basically in agreement; the assumptions they made; their qualifications; and the main areas of disagreement in the fuzzy rules. Note that the experts' assumptions were given because these have been found to drive the experts' answers (Ascher (1978), Booker and Meyer (1988)) and because decision makers may choose one expert's judgment over another, depending on whose assumptions they most agree with. The experts' qualifications, in this case, years and type of experience, were presented because this information is of interest to decision makers, particularly when the experts give differing judgments.

Finally, a table (Table 6.5) was presented that showed whether the use of one expert's judgment over another's would make a difference to the bottom-line answer for a particular condition. (Note that the performance scale for Table 6.5 is from 0.0 to 1.00, with 1.00 meaning "meets performance specifications.") Our reason for showing that there was, in this case, no difference to the bottom line is that it simplifies matters. For instance, it allows the decision makers to forgo the difficult task of selecting one expert's judgment or of mathematically combining the judgments.

Table 6.5. *Subjective performance ranges based on each expert's fuzzy rules and a hypothetical condition.*

	Median	Range
Expert 1	1.00	(0.92, 1.00)
Expert 2	1.00	(0.96, 1.00)

6.3 Summary

The driving consideration in eliciting expert judgments in fuzzy or probabilistic forms is the experts' preferences. Key points in performing the elicitation are as follows:

- Using the advisor, the expert "insider" who can advise on how to conduct the elicitation so as to fit the experts' way of thinking and doing business. Using the advisor expert embodies the principle of "asking how to ask" from cultural anthropology—the idea that the researcher may be an outsider to a culture, unaware of the special dialect and customs, and therefore may need to ask an insider how to ask the questions (Meyer and Paton (2001)).

- Pilot testing, if the judgments of more than a few experts will be elicited.

- Documenting as much as possible the experts' thinking and sources of information, as well as the results.

- Involving the experts in the review, analysis, interpretation, and presentation of the expert judgment. The experts' involvement in, or even ownership of, the process is crucial, particularly if the expert judgment must be elicited periodically to reflect the latest knowledge. If expert judgment will be repeatedly elicited, the researcher should aim to have the elicitations led by (1) the advisor, (2) a core group of trained experts, or (3) the experts themselves, through self-elicitation of their own judgments.

References

W. Ascher (1981), *Long-Range Forecasting: From Crystal Ball to Computer*, Wiley–Interscience, New York.

J. M. Booker and M. A. Meyer (1988), Sources and effects of interexpert correlation: An empirical study, *IEEE Trans. Systems Man Cybernet.*, 8, pp. 135–142.

J. M. Booker, R. E. Smith, T. R. Bement, and S. M. Parker (2000), A statistical approach to fuzzy control system methods for uncertainty distributions, in *Proceedings of the World Automation Conference*, Maui, HI, TSI Press, Albuquerque, NM.

S. French, M. D. McKay, and M. A. Meyer (2000), Critique of and limitations on the use of expert judgments in accident consequence uncertainty analysis, *Radiat. Prot. Dosim.*, 90, pp. 325–330.

R. L. Keeny and D. von Winderfeldt (1989), On the users of expert judgment on complex technical problems, *IEEE Trans. Engrg. Management*, 36, pp. 83–86.

W. J. Kerscher, J. M. Booker, T. R. Bement, and M. A. Meyer (1998), Characterizing reliability in a product/process design-assurance program, in *Proceedings of the International Symposium on Product Quality and Integrity*, Institute of Environmental Sciences and Technology, Mount Prospect, IL.

W. J. Kerscher, J. M. Booker, T. R. Bement, and M. A. Meyer (1999), *Characterizing Reliability in a Product/Process Design Assurance Program*, invited paper, Spring Research Conference on Statistics in Industry and Technology, Minneapolis, MN, ASA, Alexandria, VA.

W. J. Kerscher III, J. M. Booker, T. R. Bement, and M. A. Meyer (2000), Characterizing reliability during a product development program, in *Proceedings of ESREL 2000, SARS and SRA-Europe Annual Conference, Edinburgh, UK, 14–17 May* 2000, Vol. 2, A. A. Balkema, ed., Rotterdam, The Netherlands, pp. 1433–1440.

M. A. Meyer and J. M. Booker (2001), *Eliciting and Analyzing Expert Judgment: A Practical Guide*, ASA–SIAM Series on Statistics and Applied Probability, SIAM, Philadelphia, ASA, Alexandria, VA.

M. A. Meyer, J. M. Booker, T. R. Bement, and W. Kerscher III (1999), *PREDICT: A New Approach to Product Development*, Report LA-UR-00-543 and 1999 R&D 100 Joint Entry, Los Alamos National Laboratory and Delphi Automotive Systems, Los Alamos, NM and Troy, MI.

M. A. Meyer, J. M. Booker, R. E. Smith, and T. R. Bement (1999), Enhanced reliability update, *Weapon Insider*, 6, Los Alamos National Laboratory, Los Alamos, NM, p. 1.

M. A. Meyer and R. C. Paton (2001), Interpreting, representing and integrating scientific knowledge from interdisciplinary projects, in *Theoria Historia Scientiarum*, Nicholaus Copernicus University, Torun, Poland.

N. R. Ortiz, T. A. Wheeler, R. J. Breeding, S. Hora, M. A. Meyer, and R. L. Keeney (1991), The use of expert judgment in NUREG-1150, *Nuclear Engrg. Design*, 126, pp. 313–331.

W. J. PARKINSON, K. W. HENCH, M. R. MILLER, R. E. SMITH, T. R. BEMENT, M. A. MEYER, J. M. BOOKER, AND P. J. WANTUCK (1999), The use of fuzzy expertise to develop uncertainty distributions for cutting tool wear, in *Proceedings of the 3rd International Conference on Engineering Design and Automation* (*EDA* 1999), Vancouver, BC, Canada, Integrated Systems, Inc., Prospect, KY.

W. J. PARKINSON, R. E. SMITH, AND N. MILLER (1998), A fuzzy controlled three-phase centrifuge for waste separation, in *Proceedings of the World Automation Conference*, Anchorage, AK, TSI Press, Albuquerque, NM.

T. J. ROSS (1995), *Fuzzy Logic with Engineering Applications*, McGraw–Hill, New York.

G. M. SHIRAZ AND C. A. SAMMUT (1998), Acquiring control knowledge from examples using ripple-down rules and machine learning, in *Proceedings of the 11th Workshop on Knowledge Acquisition, Modeling and Management*, Knowledge Science Institute, University of Calgary, Calgary, AB, Canada.

R. E. SMITH, T. R. BEMENT, W. J. PARKINSON, F. N. MORTENSEN, S. A. BECKER, AND M. A. MEYER (1997), The use of fuzzy control system techniques to develop uncertainty distributions, in *Proceedings of the Joint Statistical Meetings, Anaheim, California*, American Statistical Association, Alexandria, VA.

R. E. SMITH, J. M. BOOKER, T. R. BEMENT, M. A. MEYER, W. J. PARKINSON, AND M. JAMSHIDI (1998), The use of fuzzy control system methods for characterizing expert judgment uncertainty distributions, in *Proceedings of the 4th International Conference on Probabilistic Safety Assessment and Management* (*PSAM* 4), Vol. 1, International Association for Probabilistic Safety Assessment and Management, pp. 497–502.

W. D. STANBRO AND K. BUDLONG-SYLVESTOR (2000), The role of expert judgment in safeguards, *J. Nuclear Materials Management*, Summer, pp. 17–20.

Part II

Applications

Timothy J. Ross

II.1 Chapters 7–15

In bridging the gap between probability and fuzzy theories for the quantification of uncertainty, the second part of the book illustrates this "bridging" with case studies in nine different fields, ranging from photography to signal validation:

- Chapter 7: Image Enhancement: Probability Versus Fuzzy Expert Systems;

- Chapter 8: Engineering Process Control;

- Chapter 9: Structural Safety Analysis: A Combined Fuzzy and Probability Approach;

- Chapter 10: Aircraft Integrity and Reliability;

- Chapter 11: Auto Reliability Project;

- Chapter 12: Control Charts for Statistical Process Control;

- Chapter 13: Fault Tree Logic Models;

- Chapter 14: Uncertainty Distributions Using Fuzzy Logic;

- Chapter 15: Signal Validation Using Bayesian Belief Networks and Fuzzy Logic.

In some of these chapters, a case study is described in which separate probability and fuzzy approaches are taken and the differing results—advantages and disadvantages—are compared and discussed. This approach is taken in Chapters 7, 8, 12, 13, and 15. In Chapters 9, 11, and 14, we combine probability and fuzzy models into a single, integrated approach. Finally, in Chapter 10, we show how an application using probability theory can be enhanced with fuzzy logic.

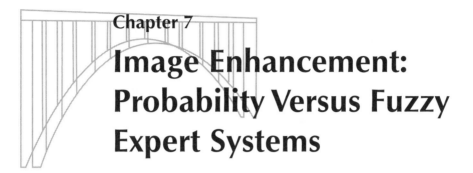

Chapter 7

Image Enhancement: Probability Versus Fuzzy Expert Systems

Aly El-Osery and Mo Jamshidi

Abstract.   The field of image enhancement is being used in various areas and disciplines. Advances in computers, microcontrollers, and digital signal processing (DSP) boards have opened new horizons to digital image processing and many avenues to the design and implementation of new innovative techniques. This chapter compares image enhancement, via modification of the probability density function of gray levels, with new techniques that involve the use of knowledge-based (fuzzy expert) systems capable of mimicking the behavior of a human expert.

7.1   Introduction

The field of image processing dates back to the 1920s, but limitations due to the speed of processing units and the size of available storage space have handicapped its progress. Nowadays, the ever-growing advances in computer technologies have reduced these limitations to a negligible level. Hence, the interest and the need for digital image processing have increased sharply. Today, image processing is being applied in many fields and disciplines.

Improving pictorial information for human interpretation and processing image information for autonomous machine perception were two main reasons behind the need for image processing (Gonzalez and Woods (1992)). There are various techniques and tools for dealing with these two categories. Traditionally, all image-processing techniques were based purely on mathematical approaches without any incorporation of human experience and knowledge that might have developed after years of field work. Currently, new techniques are being developed to mimic human reasoning and to encode it into computer programs that, in some cases, have proven superior results over conventional techniques.

Image-processing techniques may be classified into a number of categories, among them image digitization, image enhancement, image restoration, image encoding, and image segmentation and representation. In this chapter, we address only one of these categories, namely, image enhancement. The goal of image enhancement is to process a given image, resulting in a more useful image for a specific application. It is important to emphasize the words "specific application" because the choice of the processing technique is very application specific. The results obtained from one technique may be superior for one application but unacceptable for another.

Processing images for human perception is a very complicated task. The determination of what is a "good" image is a subjective matter and differs from one viewer to another based on taste and experience. On the other hand, image processing for autonomous machine perception is somewhat easier because machines do not behave subjectively. The sole goal of image processing is to provide the information needed for the machine to perform a specific task. In either case, the determination of the right technique needs experimentation.

In the first part of this chapter, we will concentrate on providing a brief understanding of a few of the key elements and methodologies of image processing. The second part will be dedicated to examining one of the conventional techniques used in image enhancement for human interpretation, known as histogram equalization, as well as the new intelligent techniques that are based on expert systems.

7.2   Background

This section describes a few of the image-processing elements necessary to understand some of the concepts introduced in this chapter. This section is not intended to provide a complete background in image-processing techniques; the sole purpose of this section is to provide a quick overview of some of the key elements used in image enhancement. Digital image-processing elements, different image formats (RGB, CMY, YIQ, HSI), spatial domain methods, and a probability density function method will be introduced.

7.2.1   Digital image-processing elements

There are four general and necessary elements of digital image processing:

- image acquisition;
- processing unit;
- displaying unit;
- storage.

Image acquisition is the means of digitizing the image and making the data available for processing. Images could be obtained using scanners, digital cameras, the Internet, etc. Next, the image information is processed to obtain the desired results. A displaying unit is necessary to view and judge the results of the applied algorithms. Finally, the results are stored on hard disks, CDs, magnetic tapes, etc.

7.2.2   Image formats

Manipulation of digital images requires a basic understanding of existing color models. Color models are used to provide a common standard in the specification of color. One of the most frequently used models is RGB (red, green, and blue) (Ballard and Brown (1982)). This model is widely used for color monitors and a broad class of color video cameras. There also exist other models, such as CMY (cyan, magenta, and yellow) for color printers and YIQ, the standard for color TV broadcast (the Y component corresponds to luminance, and I and Q are two chromatic components). HSI (hue, saturation, and intensity) is another useful color model for two main reasons: the first is that HSI is closely related to the way humans perceive color, and the second is that in the HSI model the intensity is decoupled from the color information.

7.2.2.1 RGB and CMY

In the RGB model, each color is presented in its primary spectral components of red, green, and blue. The color subspace of the RGB model is based on the Cartesian coordinate system. Figure 7.1 shows the color cube of the RGB model in which the red, green, and blue colors are at three corners, and the cyan, magenta, and yellow are at the opposite three corners. In the color cube, black is at the origin and white is at the corner farthest from it. Images in the RGB model consist of three independent image planes, one for each primary color. The presence of all colors produces white and the absence of all colors produces black.

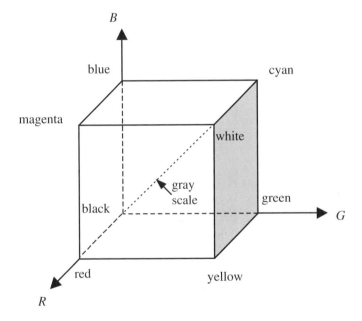

Figure 7.1. *RGB color cube.*

Cyan, magenta, and yellow are the secondary colors of light and the primary colors of pigments. The CMY model is used in color printers. The conversion from RGB to CMY is done using the following simple operation:

$$\begin{bmatrix} C \\ M \\ Y \end{bmatrix} = \begin{bmatrix} 1 \\ 1 \\ 1 \end{bmatrix} - \begin{bmatrix} R \\ G \\ B \end{bmatrix}. \tag{7.1}$$

7.2.2.2 YIQ

In the YIQ model, the Y component corresponds to luminance, and I and Q are two chromatic components called *inphase* and *quadrature*. The advantage in the YIQ model is that the luminance (Y) is decoupled from the color information (I and Q). The transformation from the RGB model to YIQ is presented below in (7.2) (Gonzalez and Woods (1992)). Due to the fact that (7.2) is a linear transformation, the computation time to obtain the YIQ values from the RGB or visa versa is inexpensive.

$$\begin{bmatrix} Y \\ I \\ Q \end{bmatrix} = \begin{bmatrix} 0.299 & 0.587 & 0.114 \\ 0.596 & -0.275 & -0.321 \\ 0.212 & -0.523 & 0.311 \end{bmatrix} \begin{bmatrix} R \\ G \\ B \end{bmatrix}. \tag{7.2}$$

Using the YIQ model, color, tint, and brightness are defined as follows:

Color. Color is defined as the average color intensity of the image under investigation. The color intensity is given by

$$\text{color} = \sqrt{Q^2 + I^2}. \tag{7.3}$$

Tint. In (7.4) below, tint is defined to be the phase of a pixel in the subject of interest (SOI) in the Q–I plane:

$$\text{tint} = \tan^{-1}\left(\frac{I}{Q}\right). \tag{7.4}$$

Brightness. Brightness is defined as the average value of Y.

7.2.2.3   HSI

HSI (hue, saturation, and intensity) is a color model closely related to the way humans perceive color. In this color model, hue represents a pure color, and saturation describes the amount by which the color is diluted with white light. Similar to the YIQ format, the intensity of the color is separated from the color information. Keep in mind that the luminance Y in the YIQ model is not directly related to the intensity I in the HSI model, and that the I in the YIQ has a different meaning than the I in the HSI model. Figure 7.2 illustrates graphically the relationship between HSI and RGB. The hue of point P is the angle of that point measured from the red axis. The saturation of point P, which is the measure of how much the color is diluted with white light, is the distance from the center of the triangle. The intensity of the color varies from white down to black by moving along the line that is perpendicular to the center triangle. Equations (7.5)–(7.7) provide the conversions from RGB to HSI:

$$H = \cos^{-1}\left\{\frac{\frac{1}{2}[(R-G)+(R-B)]}{[(R-G)^2+(R-B)(G-B)]^{\frac{1}{2}}}\right\}, \tag{7.5}$$

$$S = 1 - \frac{3}{R+G+B}[\min(R,G,B)], \tag{7.6}$$

$$I = \frac{1}{3}(R+G+B), \tag{7.7}$$

where R, G, and B are normalized between zero and one. If $B > G$, then $H = 2\pi - H$. Also if S is zero, then H is undefined, and if I is zero, then S is undefined.

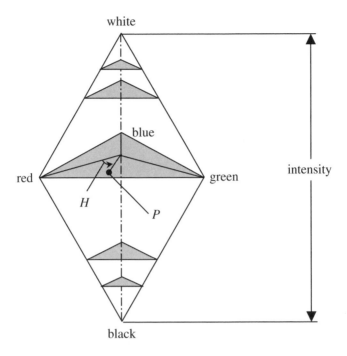

Figure 7.2. *HSI color model.*

The conversion form HSI to RGB requires the red, green, and blue to be normalized in the following fashion (Gonzalez and Woods (1992)):

$$r = \frac{R}{R + G + B}, \tag{7.8}$$

$$g = \frac{G}{R + G + B}, \tag{7.9}$$

$$b = \frac{B}{R + G + B}. \tag{7.10}$$

Using the above definitions and (7.7), the following relations can be derived:

$$R = 3Ir, \tag{7.11}$$

$$G = 3Ig, \tag{7.12}$$

$$B = 3Ib. \tag{7.13}$$

Equations (7.14)–(7.24) below represent the extent of the computations required to transform HSI to RGB.

For $0° < H \leq 120°$,

$$r = \frac{1}{3}\left[1 + \frac{S \cos H}{\cos(60° - H)}\right], \tag{7.14}$$

$$b = \frac{1}{3}(1 - S), \tag{7.15}$$

$$g = 1 - r - b. \tag{7.16}$$

For $120° < H \leq 240°$,

$$H = H - 120°, \tag{7.17}$$

$$r = \frac{1}{3}(1 - S), \tag{7.18}$$

$$g = \frac{1}{3}\left[1 + \frac{S \cos H}{\cos(60° - H)}\right], \tag{7.19}$$

$$b = 1 - r - g. \tag{7.20}$$

Finally, for the interval $240° < H \leq 360°$,

$$H = H - 240°, \tag{7.21}$$

$$g = \frac{1}{3}(1 - S), \tag{7.22}$$

$$b = \frac{1}{3}\left[1 + \frac{S \cos H}{\cos(60° - H)}\right], \tag{7.23}$$

$$r = 1 - g - b. \tag{7.24}$$

YIQ and HSI are two different models, and the use of either one is dependent on the application. YIQ has the advantage of being computationally less expensive to obtain than HSI. Essentially, YIQ is obtained by a linear transformation of the RGB, while HSI is obtained via a nonlinear transformation. The advantage of HSI is that it is a very close representation of the way humans perceive color, and the intensity is decoupled from the hue and saturation.

7.2.3   Spatial domain

Spatial domain methods are procedures that directly manipulate the pixel information. A digital image consists of a collection of pixels in the Cartesian coordinates. Thus a digital image consists of discrete elements and could be expressed as $u(x, y)$, where x and y are positive integers. The process of modifying an image in the spatial domain could be expressed as

$$h(x, y) = T[u(x, y)], \tag{7.25}$$

where T is a processing operator and $h(x, y)$ is the final processed image. Figure 7.3 is a demonstration of a digital image and its constitution of discrete elements.

7.2.4   Probability density function

The probability density function (PDF) is a key element in several image-processing techniques, among them histogram equalization, that will be covered in the following section. A PDF is defined as

$$f_X(x) = \frac{d F_X(x)}{dx}, \tag{7.26}$$

where x is a continuous random variable and $F_X(x)$ is the cumulative distribution function (CDF) and is defined as

$$F_X(x) = \text{probability that } X \leq x = P(X \leq x). \tag{7.27}$$

one pixel

Figure 7.3. *Digital image.*

The PDF has the following properties:

$$f_X(x) \geq 0, \tag{7.28}$$

$$\int_{-\infty}^{\infty} f_X(x)dx = 1, \tag{7.29}$$

$$P(x \leq a) = F_X(a) = \int_{-\infty}^{a} f_X(x)dx, \tag{7.30}$$

$$P(a \leq x \leq b) = \int_{a}^{b} f_X(x)dx. \tag{7.31}$$

The properties given in (7.28) and (7.29) show that the probability of an event occurring has to be bounded below by 0 and bounded above by 1. The properties given in (7.30) and (7.31) give some insight into how to evaluate probabilities using PDFs.

7.3   Histogram equalization

Histogram equalization is a technique of image enhancement that modifies the PDF of gray levels, and its objective is to modify the PDF of gray levels to a uniform PDF. Figure 7.4 shows the histogram of a variety of images with different brightness and contrast problems. The histogram plotted is the frequency of occurrence of a certain gray level. Brightness of the image could be determined by observing the shape of the histogram; i.e., Figure 7.4(a) is a bright image, and therefore the gray levels are concentrated in the higher end of the scale. On the other hand, Figure 7.4(b) is a dark image and therefore the gray levels are concentrated in the lower end of the scale. Figure 7.4(c) is a representation of a low-contrast image, while Figure 7.4(d) is an example of a high-contrast image. As shown by the histogram, a low-contrast image has gray levels concentrated in a small region of the scale, while a high-contrast image has a more uniform distribution. Consequently, to obtain a high-contrast image from a low-contrast one, a transformation has to be done to spread the distribution of the gray levels. This technique is histogram equalization.

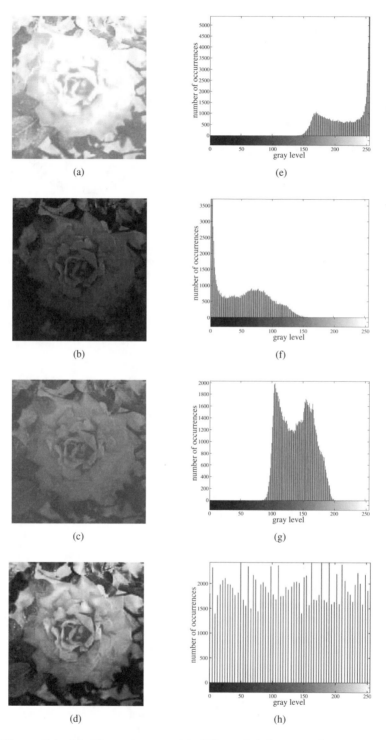

Figure 7.4. (a)–(d) *are images with different brightness and contrast problems, and* (e)–(h) *are the equivalent histogram representations.*

If we represent the normalized gray levels with the variable x which is in the interval [0, 1], where 0 represents black and 1 represents white, then our goal will be to find the new set of gray levels, to be represented by the variable y, that has a uniform distribution. The general form of such a transformation is

$$y = T(x). \tag{7.32}$$

Certain conditions need to be embossed on the transformation T to preserve the order from black to white in the gray scale and to ensure that the new gray levels will be in the allowed range. These conditions are listed below:

1. $T(x)$ is single-valued and monotonically increasing.

2. $T(x)$ is in the range [0, 1].

Since the techniques that we will develop have to be general, we have to accommodate a variety of images, and hence the gray levels of these images could be assumed to be random variables. Consequently, we need to deal with the PDFs of these variables. Let $f_X(x)$ and $f_Y(y)$ be the PDFs for the original image and the transformed image, respectively. At this point, we assume the PDFs are continuous. After we are done with the derivation of the transformation function, we will obtain the same result for the discrete case.

The goal is to find the transformation T that will transform $f_X(x)$ into $f_Y(y)$, where $f_Y(y) = 1$ for $0 \leq y \leq 1$. As a first step to achieving this goal, we need to introduce the following property (Gonzalez and Woods (1992)):

If $f_X(x)$ and $T(x)$ are known and the T^{-1} is single-valued and monotonically increasing in the range $0 \leq y \leq 1$,

then

$$f_Y(y) = \left[f_X(x) \frac{dx}{dy} \right]_{x=T^{-1}(y)}. \tag{7.33}$$

Using the above property, the transformation $T(x)$ is found to be

$$y = T(x) = \int_0^x f_X(\xi) d\xi. \tag{7.34}$$

A complete discussion of (7.34) is given in Gonzales and Woods (1992).

In digital image processing the image data are discrete; consequently, for the above equations to be useful they have to be modified for the discrete case. In the discrete case, we have

$$P(x_k) = \frac{n_k}{n}, \quad 0 \leq x_k \leq 1, \quad \text{and} \quad k = 0, 1, \ldots, L - 1, \tag{7.35}$$

where L is the number of gray levels, $P(x_k)$ is the probability of the kth level, n_k is the number of occurrences of that level in the image, and n is the total number of pixels in the image. Hence in the discrete case, (7.34) translates into

$$y_k = T(x_k) = \sum_{i=0}^{k} \frac{n_i}{n}. \tag{7.36}$$

The same concepts mentioned above also are applied for color images. For color images, the HSI format is used to decouple the hue and saturation from intensity. Then histogram equalization is applied on the intensity.

Figure 7.5 illustrates the process of enhancing color images via histogram equalization. Assuming that the original image is in RGB format, we transform the data to HSI, then apply histogram equalization to the intensity component. The final step is to convert the HSI back to RGB. The same example will be used in the next section to provide a comparison basis between this probabilistic method and the fuzzy system–based approach.

Figure 7.6 is an example of a dark image, as also indicated by the histogram shown in Figure 7.7. Therefore, in order to enhance this image using histogram equalization, we apply the procedure demonstrated in Figure 7.5. The result of the histogram equalization is shown in Figure 7.8. The brightness of the image has been improved but the integrity of the color has been lost. Figure 7.9 shows the histogram of the enhanced image of Figure 7.8.

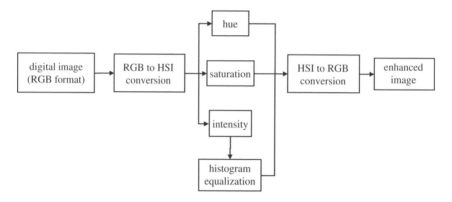

Figure 7.5. *Block diagram of the enhancement process using histogram equalization.*

Figure 7.6. *An example of a dark image.*

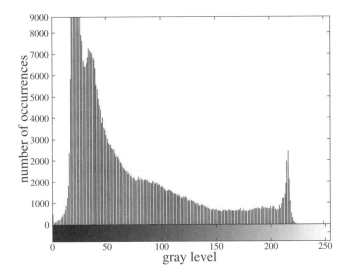

Figure 7.7. *Histogram of the photo shown in Figure 7.6.*

Figure 7.8. *Figure 7.6 after histogram equalization.*

7.4 Expert systems and image enhancement

This section presents a systematic approach to enhancing the quality of images obtained from many different sources, e.g., video, television, scanner, Internet, etc., using expert systems based on fuzzy logic. Nowadays, digital image-processing techniques are used in combination with fuzzy-based expert systems in many different applications (Jamshidi, Vadiee, and Ross (1993)). There have been numerous studies of the use of fuzzy expert systems in auto-focusing (Shingu and Nishimori (1989)), auto-exposure, auto-zooming, filtering, and edge detection (Tyan and Wang (1993), Bezdek (1981), Lim and Sang (1990), Pal (1991)). Asgharzadeh and Gaewsky (1994) outline the use of fuzzy rule-based systems in controlling video recorders. Asgharzadeh and Jamshidi (1996) received a U.S. patent on the enhancement process, which was further discussed by Asgharzadeh and Zyabakhah-Fry (1996).

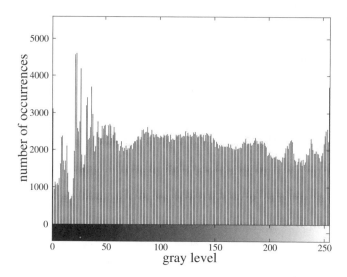

Figure 7.9. *Histogram of Figure 7.8.*

Fuzzy logic is a formalization of the human capacity for approximate reasoning (Ross (1995)). Unlike the restriction of classical propositional calculus to a two-valued logic, fuzzy logic involves concepts without clearly defined boundaries. In fuzzy logic, linguistic terms could be used to describe subjective ideas. In other words, fuzzy logic is a representation of our natural language.

For the case of enhancing images for human perception, the use of a fuzzy expert system clearly is a natural choice (Asgharzadeh, El-Osery, and Pages (1998)). Images are enhanced and subjected to human visual judgment; consequently, enhancements done with the aid of fuzzy expert systems that mimic the judgment of a human expert are expected to produce satisfactory results.

Figure 7.10 provides an overview of the necessary steps needed for enhancing an image using a fuzzy expert system (El-Osery (1998)). A digitization process is used in case the image is not digital. After obtaining the digital image, a subject of interest (SOI) has to be determined. This SOI will be used to determine the quality of the image and will be input into the expert system. The SOI then will be used to determine the attributes of the image. The attributes of the image are passed to the expert system to determine the changes necessary to obtain an optimal image. The changes are applied and are observed on a displaying unit.

7.4.1   SOI detection

SOI detection, or feature extraction, is a very crucial process in image enhancement. If the knowledge-based system is going to be developed to mimic the behavior of a human expert, then there has to be a routine that detects the same features that the human expert uses to examine the quality of that image. Several algorithms have been developed to recognize an image using edge detection, clustering (Ruspini (1969)), and pattern classification, etc. These algorithms enable the computer to understand visual information it finds in the environment. Similarly, in the process of enhancing an image for human interpretation, it is important to determine an SOI that will be used as a reference. Based on the SOI, attributes of the image are calculated. These attributes are the values that will be used by the fuzzy expert system to enhance the image.

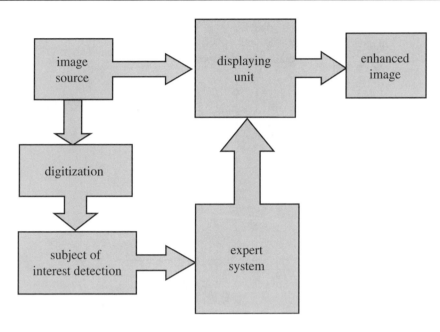

Figure 7.10. *Block diagram of the enhancement process.*

7.4.2 Fuzzy expert system development

After the SOI is detected, the next step is to use this information to develop the fuzzy expert system. There are different techniques for generating this system. By observing trends in the data, rules could be generated by intuition. Using programs like FuLDeK® (a fuzzy logic development kit) and MATLAB is another possibility. These programs give users the freedom to choose the type of rules generated and use mathematical techniques to generate the rules that will minimize the error between the desired and estimated outputs.

7.4.3 Example

In this section, the above-mentioned concepts will be demonstrated. Color images are used, and the SOI is chosen to be the human skin tone. The procedure for enhancing the image using a fuzzy expert system is illustrated in Figure 7.10. The following section will demonstrate the process of extracting the SOI (skin tone) and will provide the methods used to generate the expert system and, finally, to apply the corrections to obtain the enhanced image.

7.4.3.1 Feature extraction

When enhancing an image it is important to determine an SOI that will be used as a reference. Because human skin tone is one of the most interesting SOIs for detection, and also because images of humans are the most widely available types of images, human skin tone was chosen as the SOI for this example. The detection of the SOI occurs on the pixel level and has to include the whole spectrum of skin colors. By mapping all different combinations of red, green, and blue in the range of 0–255 to the Q–I plane, the following definition for human skin tone was obtained.

Definition   Any pixel in the Q–I plane that falls within ± 30 degrees of the I-axis and has a normalized magnitude between 10% and 55% is considered a skin pixel. Figure 7.11 is a graphical representation of this definition (El-Osery (1998)).

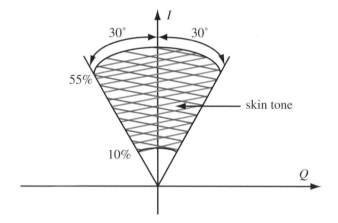

Figure 7.11. *Definition of skin tone.*

7.4.3.2   High- and low-resolution masks

In detecting the SOI in the image, two different resolution masks are used, the high-resolution mask and the low-resolution mask. The high-resolution mask uses the skin tone definition mentioned previously to detect the SOI (Asgharzadeh, El-Osery, and Pages (1998)). This mask tests every pixel. If a pixel is a skin tone, a value of 1 is stored, or 0 is stored otherwise. Figure 7.6 is used as the image under investigation. By applying the high-resolution mask, Figure 7.12 is obtained. As seen in Figure 7.12, there is some noise that needs to be eliminated. To eliminate such noise, another mask is used.

Figure 7.12. *High-resolution mask of Figure* 7.6.

Using the high-resolution mask, we evaluate the low-resolution mask. The low-resolution mask reduces noise. This process is similar to running the high-resolution mask through a low-pass filter. The low-resolution mask is evaluated by taking squares of 16×16 pixels from the high-resolution mask. If the average value inside the square is greater than a certain threshold value, then the 16×16 block is considered to have pixels with skin tone characteristics; otherwise, the entire square is not considered to have any skin tone characteristics. The default value of the threshold is 50%. Applying the low-resolution mask to Figure 7.6 with threshold 50% produces Figure 7.13, in which all the skin tone is extracted and the noise is reduced.

Figure 7.13. *Low-resolution mask of Figure* 7.6.

7.4.3.3   Expert system

The fuzzy expert system rules obtained from a human expert are similar to the following:

> If brightness is Dark, then increase brightness to Medium Bright level.

In this rule, Dark and Medium Bright are fuzzy variables (linguistic labels). This natural language was translated to computer language using statements such as

> If X is A_i and Y is B_i, then Z is C_i.

The fuzzy rule-based system is used to mimic a human expert that is capable of enhancing the image. To generate and tune the expert knowledge, a library of 300 images was used. SmartPhotoLab® was used to determine the image attributes that will be used for generating the rule-based system. With the help of a human expert, the images were enhanced manually using professional image-processing software, such as Corel® Photo-Painter™ and PhotoShop®. A database of the image attributes before and after enhancement was obtained. Using these data, fuzzy rules were generated with the help of FuLDeK.

Three different fuzzy rule-based systems—one for each image attribute (color, tint, and brightness)—were designed for enhancing the image. The shape of the membership functions of the inputs (see Figure 7.14) and outputs are triangular and singletons, respectively. Min and max operations were used for evaluating *and* and *or* of the antecedents. For rule inferencing, correlation product encoding (CPE) was employed. The centroid method was used for defuzzifications.

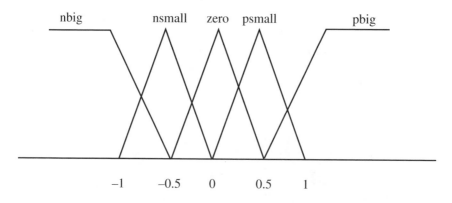

Figure 7.14. *Structure of the triangular membership function used for the inputs.*

The input of each rule-based system is normalized before fuzzification. The program then aggregates the fuzzy rules and finally defuzzifies the output. The outputs of the fuzzy system are the corrected attributes of the color, tint, and brightness. These values are averages based on the masks of the SOI. The next section will explain the process of modifying each pixel.

7.4.3.4   Image correction

The expert system generated the corrected attributes of the color, tint, and brightness using the three fuzzy rule-based systems. These values are averages based on the masks of the SOI, but to obtain the final image we need to modify each pixel. This section describes the equations used to modify each pixel.

The new color, tint, and brightness of each pixel are defined as (El-Osery (1998))

$$\text{New\_Color\_Pixel} = \text{Old\_Color\_Pixel} + \Delta_{\text{Average\_Color}}, \tag{7.37}$$

$$\text{New\_Tint\_Pixel} = \text{Old\_Tint\_Pixel} + \Delta_{\text{Average\_Tint}}, \tag{7.38}$$

$$\text{New\_Brightness\_Pixel} = \text{Old\_Brightness\_Pixel} + \Delta_{\text{Average\_Brightness}}, \tag{7.39}$$

where

$$\Delta_{\text{Average\_Color}} = \text{New\_Color(output)} - \text{Old\_Color(input)}, \tag{7.40}$$

$$\Delta_{\text{Average\_Tint}} = \text{New\_Tint(output)} - \text{Old\_Tint(input)}, \tag{7.41}$$

$$\Delta_{\text{Average\_Brightness}} = \text{New\_Brightness(output)} - \text{Old\_Brightness(input)}. \tag{7.42}$$

Using (7.3) and (7.4), we obtain the relations

$$I = \text{color} \cdot \sin(\text{tint}), \tag{7.43}$$

$$Q = \text{color} \cdot \cos(\text{tint}). \tag{7.44}$$

Taylor's series expansion was used for linearization:

$$f(x, y) = f(x_0, y_0) + \frac{\partial f}{\partial x}\Big|_{(x_0, y_0)} \cdot (x - x_0) + \frac{\partial f}{\partial y}\Big|_{(x_0, y_0)} \cdot (y - y_0)$$
$$+ \text{ higher-order terms (HOT)}, \tag{7.45}$$

where HOT ≈ 0.

Using (7.43)–(7.45) to evaluate the new values of I and Q per pixel, we obtain

$$Q_{\text{new}} = Q_{\text{old}} + \cos(\text{tint}_{\text{old}}) \cdot \Delta_{\text{Average\_Color}} - \text{Color}_{\text{old}} \cdot \sin(\text{tint}_{\text{old}}) \cdot \Delta_{\text{Average\_Tint}}, \quad (7.46)$$

$$I_{\text{new}} = I_{\text{old}} + \sin(\text{tint}_{\text{old}}) \cdot \Delta_{\text{Average\_Color}} + \text{Color}_{\text{old}} \cdot \cos(\text{tint}_{\text{old}}) \cdot \Delta_{\text{Average\_Tint}}, \quad (7.47)$$

$$Y_{\text{new}} = Y_{\text{old}} + \Delta_{\text{Average\_Brightness}}. \quad (7.48)$$

The result of applying the fuzzy system is shown in Figure 7.15. Now this result could be used to compare the fuzzy system and the histogram equalization approaches.

Figure 7.15. *Figure 7.6 enhanced via fuzzy expert system.*

7.5   Final remarks

Based on the results shown in the previous sections, it is clear that fuzzy logic is an additional and complementary solution to histogram equalization for improving the quality of digital images. This process will improve the quality of images without extensive image-processing calculations and will therefore reduce processing time. In addition, the automation will allow users to process a batch of images in a few seconds. The use of a fuzzy system for image enhancement is a natural choice since it is closely related to human subjective decision-making. Unlike the traditional method of histogram equalization, the fuzzy system incorporates human experience while modifying the image, and hence the results are more satisfactory.

We are not claiming that a fuzzy system should replace any of the conventional techniques. Rather, this chapter was intended to illustrate another tool that could be used with the appropriate application. As we mentioned earlier, it is important to experiment with different techniques to determine what is most efficient and suitable for the specific application at hand. In some cases, a combination of techniques should be used to get the best results.

References

A. ASGHARZADEH, A. EL-OSERY, AND O. PAGES (1998), Enhancement of images obtained by digital cameras using fuzzy logic, in *Advances in Soft Computing, Multimedia and Image*

Processing: Proceedings of the 3rd World Automation Congress (*WAC '98*), Vol. 8, TSI Press, Albuquerque, NM, pp. 829–835.

A. ASGHARZADEH AND J. GAEWSKY (1994), Applications of fuzzy logic in a video printer, in *Intelligent Automation and Soft Computing: Proceedings of the 1st World Automation Congress (WAC '94)*, Vol. 1, TSI Press, Albuquerque, NM, pp. 445–448.

A. ASGHARZADEH AND M. JAMSHIDI (1996), *Fuzzy Logic Controlled Video Printer*, U.S. patent no. 5,590,246.

A. ASGHARZADEH AND R. ZYABAKHAH-FRY (1996), Digital image enhancement via fuzzy logic rule-based system, in *Intelligent Automation and Soft Computing: Proceedings of the 2nd World Automation Congress (WAC '96)*, Vol. 5, TSI Press, Albuquerque, NM, pp. 25–29.

D. H. BALLARD AND C. M. BROWN (1982), *Computer Vision*, Prentice–Hall, Englewood Cliffs, NJ.

J. BEZDEK (1981), *Pattern Recognition with Fuzzy Objective Function Algorithms*, Plenum, New York.

A. EL-OSERY (1998), *Design and Implementation of Expert Systems for Digital and Analog Image Enhancement*, M.S. thesis, EECE Department and ACE Center, University of New Mexico, Albuquerque, NM.

R. C. GONZALEZ AND R. E. WOODS (1992), *Digital Image Processing*, Addison–Wesley, Reading, MA.

M. JAMSHIDI, N. VADIEE, AND T. ROSS (1993), *Fuzzy Logic and Control: Software and Hardware Applications*, Vol. 2, Series on Environmental and Intelligent Manufacturing Systems, Prentice–Hall, Englewood Cliffs, NJ.

Y. LIM AND U. SANG (1990), On the color segmentation algorithm based on the thresholding and the fuzzy C-mean techniques, *Pattern Recognition*, 23, pp. 935–952.

S. PAL (1991), Fuzzy tools for the management of uncertainty in pattern recognition, image analysis, vision and expert systems, *Internat. J. Systems Sci.*, 22, pp. 511–549.

T. ROSS (1995), *Fuzzy Logic with Engineering Applications*, McGraw–Hill, New York.

E. RUSPINI (1993), A new approach to clustering, *Inform. and Control*, 15, pp. 22–32.

T. SHINGU AND E. NISHIMORI (1989), Fuzzy-based automatic focusing system for compact camera, in *Proceedings of the 3rd International Fuzzy Systems Association Congress*, International Fuzzy Systems Association, Seattle, WA, pp. 436–439.

K. TANAKA AND M. SUGENO (1991), A study on subjective evaluation of printed color images, *Internat. J. Approx. Reasoning*, 5, pp. 213–222.

C. TYAN AND P. WANG (1993), Image Processing enhancement filtering and edge detection using the fuzzy logic approach, in *Proceedings of the 2nd IEEE International Conference on Fuzzy Systems*, Vol. 1, IEEE, Piscataway, NJ.

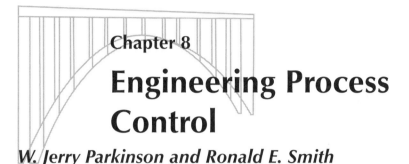

Chapter 8

Engineering Process Control

W. Jerry Parkinson and Ronald E. Smith

Abstract. Engineering process control is a reasonably mature discipline. Many books and papers have been published on the subject, and it is an integral part of many engineering curricula. Control systems are used in aircraft, chemical plants, and even robot manipulation. Although there are many other types of control, in this chapter we will discuss only feedback control, and that discussion will be quite limited. The purpose of the chapter is to provide a brief, basic introduction to classical process control. We will then introduce fuzzy process control and, finally, introduce the relatively new idea of probabilistic process control.

8.1   Introduction

In this chapter, we will present a basic introduction to engineering process control. We will look briefly at three controller types:

- the classical proportional-integral-derivative (PID) controller,

- the fuzzy logic controller, and

- the probabilistic controller discussed in detail by Laviolette et al. (1995) and Barrett and Woodall (1997).

We will look at two types of control problems:

- the setpoint tracking problem, and

- the disturbance rejection problem.

We will not discuss control system stability, which is extremely important for some types of control problems, but not an important concern for the problems discussed in this chapter. For more information on this subject, the reader is referred to Phillips and Harbor (1996), Szidarovszky and Bahill (1992), Shinskey (1988), and Ogunnaike and Ray (1994). We will discuss only feedback control systems here. Feed-forward control systems are not as common as feedback systems. A discussion of feed-forward systems can be found in Shinskey (1988), Ogunnaike and Ray (1994), and Murrill (1991, 1988). It would take an entire book or set of books to describe classical engineering process control in detail, and there are many good ones available. See, for example, Phillips and Harbor (1996), Szidarovszky and Bahill (1992), Shinskey (1988), Ogunnaike and Ray (1994), and Murrill (1991, 1988).

8.2   Background

The classical feedback control system can be described using a block flow diagram like the one shown in Figure 8.1.

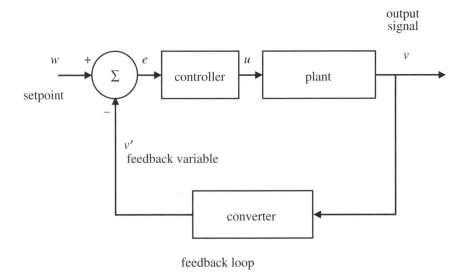

Figure 8.1. *Standard block flow diagram for a control system.*

In Figure 8.1, the first rectangular block represents our controller. The second rectangular block represents the system to be controlled, often called the plant. The block in the feedback loop is a converter. The converter converts the feedback signal to a signal useable by the summer. The summer is the circle at the far left-hand side of the diagram. The letter w represents the setpoint value or the desired control point. This is the desired value of the variable that we are controlling. The letter v represents the output signal or the current value of the variable that we are controlling. The symbol v' represents the feedback variable, essentially the same signal as the output signal but converted to a form that is compatible with the setpoint value. The letter e represents the error, or the difference between the setpoint value and the feedback variable value. The letter u represents the control action supplied by the controller to the plant. A short example will clarify this explanation.

Example 1.   Suppose that we wish to control the temperature in a cold one-room house with a variable load electric heater. The situation is pictured in Figure 8.2.

Equation (8.1) is the differential equation that describes the time (inside) temperature relation of the one-room house:

$$mC_p\frac{dT_i}{dt} = Q_i - Q_o - Q_d. \tag{8.1}$$

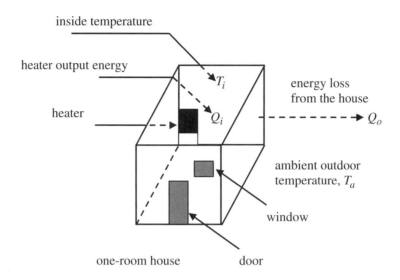

inside temperature

heater output energy

T_i

heater

Q_i

energy loss
from the house

Q_o

ambient outdoor
temperature, T_a

window

one-room house   door

Figure 8.2. *One-room house being heated.*

Equation (8.1) is the plant equation. This describes the response of the plant in the time domain. In (8.1), the letter *m* represents the mass of the air in the room. The symbol C_p represents the constant pressure heat capacity of the air in the room. The symbol T_i is the inside, or room, temperature. This is the output signal, or the controlled variable. It is equivalent to the letter *v* shown in Figure 8.1. The letter *t* represents time, and the symbol Q_i represents the heat or energy produced by the heater. The symbol Q_0 represents the normal energy loss through the walls of the house defined by

$$Q_o = U A_h (T_i - T_a). \tag{8.2}$$

The letter *U* is the overall heat transfer coefficient, described in heat transfer textbooks. For example, see Kreith (1958). This heat transfer coefficient takes into account the insulation in the walls, the heat transfer to the walls from the inside of the room, and the heat transfer from the walls on the outside of the house. The symbol A_h represents the heat transfer area of the house, essentially the surface area of all the walls and the roof. The symbol T_a represents the temperature of the outside air, or the ambient temperature. The symbol Q_d in (8.1) represents the disturbance energy loss. This is a heat loss that is incurred when a door or window is opened. It is defined by

$$Q_d = h A_{dw} (T_i - T_a). \tag{8.3}$$

The letter *h* represents a simple convective heat transfer coefficient. It is less complex than the compound heat transfer coefficient *U*; hence the use of a different letter. The symbol A_{dw} is the area of the door or window that is opened to create the disturbance energy loss.

In most cases we will want to maintain a constant room temperature, or a setpoint temperature, T_s. If we are energy conscious, this temperature might be about 68°F. We would like T_i to be equal to T_s at all times. Thus we set our room thermostat to T_s, or 68°F. In this case, T_s is equivalent to *w* in Figure 8.1. Our controller in this example manipulates the heater output by varying the resistance in the heater coil. Technically, the feedback signal to the controller will be electronic, but for our purposes, we can consider both the output signal and the feedback signal to be T_i, the inside room temperature. Now the conversion block in the feedback loop (in Figure 8.1) is equal to 1 and v' is equal to v. The error, e, is the difference between the setpoint value T_s (68°F) and the actual room temperature T_i. The control action u is the amount of resistance that the controller applies to the heating coil in the heater in order to generate the required amount of heat Q_i. For our purposes, we can assume that Q_i is equal to u, the signal that goes to the plant from the controller. The block flow diagram that accompanies this problem is shown in Figure 8.3.

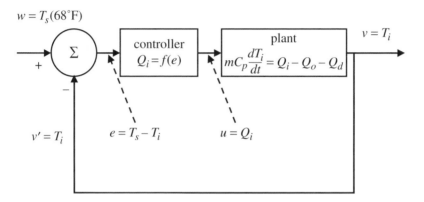

Figure 8.3. *Block flow diagram for Example 1.*

Figure 8.3 is a time domain–block flow diagram for a single-input–single-output (SISO) control system. The standard classical PID controller can be used to control this system.

8.2.1   Basic PID control

The PID control algorithm is described by

$$u = K_P e + K_I \int_0^{t^*} e\, dt + K_D \frac{de}{dt}. \tag{8.4}$$

In (8.4), the symbols K_P, K_I, and K_D represent the proportional, integral, and derivative control constants, respectively. These constants are specific to the system in question. They usually are chosen to optimize the controller performance and to ensure that the system remains stable for all possible control actions.

If we use a PID controller, or any other linear controller, for the system depicted in Figure 8.3, then the system is called a linear system. Linear systems have nice properties. The control engineer can use Laplace transforms to convert the linear equations in the blocks in Figure 8.3 to the Laplace domain. The blocks can then be combined to form a single transfer function for the entire system. The transformed version of Figures 8.1 and 8.3 is shown in Figure 8.4.

In Figure 8.4, $G(s)$ is the transfer function for the entire block flow diagram including the feedback loop. The variables $w(s)$ and $v(s)$ are the setpoint and output variables, respectively, converted to the Laplace, s, domain. The control engineer can work with the system transfer function and determine the range in which the control constants K_P, K_I, and K_D must fall in order to keep the system stable. Electrical engineers design controllers for a wide variety of systems. Many of these systems can become unstable as a result of a sudden change in the control action. An example might be an airplane control system.

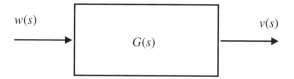

Figure 8.4. *Overall system block flow diagram or transfer function for the Laplace domain.*

Chemical engineers, on the other hand, usually design control systems only for chemical processes. These systems are not as likely to become unstable from a sudden change in the control action. An example is a liquid-level controller that controls the level in a tank full of chemicals. A sudden change in the control action, say a response to a leak in the tank, is not likely to make the system become unstable, no matter how abrupt the change. Laplace transforms and the s domain are a very important part of classical control theory, but are not necessary for the discussion in this chapter. The interested reader is referred to any fundamental text on process control. See, for example, Phillips and Harbor (1996).

The PID controller accounts for the error, the integral of the error, and the derivative of the error, in order to provide an adequate response to the error. Figure 8.5 shows a typical time-domain response curve for a PID controller, for a problem like the room-heating problem of Example 1.

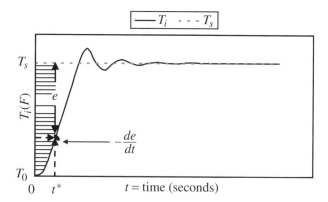

Figure 8.5. *Time temperature response for the room heating problem in Example 1.*

Figure 8.5 represents a typical response curve for a setpoint tracking problem. We will continue with Example 1 and assume that the thermostat setting has been changed from, say, $65°F(T_0)$ to $68°F(T_s)$. The control system will try to adjust as quickly as possible to the new setpoint. At time t^*, the room temperature T_i is at the point on the response curve to which the two broken arrows are pointing. The error, e, is the distance between the setpoint temperature line, T_s, and the current room temperature T_i. Since in this case

$$\frac{de}{dt} = \frac{d(T_s - T_i)}{dt} = -\frac{dT_i}{dt},$$

the derivative term in the PID control equation is shown as the negative derivative of the response curve at time t^*. The crosshatched area from the T_s line to the T_i curve between time t_0 and t^* shows the integral term. One criterion for an optimal controller is to find control constants than minimize the error integral. That is, find K_P, K_I, and K_D such that $\int_0^{t_\infty} e \, dt$ is minimized. The term t_∞ is the time at which the controlled variable (temperature in this case) actually reaches and stays at the setpoint. The proportional term in (8.4) drives the control action faster when there is a large error and slower when the error is small. The derivative term helps to home in the controller variable to the setpoint. It also reduces overshoot because of its response to the change in the sign and to the rate of change of the error. Without the integral term, however, the controlled variable would never hold at the setpoint, because at the setpoint both e and $\frac{de}{dt}$ are equal to 0. A pure proportional-derivative (PD) controller would become an on–off controller when operating about the setpoint.

8.2.2   Basic fuzzy logic control

In the simplest form, a fuzzy control system connects input membership functions, that is, functions representing the input to the controller, e, to output membership functions representing the control action, u.

A good example of a fuzzy control system is a controller that controls the room temperature in the one-room house shown in Figure 8.2. This time we will look at a disturbance rejection problem. Assume that the room is at the setpoint temperature of 68°F. This temperature is maintained because the controller is working, forcing the heater to supply enough heat to the room to maintain the setpoint temperature. If someone were to open the door to the cold outdoors, the room temperature would drop. When the door is closed again, the temperature will start to rise and head back to the setpoint, but at a pretty slow rate unless the controller and the heater are capable of dealing with this situation. A simple fuzzy control system designed for our room-heating problem consists of three rules:

1. If the temperature error is positive, then the change in control action is positive.

2. If the temperature error is zero, then the change in control action is zero.

3. If the temperature error is negative, then the change in control action is negative.

The input membership functions are shown in Figure 8.6. The reader should notice that the input membership functions as shown in Figure 8.6 are limited to a temperature error between ±20°F. It is easy, and probably useful, to extend the limits to ±∞ by using statements in the controller code such as

if   temperature error ≥ (say) 5°F,

then   the membership in positive is 1.0;

else if   temperature error ≤ (say) −5°F,

then   the membership in negative is 1.0.

The reader should also notice the "dead band" or "dead zone" in the membership function zero between ±1°F. This is optional and is a feature commonly used with on–off controllers. It is easy to implement with a fuzzy controller and is useful if the control engineer wishes to resrtict control response to small transient temperature changes. This step can save wear and tear on equipment.

The output membership functions for this controller are shown in Figure 8.7. In this figure the defuzzified output value from the controller is a fractional value representing the required heater output for the desired temperature change. It is defined by

$$\text{change in controller action} = \Delta u = \frac{Q_i - Q_{\text{sp}}}{\text{range}}. \tag{8.5}$$

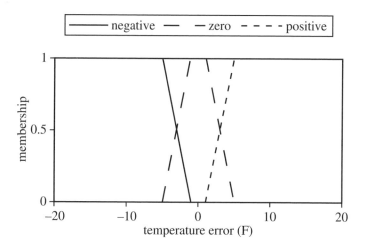

Figure 8.6. *Input membership functions for the fuzzy room temperature controller.*

In (8.5), the term Q_{sp} represents the heater output required to maintain the setpoint temperature. The term Q_i is the new heater load requested by the heater. If $\Delta u \succ 0$, then the range is defined as $Q_{max} - Q_{sp}$, where Q_{max} is the maximum heater output. If $\Delta u \succ 0$, then the range is defined as Q_{sp}. The term Q_{sp} must be calculated using a steady-state energy balance for the room or it must be estimated in some other fashion. The steady-state calculation requires only algebra. Differential equations are not needed. This is an engineering calculation quite different than calculations needed to compute K_P, K_L, and K_D for the PID controller.

The fuzzy output sets positive and negative to $+2.0$ to 0.0 and -2.0 to 0.0, respectively. Since the change in controller action is a fraction between either 0.0 and 1.0 or 0.0 and -1.0, it is clear that we will never obtain a control action outside of the range -1.0 to 1.0. Our defuzzification technique will require that we include as positive those numbers up to 2.0 in the fuzzy set or membership function and negative those numbers down to -2.0 in the fuzzy set. Even though numbers of this magnitude can never be generated by our fuzzy system,

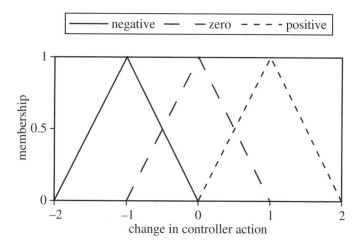

Figure 8.7. *Output membership functions for the fuzzy room temperature controller.*

we can still include them in our fuzzy sets. Users can define their fuzzy sets however they wish. The fuzzy mathematics described in earlier chapters is capable of handling objects of this type. The user has to define the fuzzy sets so that they make sense for the particular problem. In our case, we are going to use the centroid technique for defuzzification. We therefore need to extend our membership functions so that it is possible to obtain centroids of either ± 1.0. We need this capability in order for the control system to either turn the heater on "full blast" or turn it completely off.

We can describe our simple fuzzy controller as an approximation to an I or integral controller. Our rules are of the form $\Delta u = f(e)$, where Δu is the control action change for the sample time interval Δt. We can make the approximation that $\frac{\Delta u}{\Delta t} \approx \frac{du}{dt} = K_I e$ and $\int du = K_I \int e\, dt$ and $u = K_I \int_0^{t^*} e\, dt$.

We can build a fuzzy PID controller if in addition we include in our rules and membership functions the change in error with respect to time, $\frac{de}{dt}$, and the second derivative of the error with respect to time, $d(\frac{de}{dt})$. The fuzzy PID controller will be more flexible than the classical PID controller because the control constants K_P, K_I, and K_D are defined by the membership functions and can vary with the problem conditions.

Example 2. Suppose that someone does open the door to the one-room house described in Example 1 and that the room temperature drops from the setpoint value of $68°F$ to $65°F$. The error is defined as the setpoint reading, $68°F$, minus the new reading, $65°F$, or $+3°F$. Our three rules are fired, producing the following results:

1. The positive error is 0.5.

2. The zero error is 0.5.

3. The negative error is 0.0.

The results are shown graphically in Figure 8.8.

In this example, an error of $+3°F$ intersects the membership function zero at 0.5 and the membership function positive at 0.5. We say that rules 1 and 2 were each fired with a strength of 0.5. The output membership functions corresponding to rules 1 and 2 are each "clipped" at 0.5. See Figure 8.9.

The centroid of the "clipped" membership functions, the shaded area in Figure 8.9, is $+0.5$. This centroid becomes the term Δu in (8.5). Since Δu is greater than 0.0, (8.5) can be rewritten as

$$Q_i = (Q_{\max} - Q_{sp})\Delta u + Q_{sp} \quad \text{or} \quad Q_i = 0.5(Q_{\max} + Q_{sp}) \quad \text{since } \Delta u = 0.5. \quad (8.6)$$

This says that the new heater output, Q_i, should be adjusted to be halfway between the current, or setpoint, output and the maximum heater output. After an appropriate time interval, corresponding to a predetermined sample rate, the same procedure will be repeated until the setpoint temperature, $68°F$, is again achieved. The disturbance rejection response curve corresponding to this example will look something like the one shown in Figure 8.10.

8.2.3 Basic probabilistic control

The probabilistic controller (Laviolette et al. (1995); Barrett and Woodall (1997)) is very similar to the fuzzy controller. The probabilistic controller can use the same rules as the fuzzy controller, but the rules tie together input conditional probabilities and output probability

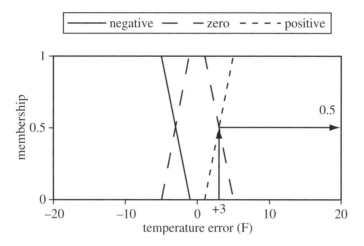

Figure 8.8. *Resolution of input for Example* 2.

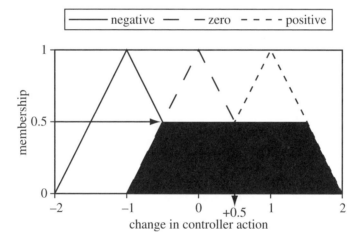

Figure 8.9. *Resolution of output for Example* 2.

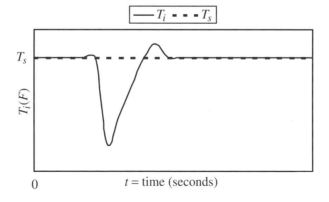

Figure 8.10. *Typical disturbance rejection response curve.*

density functions (PDFs) instead of membership functions. If we use the same three rules that we used for the fuzzy controller in section 8.2.2, then we can use input conditional probabilities that look similar to the fuzzy input membership functions shown in Figure 8.6. The input conditional probabilities are shown in Figure 8.11 and described by the following equations:

$$P[\text{``negative''}|x] = 1, \qquad\qquad x \le -5,$$
$$= -\frac{x+1}{4}, \quad -5 \le x \le -1,$$
$$= 0 \qquad\qquad \text{otherwise}, \qquad\qquad (8.7a)$$
$$P[\text{``zero''}|x] = \frac{x+5}{4}, \qquad -5 \le x \le -1,$$
$$= 1, \qquad\qquad -1 \le x \le 1,$$
$$= \frac{5-x}{4}, \qquad\quad 1 \le x \le 5$$
$$= 0 \qquad\qquad \text{otherwise}, \qquad\qquad (8.7b)$$
$$P[\text{``positive''}|x] = \frac{x-1}{4}, \qquad\quad 1 \le x \le 5,$$
$$= 1, \qquad\qquad\quad x \ge 5,$$
$$= 0 \qquad\qquad \text{otherwise}. \qquad\qquad (8.7c)$$

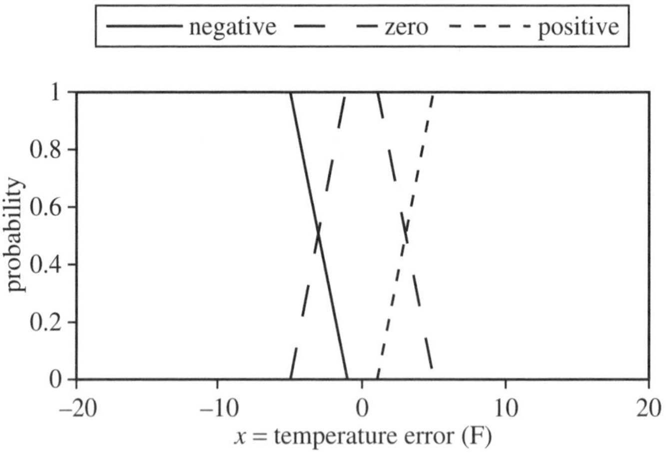

Figure 8.11. *Probabilities for temperature error classes.*

Figures 8.6 and 8.11 look very much alike, but they represent different concepts. Even though probability functions and membership functions are different concepts, they are treated in a similar fashion when solving control problems. A major difference in the way that these functions are used is that probabilities *must* add to 1. All memberships computed based on Figure 8.6 will also add to 1, but this is not a requirement for membership functions. The user or control system developer is free to choose any fuzzy set that might fit the particular problem. The sum of the set memberships is not restricted to equal 1. In the case of the fuzzy controller, the control system designer defines the membership functions, as in Figure 8.6, and knows exactly where the set boundaries are. In Example 2 (Figure 8.7), our error is +3°F. From Figure 8.7, we can determine that the membership in the fuzzy set zero is 0.5 and the membership in the fuzzy set positive is also 0.5. In the case of the probabilistic controller, the error is the same, $+3°F$. The difference is that the probabilist uses Figure 8.11 and/or (8.7b) and (8.7c) to determine the probabilities of the error being in either the crisp set zero or the crisp set positive. The probabilist assumes that the sets negative, zero, and positive are crisp sets with definite boundaries, but that these boundaries are unknown. The measured error belongs to only one of these sets and belongs with a full membership of one. Figure 8.11 and (8.7a)–(8.7c) were constructed to help the user estimate the probability of whether the error belongs to a given set. If we rework Example 2 for a probabilistic controller, with an error of $+3°F$, we would say that there is a probability of 0.5 that the error belongs to the set zero and a probability of 0.5 that the error belongs to the set positive. As with the fuzzy example, rules 1 and 2 are each fired with a probability (strength) of 0.5. The output, however, is computed differently than in the fuzzy example. In the probabilistic case, the output is described by PDFs, rather than output membership functions. For the probabilistic continuation of Example 2, the output PDFs are shown in Figure 8.12 and described by the following equations:

$$
\begin{aligned}
\text{"negative"}: f_1(z) &= -2z, & -1 \leq z \leq 0, \\
&= 0 & \text{otherwise,} & \quad (8.8a) \\
\text{"zero"}: f_2(z) &= 1+z, & -1 \leq z \leq 0, \\
&= 1-z, & 0 \leq z \leq 1, \\
&= 0 & \text{otherwise,} & \quad (8.8b) \\
\text{"positive"}: f_3(z) &= 2z, & 0 \leq z \leq 1, \\
&= 0, & \text{otherwise.} & \quad (8.8c)
\end{aligned}
$$

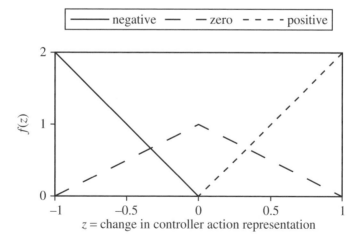

Figure 8.12. *Change in controller action PDFs.*

In the probabilistic case, the change in controller action, Δu, is a function of a representative value of the PDFs that are used to define the rule outputs. The representative value of the PDF can be one of several properties of that PDF. The mean, the mode, and the median are examples of the PDF representative values. For notational purposes we will use the letters RV to denote the general representative value.

The area under the PDF curve must equal 1, and the bounds must include only numbers that belong to the distribution. In our example, the lower bound of the PDF that represents the set negative can be no lower than -1.0, and the upper bound for the PDF that represents the set positive can be no larger than $+1.0$. The rules governing the construction of the output PDFs are more restrictive than those governing the construction of the fuzzy output membership functions. However, the resolution of the probabilistic rules is normally easier than the resolution of the fuzzy rules. The resolution formula that will determine the change in controller action for our probabilistic SISO control problem is given by

$$\Delta u = \sum_{i=1}^{3} P_i (\text{PDF–RV})_i. \tag{8.9}$$

In (8.9), P_i is the input probability for rule i, and $(\text{PDF–RV})_i$ is the representative value for the output PDF for rule i. Table 8.1 gives the RVs of the mean and the mode for Figure 8.12.

Table 8.1. *RVs of the mean and the mode for Figure* 8.12.

RV/PDF	Negative	Zero	Positive
mean	$-\frac{2}{3}$	0	$+\frac{2}{3}$
mode	-1	0	$+1$

If the mean is the RV, then in our example, with a probability of 0.5 that the error is equal to zero and a probability of 0.5 that the error is equal to positive, from (8.9) Δu is

$$\Delta u = (0) * \left(-\frac{2}{3} \right) + (0.5) * (0) + (0.5) * \left(\frac{2}{3} \right) = \frac{1}{3}.$$

If the mode is the RV, (8.9) produces $\frac{1}{2}$ for the value of Δu:

$$\Delta u = (0) * (-1) + (0.5) * (0) + (0.5) * (1) = \frac{1}{2}.$$

This is the same value that was produced by the fuzzy controller. The actual controller output is still produced by using equation (8.5).

8.3   SISO control systems

The probabilistic control problem described by Laviolette et al. (1995) and Barrett and Woodall (1997) is an interesting one. It is somewhat like the room-heating problem of Examples 1 and 2. It is different enough, however, to demonstrate some important ideas. This probabilistic controller was constructed by Barrett and Woodall (1997) based on a fuzzy control problem described by Kosko (1983). Unfortunately, there is no model or real system on which to test either this fuzzy control problem or this probabilistic control system. All one can tell is that both control systems will provide similar control actions for the same input (Barrett and Woodall (1997), Laviolette et al. (1995)). It is quite instructive to provide a model of the system for the controllers and compare the system response to these controllers and also to a PID controller. A model was constructed and we attempted to fit this model to the existing controller. It is important to note that this procedure usually is done the other way around. A control system usually is designed to conform to some model of the real

Table 8.2. *Fuzzy and probabilistic rules for the room comfort problem.*

Rule number	Room Temperature	Fan Motor Speed
1	Cold	Stop
2	Cool	Slow
3	Just Right	Medium
4	Warm	Fast
5	Hot	Blast

world. Consequently, the model used for this example is only a first approximation of the real world, but it is good enough to demonstrate several important points.

The membership functions used by Barrett and Woodall (1997) and Laviolette et al. (1995) are slightly different than those used by Kosko (1983). We will use the membership functions of Barrett and Woodall (1997) and Laviolette et al. (1995) since they developed the probabilistic controller. On the surface, the problem appears to be similar to temperature control of a room as in Examples 1 and 2. It is actually comfort control of the room. The sole control of the comfort is the motor speed of the fan. The five rules for both the fuzzy and the probabilistic controllers are given in Table 8.2. The rules for this problem are of the form

if the "room temperature" is..., then set the "fan motor speed" to....

For the fuzzy controller, the membership functions for the input variable "room temperature" are shown in Figure 8.13. The membership functions for the output variable "fan motor speed" are shown in Figure 8.14. Note that we have extended the output membership functions from −30RPM to 130RPM so that we can use centroid defuzzification and still stop the fan completely and also get speeds of 100RPM.

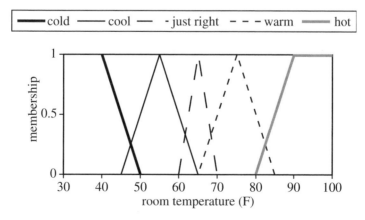

Figure 8.13. *Input membership functions for the room comfort problem.*

For the probabilistic controller, the input conditional probabilities for temperature classes are shown in Figure 8.15. The rule outputs for the probabilistic controller are motor speed PDFs. They are shown in Figure 8.16.

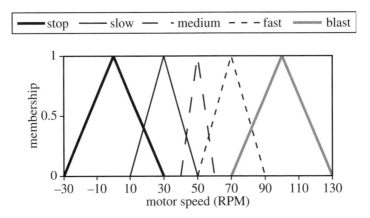

Figure 8.14. *Output membership functions for the room comfort problem.*

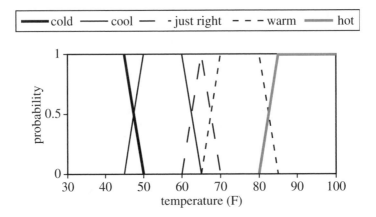

Figure 8.15. *Input conditional probabilities for the room comfort problem.*

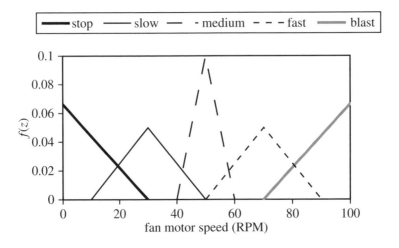

Figure 8.16. *Output probability density functions for the room comfort problem.*

The conditional probabilities shown in Figure 8.15 are described in the following equations:

$$P[\text{``cold''}|x] = 1, \qquad x \leq 45,$$
$$= \frac{50 - x}{5}, \quad 45 \leq x \leq 50,$$
$$= 0 \qquad \text{otherwise,} \qquad (8.10a)$$

$$P[\text{``cool''}|x] = \frac{x - 45}{5}, \quad 45 \leq z \leq 50,$$
$$= 1, \qquad 50 \leq x \leq 60,$$
$$= \frac{65 - x}{5}, \quad 60 \leq x \leq 65,$$
$$= 0 \qquad \text{otherwise,} \qquad (8.10b)$$

$$P[\text{``just right''}|x] = \frac{x - 60}{5}, \quad 60 \leq x \leq 65,$$
$$= \frac{70 - x}{5}, \quad 65 \leq x \leq 70,$$
$$= 0 \qquad \text{otherwise,} \qquad (8.10c)$$

$$P[\text{``warm''}|x] = \frac{x - 65}{5}, \quad 65 \leq x \leq 70,$$
$$= 1, \qquad 70 \leq x \leq 80,$$
$$= \frac{85 - x}{5}, \quad 80 \leq x \leq 85,$$
$$= 0 \qquad \text{otherwise,} \qquad (8.10d)$$

$$P[\text{``hot''}|x] = \frac{x - 80}{5}, \quad 80 \leq x \leq 85,$$
$$= 1, \qquad x \geq 85,$$
$$= 0 \qquad \text{otherwise.} \qquad (8.10e)$$

The PDFs shown in Figure 8.16 are described by the following equations:

$$\text{``stop''} : f_1(z) = \frac{30 - z}{450}, \quad 0 \leq z \leq 30,$$
$$= 0 \qquad \text{otherwise,} \qquad (8.11a)$$

$$\text{``slow''} : f_2(z) = \frac{z - 10}{400}, \quad 10 \leq z \leq 30,$$
$$= \frac{50 - z}{400}, \quad 30 \leq z \leq 50,$$
$$= 0 \qquad \text{otherwise,} \qquad (8.11b)$$

$$\text{``medium''} : f_3(z) = \frac{z - 40}{100}, \quad 40 \leq z \leq 50,$$
$$= \frac{60 - z}{100}, \quad 50 \leq z \leq 60,$$
$$= 0 \qquad \text{otherwise,} \qquad (8.11c)$$

$$\text{``fast''} : f_4(z) = \frac{z - 50}{400}, \quad 50 \leq z \leq 70,$$

$$= \frac{90 - z}{400}, \quad 70 \leq z \leq 90,$$

$$= 0 \qquad \qquad \text{otherwise,} \tag{8.11d}$$

$$\text{"blast"} : f_5(z) = \frac{z - 70}{450}, \quad 70 \leq z \leq 100,$$

$$= 0 \qquad \qquad \text{otherwise.} \tag{8.11e}$$

We notice that the probabilities shown in Figure 8.15 and described by (8.10a)–(8.10e) always add to 1. The PDFs shown in Figure 8.16 and described by (8.11a)–(8.11e) are created so that the area under each PDF is equal to 1 and the base of each PDF is the range of fan motor speed (z) that can be described by the PDF labels "stop," "slow," etc. The input values (room temperatures) are used with the probabilities just like they are used with fuzzy logic input membership functions. The same five rules are used and the appropriate rules are fired. The output from the method is a PDF, and the crisp value required for the fan speed is computed by (8.9).

Example 3a: Probabilistic controller.   Suppose that we have a room temperature of 100°F. One rule is fired, rule number 5: "If the room is hot, then blast." In this case, the output value requested by the controller is the mean value for the PDF that represents "blast." This value is 90RPM. Although the fan is capable of 100RPM, the highest value that the probabilistic control system can request is 90RPM. If the mode were used instead of the mean to represent the PDF, the controller would work better (at least in this case).

Example 3b: Fuzzy controller.   The room still is at 100°F and rule number 5 is fired. With the fuzzy control system, one technique used with this type of problem is to extend the fuzzy set "blast" to 130RPM. Even though the fan cannot reach these speeds, there is nothing to preclude these numbers from being in that fuzzy set. If "blast" is now a triangle that extends from 70RPM to 130RPM, the centroid of that triangle is 100RPM. This is the fan maximum speed and the desired output value for the control system.

8.3.1   A system model and a PID controller

If we write a computer model for the room and the fan, we can track the response of the model to the control actions over a period of time. The model that we chose is somewhat crude but is a reasonably good "first" approximation of the system. We assumed that the model was a small one-room house with poor insulation. We assumed an overall R-value of approximately 10. The heat sources are heat coming through the walls and the ceiling from the outside world, a possible heat leak from an open window or door, and heat transferred from one human body, at a temperature of 98.6°F, to the room. The differential equation describing the room temperature as a function of time is given by

$$\rho C_p V \frac{dT_i}{dt} = (h_o A_o + U_h A_h)(T_a - T_i) + h_b A_b(98.6 - T_i). \tag{8.12}$$

In (8.12), the terms are as follows:

ρ is the density of the air in the room.

C_p is the heat capacity of the air in the room.

V is the volume of air in the room.

T_i is the room temperature.

t is time.

h_o is the heat transfer coefficient for the opening (window or door).

A_o is the area of the opening, normal to the direction of heat transfer.

U_h is the average overall heat transfer coefficient of the house.

A_h is the heat transfer area for the house.

T_a is the outside, ambient temperature.

h_b is the heat transfer coefficient for the human body.

A_b is the surface area of the human body.

Equation (8.12) is obviously a simplified model because the mass of the house, among other things, is ignored. Values and models for the heat transfer coefficients in (8.12) can be obtained from any engineering heat transfer text. For example, see Kreith (1958).

Example 4a: Probabilistic controller. In this problem, we imposed an ambient temperature profile on the house that went from 45°F to nearly 103°F then dropped to nearly 90°F in a period of four hours. At 90 minutes into the run, we imposed a disturbance on the system, by opening the door for 30 minutes. The ambient temperature profile and the computed room temperature response (8.12) are shown in Figure 8.17.

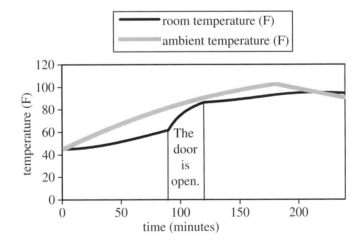

Figure 8.17. *Room and ambient temperature profiles for Example* 4.

The fan speed response from the probabilistic controller, with the room temperature, is plotted in Figure 8.18. These plots demonstrate that the probabilistic controller causes the fan speed to track the temperature reasonably well. The controller also adjusts the fan speed quite well to compensate for the disturbance caused by leaving the door open. We can say that this probabilistic controller can be used for both setpoint tracking and disturbance rejection. This controller probably is adequate for keeping a human reasonably comfortable in a hot room. It probably mimics the human manual controller quite well. We should notice, however, the flat regions and abrupt changes in the response curve (fan speed). The flat regions are caused by conditional probabilities that are equal to 1 for a wide range. In this case, it is the probabilities of cool and warm shown in Figure 8.15. For example, between the temperatures of 50°F and 60°F, the probability that the temperature is cool is 1. Rule 2, Table 8.2, is the only rule that is fired as long as the temperature is in this range. The output of that rule is a fan speed of slow, which corresponds to 30RPM, the mode (and the mean value) of the PDF that represents a fan speed of slow in Figure 8.16. Flat spots and abrupt changes can easily be avoided with fuzzy and probabilistic control systems. Smooth control transitions usually are one of the benefits of these control systems. This is not meant as a criticism of the control system. The original problem (Kosko (1983)) was set up to demonstrate how to build control systems using fuzzy logic. It was not meant to show how to build the best control systems. The systems shortcomings carried over to the probabilistic treatment.

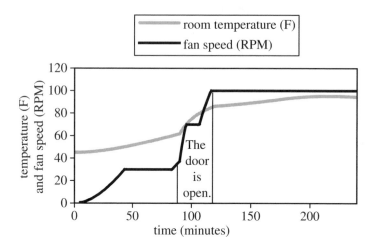

Figure 8.18. *Room temperature and fan speed profiles for Example* 4a.

Example 4b: Fuzzy controller.   For the fuzzy problem, we imposed the same ambient temperature profile and same disturbance on the house. The ambient temperature profile and the computed room temperature response were the same as those for the probabilistic problem shown in Figure 8.17. The fan speed response for the fuzzy controller was similar also. It is shown in Figure 8.19.

The flat regions and abrupt changes in the response curve shown in Figure 8.19 are due to the portions of the input membership functions shown in Figure 8.13, where only one membership function is used and only one rule is fired. It is similar to the problem with the probabilistic controller. It is evident from Figures 8.18 and 8.19 that, at least for this problem, probabilistic controllers can be written to behave almost exactly like fuzzy controllers, or visa versa.

Example 4c: PID controller.   To construct a PID controller as described by (8.4), we need to find an error term, *e*. If we look at Figures 8.18 and 8.19, we see that these response curves

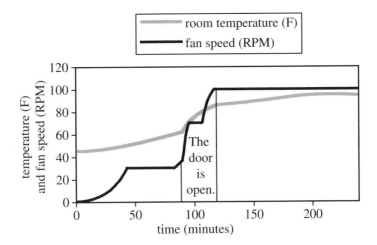

Figure 8.19. *Room temperature and fan speed profiles for Example* 4b.

do not really show an error. This is because we are not really controlling the temperature with the fan. We are controlling human comfort. We cannot choose a desired temperature as a setpoint and control that temperature with the fan. Since the controller, and therefore the fan, does not determine the difference between some setpoint and the "controlled" room temperature (the error), an error term, for this situation, supplied to a PID controller is meaningless. To use a PID controller, we have to reformulate the problem.

If we assume that the room comfort is totally defined by the heat transfer rate from a single human body, we can build a "soft-sensor" for the control system using the second portion of the right-hand side of (8.12):

$$Q_b = h_b A_b (98.6 - T_i). \tag{8.13}$$

In (8.13), Q_b is the heat transfer rate from a single human body in the room, in BTU/minute. Equation (8.13) is called a soft-sensor because we cannot use an actual hardware sensor to determine the heat transfer rate. We need to measure the room temperature and compute the heat transfer rate using software. To determine a setpoint for our PID control system, we need to estimate a value for a comfortable heat transfer rate. For this example, we assumed that the human body produces 2000Kcal per day of heat at a temperature of 98.6°F. We further assumed that this energy should be dissipated at an even rate of 5.51BTU/min for the entire 24-hour period. For this example, our setpoint is going to be 5.51BTU/min, although a skilled biologist might come up with a better answer.

In (8.13), the convective heat transfer coefficient, h_b, depends on both free convection and forced convection. When the fan is not turning, the heat is transferred from the body by natural or free convection. In this mode air, and consequently heat, movement is caused by the difference in the air density at the body surface and in the room. This density difference is in turn caused by the differences in temperature at these points. Consequently, most free convection heat transfer coefficient correlations are of the following form:

$$h_b(\text{free}) = f((98.6 - T_i)^{0.4}). \tag{8.14}$$

When the fan is turning, the convective heat transfer coefficient is a function of the velocity of the air leaving the fan blade. Several models are available for this situation (Kreith (1958)).

One model could be that this also is a free convection heat transfer mode, with gravity force air movement replaced with the fan centrifugal force movement. This is much like the way a centrifuge replaces gravitation force with centrifugal force. No matter the exact model, the heat transfer coefficient correlation is of the following form:

$$h_b(\text{forced}) = f((\text{fan\_speed})^{0.8}). \qquad (8.15)$$

Obviously, increasing the fan speed is going to increase the heat transfer rate. The human body heat transfer coefficient, h_b, is going to be based on both free convection and forced convection depending upon the fan speed and the observer's position in the room. As a first approximation, for this example we assumed that h_b is a linear combination of forced convection and free convection. (Note that this method of compacting h_b was also used with the fuzzy and probabilistic controller examples.)

Now we can reformulate Example 4c so that we can control the system using a PID controller. The block flow diagram is shown in Figure 8.20. The heat loss response curve for the example is shown in Figure 8.21. The room temperature and the fan speed profiles are shown in Figure 8.22.

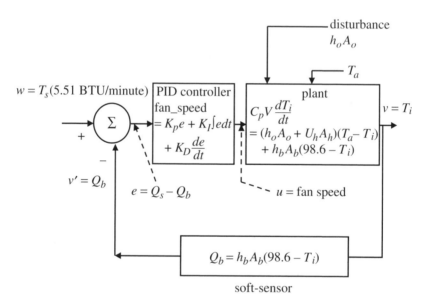

Figure 8.20. *Block flow diagram for a PID controller(Example 4c).*

Observing Figure 8.21, we can see that the PID controller does not turn the fan on until the body heat loss is less than or equal to the setpoint. This makes sense, because if the temperature is too low, turning the fan on will increase the heat loss from the body, making the person in the room less comfortable. Further, the controller is able to hold the comfort level at the setpoint until the door is opened. Observing Figure 8.22, we can see that the PID controller tries very hard to reject this disturbance by turning on the fan full blast. Unfortunately, the fan is not capable of increasing the heat transfer rate enough to control the extra heat load. The PID controller does a reasonably good job of tracking the "comfort" setpoint until the room temperature becomes too close to the body temperature of 98.6°F and the fan can no longer remove heat fast enough to keep the human body comfortable. The PID controller makes a strong attempt at disturbance rejection, but the fan is not adequate for the

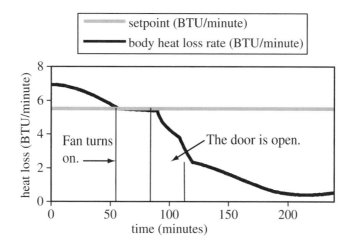

Figure 8.21. *Response curve for the PID controller (Example 4c).*

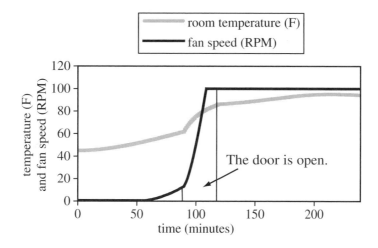

Figure 8.22. *Room temperature and fan speed profiles for Example 4c.*

task. The PID controller constants were assigned based on past experience. No attempt was made to optimize the controller. Optimization was not important in this problem because the fan was not adequate for the task.

Figures 8.23 and 8.24 are the comfort response curves for the same problem using the probabilistic controller and the fuzzy controller, respectively. Since these controllers were not programmed to track the comfort setpoint, they did not do a very good job of that. Similarly, the PID controller did not track the room temperature very well because it was not programmed to do so. In Examples 4a and 4b, the probabilistic controller and the fuzzy controller produced response curves that were nearly identical. The PID controller is different. It required a model (the soft-sensor), a calculated error, and a setpoint. On the surface, it appears that the PID controller might be superior to both the probabilistic controller and the fuzzy controller. It did, after all, track and temporarily maintain the real variable—comfort—better than either of the other two controllers.

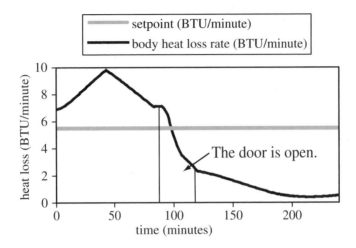

Figure 8.23. *Comfort response curve for the probabilistic controller* (*Example* 4a).

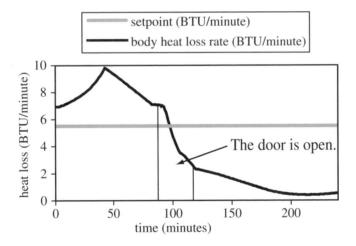

Figure 8.24. *Comfort response curve for the fuzzy controller* (*Example* 4b).

Several observations need to be made about Example 4c:

• We are controlling a model, not a real-world situation.

• The model for the soft-sensor for the PID controller is identical to the human heat loss equation in the model. This will never be the case in a real-world problem.

• To compute an error term for the PID controller, an estimate needed to be made for the setpoint and a model had to be constructed for the soft-sensor.

• The accuracy of the setpoint estimation and the quality of the model will determine the quality of the PID controller.

All three techniques are available to be used. Each technique has weak and strong points. The controller choice will probably depend upon the problem. For this problem, the control engineer might ask the following questions:

Can I design a better control system by doing more research and possibly some experimentation in order to build an accurate soft-sensor model and determine an accurate setpoint? Or is it better to sit in a room with a thermometer and a tachometer for the fan in order to determine the best rules and membership functions (or conditional probabilities and PDFs) for the fuzzy or probabilistic controllers?

The goal, after all, is to build the best control system possible.

8.4 Multi-input–multi-output control systems

The multi-input–multi-output (MIMO) control system is a little bit more complicated. The problem is best described by the following examples.

Example 5a: Probabilistic controller.   In this example, we extend the work of Barrett and Woodall (1997). In their example, they add humidity measurements to the controller. The controller becomes a multi-input (binary)–single-output (MISO) controller. If we assume five input classes for humidity, then we need 25 rules for a complete control set. The number of rules equals the product of the number of membership functions for each input variable. The rules are of the form

> if the "room temperature" is. . . and if the "room humidity" is. . . , then set the "fan motor speed" to. . . .

The input conditional probabilities for humidity classes could be represented as those shown in Figure 8.25. The other inputs, temperature classes, are still the same (Figure 8.15). The rule outputs for the probabilistic controller are the same motor speed PDFs shown in Figure 8.16. For this example and the next one (the fuzzy controller), we will use Figure 8.25 for the conditional probabilities of both the probabilistic controller input and membership functions input for the fuzzy controller.

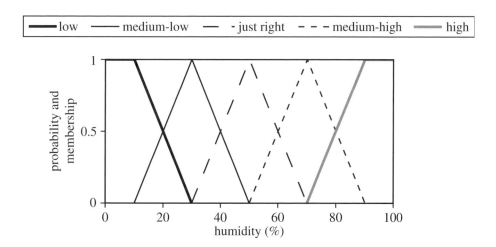

Figure 8.25. *Input conditional probabilities and membership functions for humidity for Examples* 5a *and* 5b.

Table 8.3. *Fuzzy and probabilistic rules for the two-input room comfort problem* (*abbreviated*).

Rule number	Room temperature	Room humidity	Fan motor speed
1	cold	low	stop
⋮	⋮	⋮	⋮
13	just right	just right	medium
14	just right	medium-high	medium
⋮	⋮	⋮	⋮
18	warm	just right	medium
19	warm	medium-high	fast
⋮	⋮	⋮	⋮

Table 8.3 is an abbreviated set of the 25 rules required for this controller.

If the room temperature is 68°F and the room humidity is 55%, then from Figure 8.15 the probability that the temperature is warm is 0.6 and the probability that the temperature is just right is 0.4. From Figure 8.25, the probability that the humidity is just right is 0.75 and the probability that the humidity is medium-high is 0.25. Since the 25 rules used by the controller represent each possible combination of humidity and temperature, only the rules 13, 14, 18, and 19 will be fired. The formula for computing the fan motor speed is given by (8.16), a slightly modified version of (8.9):

$$u = \sum_{i=1}^{25} P_{ij} P_{ik} \, (\text{PDF–RV})_i \,. \tag{8.16}$$

In (8.16), P_{ij} is the input probability for rule i and input j (temperature), P_{ik} is the input probability for rule i and input k (humidity), and $(\text{PDF–RV})_i$ is the representative value for the output PDF for rule i. Equation (8.16) can easily be generalized to the case of more than two inputs and more than 25 rules. Again, in a problem like this, where we need to actually stop the fan or bring it to a full 100RPM, we like to use the mode for the RV value. From Figure 8.16, the mode for the medium PDF is 50RPM and the mode for the fast PDF is 70RPM. For this example, from (8.16), the controller output u is

$$u = ((0.4) * (0.75) + (0.4) * (0.25) + (0.6) * (0.75)) * 50 + (0.6) * (0.25) * 70$$
$$= 53\text{RPM}.$$

Example 5b: Fuzzy controller. The same problem using a fuzzy logic controller is solved a little bit differently. The input membership functions for temperature and humidity are shown in Figures 8.13 and 8.25, respectively. For a temperature of 68°F, the membership (from Figure 8.13) in just right is 0.4 and the membership in warm is 0.3. For a humidity of 55%, from Figure 8.25 the membership in just right is 0.75 and the membership in medium-high is 0.25. The same four rules as in Example 5a—rules 13, 14, 18, and 19—are fired. With the fuzzy controller, we must determine the strength with which each rule is fired. There are several techniques available for resolving this problem. We like the min–max rule. The rules and their values shown in Table 8.4 will help to demonstrate this technique.

From Table 8.4, we can see that each rule is fired with a strength equal to its minimum antecedent value. For this example, the consequent of three of the rules is the membership

Table 8.4. *Demonstration of the min–max using fuzzy rules.*

Rule number	Room temperature (membership)	Humidity (membership)	Fan motor speed (strength)
13	just right (0.4)	just right (0.75)	medium (0.4)
14	just right (0.4)	medium-high (0.25)	medium (0.25)
18	warm (0.3)	just right (0.75)	medium (0.3)
19	warm (0.3)	medium-high (0.25)	fast (0.25)

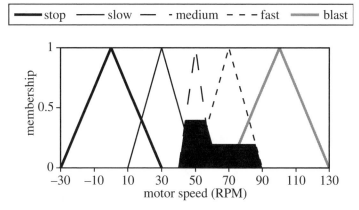

Figure 8.26. *Resolution of output for Example 5b.*

function medium. Medium has three consequent values 0.4, 0.25, and 0.3. We choose the maximum value of 0.4 and truncate the medium membership function at that strength. The maximum portion of the min–max rule does not have to be invoked for rule 19 since it is the only rule that refers to the output membership function fast. In this example, fast is truncated at 0.25. The output membership function corresponding to this example is shown in Figure 8.26.

The centroid of the output membership function shown in Figure 8.26 is 62.25RPM. This is significantly different than 53RPM computed by the probabilistic controller in Example 5a. There are two reasons for this:

- The input memberships for temperature sum to less than 1 for this problem.

- The centroid defuzzification technique is quite different than the technique used to resolve the probabilistic controller output. The difference is most apparent when more than one of the fired rules point to the same membership function. Tables 8.3 and 8.4 suggest that the fan motor speed should be medium. We do not run into this problem very often in SISO problems. The centroid technique counts the maximum value of medium, but counts it only once.

In this case, the fact that the input memberships for temperature sum to less than 1 makes little difference. If the temperature membership in warm were 0.6 instead of 0.3, then rule 18 would provide a maximum value for fan motor speed = medium of 0.6. The centroid defuzzification technique would provide a new fan speed of 60.7RPM instead of 62.25RPM. Another defuzzification technique that provides solutions closer to those produced by the probabilistic controller is the correlation minimum encoding technique. This technique uses the min portion of the min–max rule to determine the rule strength and truncates the output membership functions at that value. The controller output is then determined by the following:

$$u = \frac{\sum \text{area} * \text{centroid}}{\sum \text{area}}. \tag{8.17}$$

For this example, the controller output from (8.17) would be

$$u = \frac{(4.375 + 5.1 + 6.4) * 50 + 8.75 * 70}{(4.375 + 5.1 + 6.4 + 8.75)} = 57.1\text{RPM}.$$

If the membership for temperature in warm were 0.6 instead of 0.3, (8.17) would yield a value of 56.25RPM, much closer to the value produced by the probabilistic controller.

The choice of the defuzzification technique is up to the control engineer. The membership functions and the rules will be adjusted to fit the problem. This involves understanding the problem and not just blindly using mathematical formulas to solve the problem.

For both the fuzzy and the probabilistic controllers, the MISO problem described above can be easily extended to a MIMO problem. If we had one other output from the controller, say a room humidifier with several different settings, the rules would be of the form

> if the "room temperature" is... and if the "room humidity" is..., then set the "fan motor speed" to... and set the "humidifier" to....

One more column would have to be added to Table 8.3 for humidifier setting, but no more rules would normally need to be added. Another set of PDFs and/or membership functions would be added for the humidifier settings. The resolution of this output would be the same as for the fan motor speed. The probability multipliers or rule strengths would be the same for each rule. The final control value for the humidifier setting would depend upon the shape of the output PDFs or membership functions.

Example 5c: PID controller. For the PID approach, we will shift to a different, more manageable problem. The example used in this problem will be control of the liquid levels in two connected tanks. Two pumps that supply liquid to each tank will maintain the liquid level. Figure 8.27 is a schematic diagram of this plant.

In this problem, the object is to maintain the liquid level in tank 1, h_1, at a setpoint level that we call w_1. We want to simultaneously control the liquid level in tank 2, h_2, at a setpoint level we will call w_2. Liquid flows from tank 1 to tank 2 by means of a small pipe connection between the bottoms of the two tanks. This pipe has a valve in it to restrict the flow, q_{12}, between the two tanks. Similarly, liquid flows from tank 2 into a sump whose level is below the level of tank 2. This flow is labeled q_{20}. To maintain the levels in the tanks the controller must continually supply the proper power to pumps 1 and 2 so that they can maintain the correct flow to the individual tanks. Equations (8.18) and (8.19) are the differential equations that describe this system, or plant:

$$A\frac{dh_1}{dt} = Q_1 - q_{12}, \tag{8.18}$$

$$A\frac{dh_2}{dt} = Q_2 + q_{12} - q_{20}. \tag{8.19}$$

The symbol A represents the cross-sectional area of the two tanks. In this problem A is the same for both tanks. The flows q_{12} and q_{20} are further defined by the following equations:

$$q_{12} = \Phi_1 A_p \, \text{Sign}(h_1 - h_2)\sqrt{2g|h_1 - h_2|}, \tag{8.20}$$

$$q_{20} = \Phi_2 A_p \sqrt{2gh_2}. \tag{8.21}$$

The Φ_i symbols in the above equations represent friction coefficients for the flow in the pipes, with the restrictions, between tanks 1 and 2 and leaving tank 2. The symbol A_p represents the cross-sectional area of these pipes. In this problem, the cross-sectional area is the same for both pipes. The symbol g is the acceleration of gravity. If we substitute (8.20) and (8.21) into (8.18) and (8.19), the differential equations become nonlinear and more difficult to deal with. For this example, let us make the following assumptions:

- The cross-sectional area, A, is 1m^2.

- h_1 will always be greater than h_2.

- The setpoint for tank 1, w_1, is 2m, and h_1 will always remain very close to that setpoint.

- The setpoint for tank 2, w_2, is 1m, and h_2 will always be very close to that setpoint.

- The portion of (8.20) represented by $\Phi_1 A_p \sqrt{2g}$ is approximately equal to 1.

- The portion of (8.21) represented by $\Phi_2 A_p \sqrt{2g}$ is approximately equal to 2.

With these assumptions, (8.19) and (8.20) become the following:

$$\frac{dh_1}{dt} = Q_1 - \sqrt{h_1 - h_2}, \tag{8.22}$$

$$\frac{dh_2}{dt} = Q_2 + \sqrt{h_1 - h_2} - 2\sqrt{h_2}. \tag{8.23}$$

Since h_1 will always be close to 2 and h_2 will always be close to 1, $\sqrt{h_1 - h_2} \approx 1 \approx h_1 - h_2$ and $\sqrt{h_2} \approx 1 \approx h_2$. Therefore, we can write (8.22) and (8.23) as follows:

$$\frac{dh_1}{dt} = Q_1 - h_1 + h_2, \tag{8.24}$$

$$\frac{dh_2}{dt} = Q_2 + h_1 - 3h_2. \tag{8.25}$$

These equations are relatively easy to deal with. We should mention at this point that there are more elegant ways of linearizing differential equations. These techniques usually involve

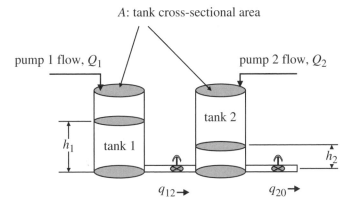

Figure 8.27. *Schematic diagram for two-tank problem of Example* 5c.

expanding the nonlinear function in a Taylor series about the setpoint and truncating the series after the first derivative term. What we have done here is rather crude, but the idea is the same. We assume that the small operating region near the setpoint, where we plan on operating, is linear. Many controllers are designed this way; for a more thorough discussion on the subject see Phillips and Harbor (1996), Ogunnaike and Ray (1994), and Luyben (1989).

To be in tune with the control literature we refer to h_1 and h_2 as x_1 and x_2, respectively. The "x's" are called state variables and represent the state of the plant or system. We will refer to Q_1 and Q_2 as $u_1(t)$ and $u_2(t)$, the controller output. Equations (8.24) and (8.25) become

$$\frac{d\bar{x}}{dt} = \underline{A}\bar{x} + \underline{B}\bar{u}(t). \tag{8.26}$$

In (8.26), $\frac{d\bar{x}}{dt}$ is the vector

$$\begin{bmatrix} \dfrac{dx_1}{dt} \\[2mm] \dfrac{dx_2}{dt} \end{bmatrix},$$

$\underline{A}$ is the matrix $\begin{bmatrix} -1 & 1 \\ 1 & -3 \end{bmatrix}$, $\underline{B}$ is the matrix $\begin{bmatrix} 1 & 0 \\ 0 & 1 \end{bmatrix}$, and $\bar{u}(t)$ is the vector $\begin{bmatrix} u_1 \\ u_2 \end{bmatrix}$ or $\begin{bmatrix} Q_1 \\ Q_2 \end{bmatrix}$. Often, the state variable x is not the control variable or sensed variable. In this case, the controlled variable y will be associated with the state variable by the matrix $\underline{C}$. In the literature, the following equation will often be seen associated with (8.26):

$$\bar{y} = \underline{C}\bar{x}. \tag{8.27}$$

In this example, the state variable is the controlled variable, and there is no need for (8.27).

A block flow diagram for the plant and controller is shown in Figure 8.28.

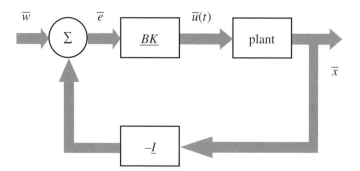

Figure 8.28. *Block flow diagram for two-tank problem of Example 5c.*

It is common practice to use heavy arrowed lines in MIMO block flow diagrams to show the flow of vector variables as opposed to single arrowed lines that denote the flow of scalar variables.

In Figure 8.28, the setpoint vector $\bar{w}$ is $\begin{bmatrix} w_1 \\ w_2 \end{bmatrix}$ or, in our case, $\begin{bmatrix} 2 \\ 1 \end{bmatrix}$. The error vector $\bar{e}$ is $\begin{bmatrix} w_1 - x_1 \\ w_2 - x_2 \end{bmatrix}$. The $\underline{BK}$ matrix is the product of the $\underline{B}$ matrix and the controller matrix $\underline{K}$, $\begin{bmatrix} k_{11} & k_{12} \\ k_{21} & k_{22} \end{bmatrix}$. In our definition of this system, $\underline{B}$ is the identity matrix $\underline{I}$ equal to $\begin{bmatrix} 1 & 0 \\ 0 & 1 \end{bmatrix}$. Thus the $\underline{BK}$ is equal to $\underline{K}$. The controller matrix represents a proportional-only controller. This controller represents only the P portion of the PID control algorithm. The control constants k_{12} and k_{21} are employed to modify or decouple the control actions for pumps 1 and 2. If $k_{12} = k_{21} = 0$, the control actions become those of two SISO proportional controllers. For simplicity's sake, we will continue the example using the assumption that two SISO controllers are adequate to control this system. Our closed-loop system can now be described by the following:

$$\begin{bmatrix} \dfrac{dx_1}{dt} \\ \dfrac{dx_2}{dt} \end{bmatrix} = \left[\begin{bmatrix} -1 & 1 \\ 1 & -3 \end{bmatrix} - \begin{bmatrix} k_{11} & 0 \\ 0 & k_{22} \end{bmatrix} \right] \begin{bmatrix} x_1 \\ x_2 \end{bmatrix} + \begin{bmatrix} k_{11} & 0 \\ 0 & k_{22} \end{bmatrix} \begin{bmatrix} w_1 \\ w_2 \end{bmatrix}. \qquad (8.28)$$

The term

$$\left[\begin{bmatrix} -1 & 1 \\ 1 & -3 \end{bmatrix} - \begin{bmatrix} k_{11} & 0 \\ 0 & k_{22} \end{bmatrix} \right]$$

is known as the closed-loop A matrix or $\underline{A}_{\text{closed\_loop}}$. To have a control system that provides stable closed-loop control, we must use control constants that will provide our closed-loop system with eigenvalues that are negative if they are real or on the left-hand side of the complex plain if they are complex. The eigenvalues λ_1 and λ_2 of the closed-loop system are determined by

$$\lambda_1 \text{ and } \lambda_2 = \text{Det}[\lambda \underline{I} - \underline{A}_{\text{closed\_loop}}]. \qquad (8.29)$$

The determinant Det in (8.29) expands into the quadratic equation for a two-by-two system. The eigenvalues λ_1 and λ_2 can be forced into the left-hand region of the complex plane by choosing the proper values of k_{11} and k_{22}. This will often be a trial-and-error procedure. Once useable values of the control constants are found, only stability of the system is guaranteed; nothing can be said for performance.

Stability is important for many control problems. Most control texts spend a great deal of time discussing stability. The problem here is that we have made a great many assumptions in order to simplify our model. All that we can really guarantee with our chosen control constants is that the model will be stable under closed-loop control. We cannot say for sure whether the real system, which we have only approximated, will be stable under closed-loop control. If this were control of an airplane, the control engineer might do well to find a better model. The two-tank system described here, however, is inherently quite stable. Compared to an airplane, it is highly damped and sluggish. Finding control constants that keep the system stable should be no problem. Finding high-performance control constants is another question. This procedure will probably involve some trial and error, and the control engineer will probably want to add at least an integral term to the controllers, making them PI controllers.

Example 5c should help make clear that as control systems become more complex, standard control system design and analyses can become much more complex. The mathematical models for these large systems can become very difficult to use without simplifying assumptions that may actually invalidate them. This is not to say that we should not use these tools; they can be very powerful. The idea is to use the right tool for the right job. Fuzzy and probabilistic controllers also can become quite complex. As the number of system variables increases, the number of rules that are needed will increase rapidly. Large rule systems are hard to manage and verify. When experts are used to determine the rules, the number of rules that an expert will actually know is limited.

These examples are meant to show some of the differences and similarities between standard control techniques, fuzzy techniques, and probabilistic techniques. They are not meant to show that one technique is better than another. At this point, we have covered all the basic material that we are going to in this chapter. The next section provides the description of the development of three control systems for a real experimental problem. It is intended to give a little more insight into some of the "real-world" problems that might be encountered in control system design. The experiment was based upon a three-tank problem similar to the two-tank problem described in Example 5c. The three control systems tested were a PI controller, a fuzzy controller, and a probabilistic controller.

8.5   The three-tank MIMO problem

The three-tank system used in this experiment is similar to the two-tank system discussed in Example 5c. The tanks are smaller, however. The experimental apparatus consists of three Lucite tanks, in series, each holding slightly less than 0.01m^3 of liquid.

The system is shown in Figure 8.29. The tanks are numbered from left to right as tank 1, tank 3, and tank 2. All three tanks are connected, with the third tank in the series, tank 2, draining into the system exit. Liquid is pumped into the first and the third tanks to maintain their levels. The levels in the first and third tanks control the level in the middle tank. The level in the middle tank affects the levels in the two end tanks. This problem, three tanks in series, where the middle tank interacts with the two tanks at either end of the system, is an analogue of many other chemical process industry control problems. Several chemical plant systems can be represented in this manner. For example, many distillation column systems consist of a total condenser with liquid-level control, a reboiler with liquid-level control, and an interactive column in between. In short, the solution to the three-tank system provides insight into many of the nonlinear control problems found in chemical process industries.

The differential equations (8.30)–(8.32) describe this experimental system. The tank flows are described by (8.33)–(8.35). The symbols used in these equations have the same meaning as those used in Example 5c, (8.18)–(8.21). The subscripts, however, pertain to Figure 8.29.

$$A \frac{dh_1}{dt} = Q_1 - q_{13}, \tag{8.30}$$

$$A \frac{dh_2}{dt} = Q_2 + q_{32} - q_{20}, \tag{8.31}$$

$$A \frac{dh_3}{dt} = q_{13} - q_{32}, \tag{8.32}$$

$$q_{13} = \Phi_1 A_p \, \text{Sign}(h_1 - h_3) \sqrt{2g|h_1 - h_3|}, \tag{8.33}$$

$$q_{32} = \Phi_3 A_p \, \text{Sign}(h_3 - h_2) \sqrt{2g|h_3 - h_2|}, \tag{8.34}$$

$$q_{20} = \Phi_2 A_p \sqrt{2gh_2}. \tag{8.35}$$

8.5.1   The fuzzy and probabilistic control systems

The designs of the fuzzy and probabilistic controllers are similar so they can be described in parallel. Both of these controllers were designed with two modules. One module is for setpoint tracking and the other for disturbance rejection. The setpoint tracking portion is the workhorse for both systems. The disturbance rejection module merely makes corrections and then returns control to the setpoint tracking module. Figure 8.30 shows the structure of both controllers.

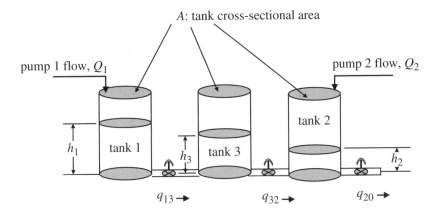

Figure 8.29. *The experimental three-tank system.*

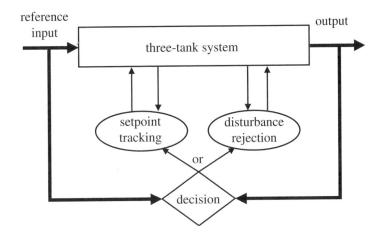

Figure 8.30. *Schematic for the fuzzy and probabilistic controllers for the three-tank system.*

We will first describe the setpoint tracking portion of each controller and the results for the setpoint tracking experiments. We then describe the disturbance rejection portion of the controllers and the results of the disturbance rejection experiments.

The fuzzy and probabilistic rules for this setpoint tracking module are given in Table 8.5. The rules are of the form

$$\text{if Error}(i) \text{ is}\ldots, \text{ then Flow\_Change}(i) \text{ is}\ldots.$$

The term Error(i) is defined as follows for both tanks 1 and 2:

$$\text{Error}(i) = \frac{w_i - h_i}{\text{Range } h(i)} \quad \text{for } i = 1 \text{ or } 2. \tag{8.36}$$

If $w_i \succ h_i$, then Range $h(i)$ equals w_i. Otherwise, Range $h(i)$ equals $h_{i\,\max} - w_i$ for $i = 1$ or 2.

The variable h_i is the current level for tank i and w_i is the setpoint for tank i. The term $h_{i\,\max}$ is the maximum level for tank i, or the level at which the tank is full. The

Table 8.5. *Fuzzy and probabilistic rules for setpoint tracking.*

Rule number	Error(i)	Flow_Change(i)
1	Error(1) = negative	Flow_Change(1) = negative
2	Error(1) = zero	Flow_Change(1) = zero
3	Error(1) = positive	Flow_Change(1) = positive
4	Error(2) = negative	Flow_Change(2) = negative
5	Error(2) = zero	Flow_Change(2) = zero
6	Error(2) = positive	Flow_Change(2) = positive

Flow_Change(i) variable is defined by

$$\text{Flow\_Change}(i) = \frac{q_i - q_{i\text{SS}}}{\text{Range } q(i)}. \tag{8.37}$$

If Flow_Change(i) $\succ$ 0, then Range $q(i)$ equals $q_{i\text{SS}}$. Otherwise, Range $q(i)$ equals $q_{i\max} - q_{i\text{SS}}$ for $i = 1$ or 2.

The term $q_{i\max}$ is the maximum possible flow from pump i, which is 6.0 liters/min for each pump. The variable q_i is the current pump flow from pump i, and $q_{i\text{SS}}$ is the steady-state, or setpoint, flow for pump i, computed from a mass balance calculation. The mass balance calculation is a simple algebraic calculation. It is based on (8.30)–(8.35) with derivatives set to zero and tank levels set at the setpoints. The calculation simplifies to three equations with three unknowns. The three unknowns obtained from the solution are the two steady-state pump flows and one independent steady-state tank level. One difficulty is determining the flow coefficients Φ_i. These coefficients can be easily measured or calculated using textbook values. Measurement of these values is best. If the coefficients are calculated and the calculation is not correct, the disturbance rejection mode will be invoked automatically and the values will be determined by the control system.

The membership function universes and conditional probabilities for Error(i), and membership function universes and PDFs for Flow_Change(i), have been normalized from -1 to 1. This was done in order to construct generalized functions. In this type of control problem, we want one set of rules to apply to all tank level setpoints. Generalizing these functions makes it possible to use only six rules to handle all tank levels and all tank level changes.

Figure 8.31 shows the input membership functions and conditional probabilities for both the fuzzy and probabilistic controllers used to solve this three-tank problem. Figure 8.32 shows the output membership functions for the fuzzy control system. These membership functions have been expanded to limits -2 and 2 so that the pumps can be turned on full blast if needed. Figure 8.33 shows the output PDFs for the probabilistic controller. These inputs and outputs are connected by the rules shown in Table 8.5.

Example 6. For this example, we chose the setpoints w_1 and w_2 for tanks 1 and 2 to be 0.40 and 0.20m, respectively. These are the same setpoints that were used in the actual experiment. We start with three empty tanks. The steady-state flows required to maintain these levels are computed from the mass balance calculation to be $q_{1\text{SS}} = 1.99346$ liters/min and $q_{2\text{SS}} = 2.17567$ liters/min. The maximum pump flows are $q_{1\max} = q_{2\max} = 6.0$ liters/min. For the example, we assume that the tank levels have nearly reached their setpoint values. The level in tank 1, h_1, is 0.38m and the level in tank 2, h_2, is 0.19m. The maximum tank levels are $h_{1\max} = h_{2\max} = 0.62$.

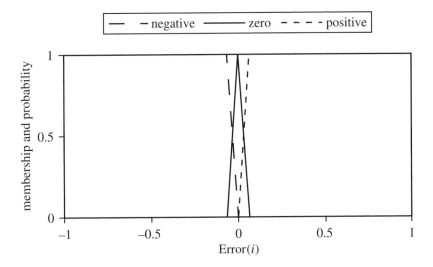

Figure 8.31. *Membership functions and conditional probabilities for input* Error(i) *for i = both* 1 *and* 2 (*pumps* 1 *and* 2).

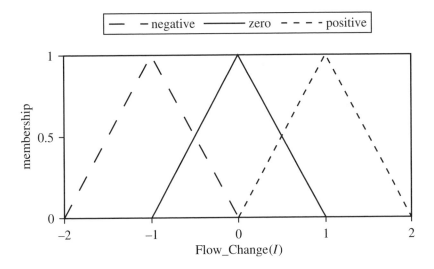

Figure 8.32. *Membership functions for* Flow_Change(i) *for i = both* 1 *and* 2 (*pumps* 1 *and* 2).

Both Error(1) and Error(2) are calculated to be -0.05 from (8.36). In both cases, the memberships and probabilities from Figure 8.31 are computed to be negative = 0.775 and zero = 0.225. These values cause rules 1 and 4 to be fired with a strength of 0.775, and rules 2 and 5 to be fired with a strength of 0.225. The output membership functions shown in Figure 8.32 are truncated at these values. The defuzzification method used with the setpoint tracking control module is the correlation minimum encoding (CME) technique. This method computes the areas and centroids of the entire truncated triangles. That is, the negative and positive Flow_Change membership functions are extended to -2 and $+2$, respectively, in order to compute the centroids and the areas. This is one of many techniques

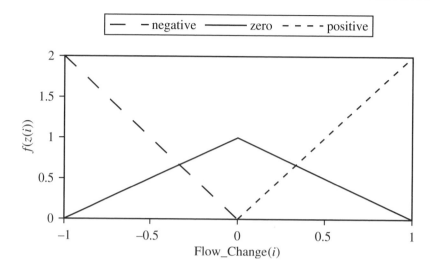

Figure 8.33. *Output PDFs for the three-tank problem, pumps* 1 *and* 2 *(i = 1 and 2).*

used with fuzzy output membership functions when it is important to use the centroid of the membership function to designate complete "shut off" or "go full blast." The crisp, or defuzzified, value used in the control equations, and obtained from firing these rules, is defined by (8.17).

Both Flow_Change(1) and Flow_Change(2) are computed to be -0.704. For both pumps, Flow_Change(i) is less than 0, so Range $q(i)$ is equal to $q_{i\,max} - q_{iss}$. Range $q(1)$ is then 4.00654 and Range $q(2)$ is equal to 3.82433. By manipulating (8.37), we obtain

$$q_i = q_{iss} - \text{Flow\_Change}(i) * \text{Range}\, q(i). \tag{8.38}$$

From (8.38), the value of q_i is computed as

$$q_1 = 1.99346 + (0.704)(4.00654) = 4.8141,$$

and q_2 is

$$q_2 = 2.17567 + (0.704)(3.82433) = 4.8680.$$

For the probabilistic controller, using (8.9) and the mode for PDF–RV, we obtain a value of -0.775 for both Flow_Change(1) and Flow_Change(2). From (8.38), we obtain the following for the new pump flows:

$$q_1 = 1.99346 + (0.775)(4.00654) = 5.0985,$$
$$q_2 = 2.17567 + (0.775)(3.82433) = 5.1395.$$

8.5.2 The PI controller

For the three-tank problem, we used two separate PI controllers and used only the first two terms of (8.4) for the control algorithm. This is common practice for situations where the D portion or derivative of the error can do more harm than good. In the case of tank flow problems like this one, the liquid level in the tanks will oscillate, or make waves, as the liquid from the pumps hits the surface. The derivative portion of a PID controller will respond unnecessarily to this oscillation. The PI controller is quite simple. It does not need separate loops for setpoint tracking and disturbance rejection. The difficulty in using this controller is coming up with optimal control constants K_P and K_I. There are several empirical techniques that can be used to estimate the value of good or optimal control constants for PI and PID controllers. These techniques usually involve perturbing the system and observing responses. We used the Ziegler–Nichols approach, described by Ogunnaike and Ray (1994), for determining a first estimate for the control constants for our PI controllers. We then used a computer model of the three-tank system and a great deal of trial and error to determine the optimal control constants. The optimal constants were those that produced the smallest error integral for the controller response for the setpoint tracking test described in the next section. The error integral is the area between the setpoint line and the controller response for a given test. The crosshatched area in Figure 8.5 shows the integral error. For our optimization tests, the area was measured both above and below the setpoint line and was measured to the end of the test time. The PI constants determined by the trial-and-error test were $K_P = 0.1365$ and $K_I = 0.000455$ seconds^{-1}. The value for K_P was approximately the same as the one determined by the Ziegler–Nichols calculation, but the value used for K_I was significantly smaller than that determined by Ziegler–Nichols. This produced a rather highly damped controller, but it reduced the amount of overshoot significantly. It is important to reduce overshoot in tank-filling problems in order to eliminate the possibility of spilling liquid. The same control constants were used for each controller, pump 1 and pump 2. These control constants worked quite well for the setpoint tracking test described in the next section.

8.5.3   Setpoint tracking: Comparison between the controllers

This setpoint tracking test case was a tank-filling problem. The setpoint level was changed from 0, an empty condition, to 0.4m for tank 1 and 0.2m for tank 2. The level for tank 3 cannot be set independently. For this test, tank 3 found its own level at about 0.3m. The run time was set to 7.5 minutes, or 450 seconds. Figure 8.34 shows the tank response to the probabilistic controller for the experiment described above. The results for the fuzzy controller are shown in Figure 8.35.

If Figures 8.34 and 8.35 are superimposed upon one another, the controller response curves cannot be differentiated. The integral error test for each set of curves produces a value that is the same for each controller and that is within the error tolerances of the numerical algorithm used for the integration. This result shows that for this problem, even though the controller algorithms can produce slightly different values for individual points, as shown in Example 6, the response curve for the entire problem is essentially the same.

The response curve for the PI controller looks very much like the curves in Figures 8.34 and 8.35. The integral error test, however, produces slightly different results for the PI controller. Table 8.6 shows integral error and rise time for the PI controller relative to the fuzzy and probabilistic controllers. The definition of rise time T_r used here is the time required for the system response, to a step input, to move from 10% of the final setpoint to 90% of that setpoint. In this case, the tank 1 step input is the change in the setpoint from 0 to 0.4 (m). The rise time is the amount of time it takes for the level in tank 1 to go from 0.04 to 0.36 (m).

In addition to the differences shown in Table 8.6, the PI controller produced a very small amount of overshoot. Neither the fuzzy controller nor the probabilistic controller produced any overshoot. In spite of the differences shown in Table 8.6, the PI controller did a pretty

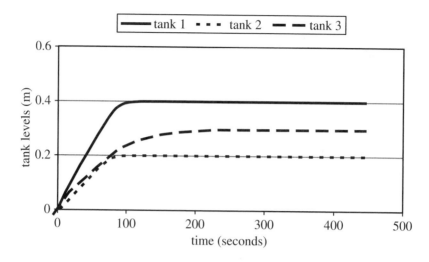

Figure 8.34. *Tank levels as a function of time for the probabilistic controller.*

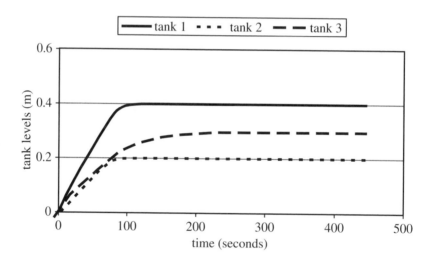

Figure 8.35. *Tank levels versus time for the fuzzy controller.*

Table 8.6. *Test results for the PI controller relative to the fuzzy and probabilistic controllers.*

Tank	Integral error	Rise time T_r
1	$+ \prec 1\%$	+1 seconds
2	+41.1%	+7 seconds
3	+7.8%	+11 seconds

good job on this control problem. Even the additional 41% error, exhibited in the response to the change in the tank 2 setpoint, is difficult to see in examples like Figures 8.34 and 8.35. It seems obvious that if the control constants for the pump 2 controller were different than those for the pump 1 controller, then the PI controller would have produced an even better response. One problem with the PI controller is the amount of time devoted to finding

Table 8.7. *Rules for disturbance rejection for both tanks* 1 *and* 2.

Rule number	If Error(i) is . . .	. . . and Slope(i) is . . .	. . . then Help_Factor(i) is . . .
1	negative	negative	small negative
2	negative	zero	small negative
3	negative	positive	large negative
4	zero	negative	zero
5	zero	zero	zero
6	zero	positive	zero
7	positive	negative	large positive
8	positive	zero	small positive
9	positive	positive	small positive

optimal control constants. In recent years, automated techniques have been developed that significantly reduce the amount of time to find control constants (Smith (1994)).

8.5.4 Disturbance rejection: The fuzzy and probabilistic controllers

Although there are other methods for designing the fuzzy and probabilistic controllers for this problem, we chose the technique described below. We chose this technique because an engineering understanding of the problem made this approach easy and efficient. Another approach would be to use more rules to cover all situations.

The fuzzy logic and probabilistic controllers are actually hierarchical controllers, as shown in Figure 8.30, with two sets of rules. One set of rules covers setpoint tracking, as already discussed. The other set covers disturbance rejection. Disturbance rejection is broken into two parts, decision-making and disturbance rejection corrections. The same rules are used for both parts. The input membership functions/conditional probabilities are slightly different when the system is trying to decide if it should be in disturbance rejection mode than when it is actually in that mode. The rules determine if the controller should be in disturbance rejection mode and also determine the size of what we will call a help factor.

In disturbance rejection mode, the original material balance is no longer valid since the disturbance will be either a tank leak or a plug in between the tanks. These conditions were not in the original material balance equations used to compute the steady-state pump flows $q_{i\text{SS}}$ used in (8.37) and (8.38). If we knew the friction coefficient ϕ_i for the plug or the leak, a rapid material balance calculation would give us the new $q_{i\text{SS}}$ and we would need only the setpoint tracking portion of the fuzzy algorithm. Unfortunately, in practice we very seldom know this friction coefficient so we must find our new steady-state pump flows, $q_{i\text{SS}}$, by using a root-finding technique.

There are 18 rules: nine for tank 1 and pump 1 and nine for tank 2 and pump 2. The rules are exactly the same for each tank so only nine rules are listed in Table 8.7. The rules are of the form

if Error(i) is. . . and if Slope(i) is. . . , then Help_Factor(i) is. . . ,

where Error(i) is defined by (8.36), Slope(i) is defined by (8.39) below, and Help_Factor(i) is a fuzzy or probabilistically determined factor used to speed the convergence of the root-finding algorithm

$$\text{Slope}(i) = \frac{h_i^{j+1} - h_i^j}{\Delta t}, \tag{8.39}$$

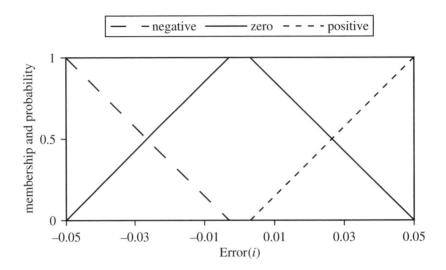

Figure 8.36. *Membership functions/conditional probabilities for* Error(*i*) *for i =* both 1 *and* 2 *(disturbance rejection—decision).*

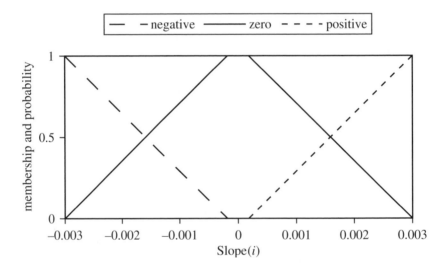

Figure 8.37. *Membership functions/conditional probabilities for* Slope(*i*) *for i =* both 1 *and* 2 *(disturbance rejection—decision and correction).*

where Δt is the sample interval equal to time $t(j+1)$ minus time $t(j)$ and h_i^{j+1} and h_i^{j} are the level measurements of tank i at times $t(j+1)$ and $t(j)$, respectively.

The membership functions/conditional probabilities for the rule input Error(i) for the decision-making portion of the disturbance rejection method are given in Figure 8.36. The membership functions/conditional probabilities for rule input Slope(i) for both the decision making and the actual correction to the root finder are given in Figure 8.37. The rule input Error(i) membership functions/conditional probabilities for the correction mode are given in Figure 8.38. The rule output Help_Factor(i) membership functions for the fuzzy correction are given in Figure 8.39. The rule output Help_Factor(i) PDFs for the probabilistic correction

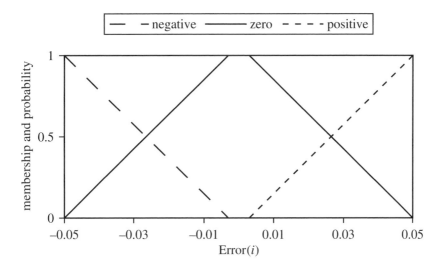

Figure 8.38. *Membership functions/conditional probabilities for* Error(i) *for* $i =$ *both* 1 *and* 2 (*disturbance rejection—correction*).

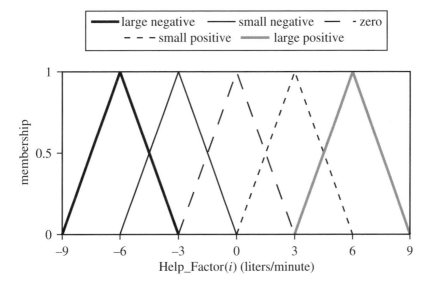

Figure 8.39. *Membership functions for* Help_Factor(i) *for* $i =$ *both* 1 *and* 2 *in liters/min* (*disturbance rejection—fuzzy correction*).

are given in Figure 8.40.

In Figures 8.36 and 8.38 for $0.05 \leq$ Error(i) the membership of Error(i) in Positive is 1.0 (or the probability that Error(i) is Positive is 1.0), and for Error(i) ≤ -0.05 the membership of Error(i) in Negative is 1.0 (or the probability that Error(i) is Negative is 1.0). In Figure 8.37, for $0.003 \leq$ Slope(i) the membership of Slope(i) in Positive is 1.0 (or the probability that Slope(i) is Positive is 1.0), and for Slope(i) ≤ -0.003 the membership of Slope(i) in Negative is 1.0 (or the probability that Slope(i) is Negative is 1.0).

The major difference between Figures 8.36 and 8.38 is the true zero range, that is, the

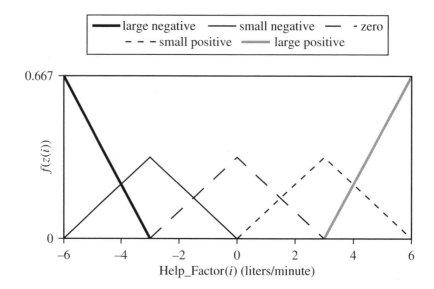

Figure 8.40. *PDFs for* Help_Factor(i) *for* i = *both* 1 *and* 2 *in liters/min* (*disturbance rejection—probabilistic correction*).

flat portion of the membership function/probability function zero for the variable Error(i) in these figures. If we use Figure 8.36, the variable Error(i) has a membership value 1.0 in zero or a probability of 1.0 that it is zero if $-0.0025 \leq$ Error(i) ≤ 0.0025. If we use Figure 8.38, the variable Error(i) has a membership value 1.0 in zero or a probability of 1.0 that it is zero if $-0.001 \leq$ Error(i) ≤ 0.001. This is equivalent to putting a dead band around the liquid levels in the tanks, where no action is taken as long as the liquid level remains within the band. The band is two and one-half times larger when we are trying to decide if the system needs disturbance rejection help than when the system is trying to find the values of the steady-state pump flows q_{iSS}. The system works as follows:

- At sample time, liquid levels h_i are measured and Error(i), the normalized difference between the liquid level and the setpoint level, is computed from (8.36).

- At the same sample time, the Slope(i) is computed from (8.39).

- The rules shown in Table 8.7 are fired using the values for Error(i) and Slope(i) and the membership functions or conditional probabilities shown in Figures 8.36 and 8.37.

- If the value for Help_Factor(i) is zero, then the controller stays in the setpoint tracking mode. If the value for Help_Factor(i) $\neq$ zero, the controller goes into disturbance rejection mode. This mode combines setpoint tracking and the disturbance rejection root-finding technique to compute the value of q_{iSS} in (8.37).

- If the controller goes into disturbance rejection mode, the rules in Table 8.7 are fired again using the values of Error(i) and Slope(i) and using the membership functions or probabilities from Figures 8.37 and 8.38. The Help_Factor(i) is used in (8.40) to compute a new q_{iSS}:

$$q_{iSS}(\text{new}) = q_i + \text{Help\_Factor}(i). \tag{8.40}$$

- The term q_i is the current pump flow from (8.38), and at this point it usually is different from q_{iSS}(old) because the setpoint tracking algorithm has already started the correction process.

- The procedure is repeated until Help_Factor(i) becomes zero again ($-0.001 \leq$ Error(i) ≤ 0.001) and stays there for five sample steps.

- The new q_{iSS} is chosen as the average of the last two values of q_{iSS}(new).

- The control system then returns to setpoint tracking mode.

This portion of the controller can work with either tank 1 or tank 2 alone or both tanks together. It usually works well with both tanks together because a reasonable-sized disturbance in one tank will affect both tanks. A one-step example for the fuzzy controller will help clarify the above procedure. The procedure for the probabilistic controller is analogous.

Example 7.   If we look at tank 1 only, the setpoint level, w_1, is 0.4m and the steady-state flow rate for pump 1, q_{1SS}, is 1.99346 liters/min. Suppose that a leak occurs in tank 1 and at the next sample time (sample interval is two seconds) the liquid level is just below 0.399 meters. The Error(1) is computed from (8.36) to be just over 0.0025. The Slope(1) is computed from (8.39) to be $\approx -0.0005 = \frac{0.399-0.4}{2.0}$. The negative Slope(1) and positive Error(1) indicate a leak and the Help_Factor(1) is computed to be slightly above zero. (This calculation does not have to be highly accurate. We just need to show that the Help_Factor(1) $\neq$ zero.) If Error(1) is 0.0025+, from Figure 8.36, the Error(1) membership is ≈ 0.99 in zero and is ≈ 0.01 in positive. For Slope(1) ≈ -0.0005, from Figure 8.37, the Slope(1) membership is ≈ 0.1 in negative and ≈ 0.9 in zero. Rule 4 will be fired with the minimum value of 0.1. Rule 5 will be fired with the minimum value of 0.9. Rule 7 will be fired with the minimum value of 0.01, and rule 8 will be fired with a minimum value of 0.01. The output membership functions from the maximum part of the min–max rule will have the following values: zero $= 0.9$, large positive $= 0.01$, and small positive $= 0.01$.

From Figure 8.39, it is obvious that the centroid of clipped output membership functions will be greater than zero. Therefore, Help_Factor(1) is $\succ$ zero. This is enough to set the disturbance rejection flag. This last step is now repeated using Figures 8.37 and 8.38 in order to determine a useful Help_Factor(1). The membership values for Slope(1) remain the same, but for Error(1) $= 0.0025+$, from Figure 8.38, the Error(1) membership is ≈ 0.9694 in zero and is ≈ 0.0306 in positive. Rule 4 will be fired with the minimum value of 0.1. Rule 5 will be fired with the minimum value of 0.9. Rule 7 will be fired with the minimum value of 0.0306, and rule 8 will be fired with a minimum value of 0.0306. The output membership functions from the maximum part of the min–max rule will have the following values: zero $= 0.9$, large positive $= 0.0306$, and small positive $= 0.0306$. The centroid from clipped output membership functions from Figure 8.39 is about 0.35 liters/min. Therefore, Help_Factor(1) is 0.35 liters/min for this iteration. We assume that the setpoint algorithm is already starting to correct itself because the level has moved away from the setpoint. For convenience, let us assume that the current value of q_1 is 2.0 liters/min. From (8.40), the new steady-state pump flow for pump 1, q_{1SS}(new), is 2.35. At the next sample time, the level still will be below the setpoint. Thus Error(1) will be positive and, from the setpoint tracking rules (Table 8.5), rule 3 will dominate. Flow_Change(1) will be positive. From (8.38), the new pump 1 flow, q_1, will be $= 2.35 * ($Flow_Change(1)$) + 2.35$ instead of $= 1.99346 * ($Flow_Change(1)$) + 1.99346$, which is the pump flow that would have been requested without the disturbance rejection correction.

This procedure is repeated (from the point that the disturbance rejection flag is set) until Help_Factor(1) is zero, or $-0.001 \leq \text{Error}(i) \leq 0.001$, for five consecutive time sample steps. At this point, the new q_{1SS} is set to be the average of the final *two* steady-state pump flow calculations. The disturbance rejection flag is then reset to zero and the controller returns to the setpoint-tracking-only mode.

The root-finding technique is just the simple iterative method defined by (8.41), found in most numerical methods texts (for example, see Henrici (1964)):

$$x^{j+1} = f(x^j), \tag{8.41}$$

where j is the iteration number and the time between iterations is the sample interval. This technique is quite stable but also quite slow. The help factor speeds it up significantly. The technique represents an approximation of the human approach to solving the disturbance rejection problem.

For the disturbance rejection module, the fuzzy method uses the min–max centroid technique for defuzzification. This is different from the CME defuzzification technique used in the setpoint tracking module. The point is that it does not matter what defuzzification technique is used as long as the user can make the rules map the input to the correct output. The probabilistic controller uses a generalized version of (8.16), with the mode for the PDF–RV, to resolve the disturbance rejection problem.

The disturbance rejection module also can serve as a correction module for the setpoint tracking mode when q_{iSS} is computed incorrectly. The disturbance rejection module quickly recognizes an incorrect steady-state flow and computes the correct value using the algorithm described above.

The PI controller used for disturbance rejection is the same controller that was used for the setpoint tracking problem. With the standard PI controller, one set of control constants is used in all situations for each controller.

8.5.5 Disturbance rejection: Comparison between the controllers

For this comparison, two tests were run for each of the controllers. Both tests were run for 1000 seconds. At 200 seconds a disturbance was introduced. At 700 seconds the disturbance was removed. The disturbance for the first test was a leak in tank 1. The leak was opened at 200 seconds and found and closed at 700 seconds. The disturbance for the second test was a partial plug between tanks 1 and 3. The plugged condition began 200 seconds into the run and was removed 700 seconds into the run. The initial conditions, or steady-state conditions, for these runs were the final setpoint conditions from the setpoint tracking problem. That is, the setpoint level was 0.4m for tank 1 and 0.2m for tank 2. Tank 3 finds its own setpoint near 0.3m.

For these tests, a computer model of the three-tank system was used. This same model was used to find the control constants for the PI controller. A model was used for these tests to ensure that the leaks and plugs were identical in each test for each controller. This identical condition is very difficult to control with the experimental apparatus. The model in this case is very good. The results were compared to the actual experimental results for several cases and the agreement was excellent. In the model, we can set the friction coefficient ϕ_i for the plug and the leak. In these tests, the friction coefficient for the leak was set at 0.65 and the friction coefficient for the plug was set at 0.4. The response curves for test 1 (the leak test) for the three controllers fuzzy, probabilistic, and PI are shown in Figures 8.41–8.43, respectively. The response curves for test 2 (the plug test) for the same three controllers are shown in Figures 8.44–8.46, respectively.

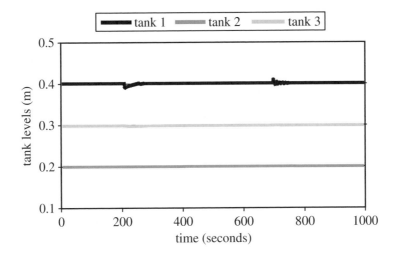

Figure 8.41. *Fuzzy disturbance rejection for a leak in tank* 1 *with a friction coefficient of* 0.65 (*test* 1).

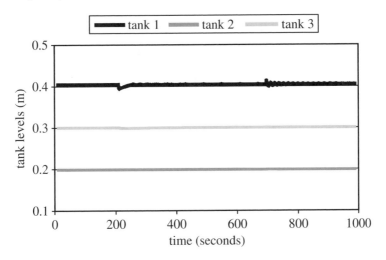

Figure 8.42. *Probabilistic disturbance rejection for a leak in tank* 1 *with a friction coefficient of* 0.65 (*test* 1).

The integral error was computed in each test for each tank for each controller. These values are listed in Table 8.8.

Figures 8.41–8.46 and Table 8.8 demonstrate that the fuzzy and probabilistic controllers are very much alike in their responses. The fuzzy controller does a little bit better on test 1 and the probabilistic controller does a little bit better on test 2. Both controllers beat the PI controller quite handily. On reason is because the PI controller was "tuned" for the setpoint tracking problem and is highly damped to reduce overshoot. This made it less responsive to the disturbance rejection problem. However, with the standard PI controller, we have only one set of control constants for all situations. Control constants that improve the disturbance rejection response will hurt the setpoint tracking response. Note that none of the controllers can do much with the level in tank 3 when there is a plug between tanks 1 and 3 since there is no pump to supply input to tank 3.

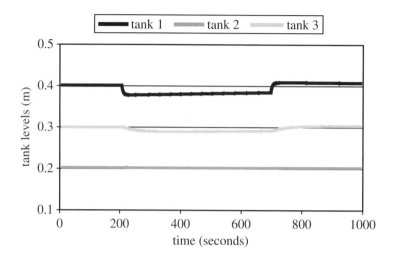

Figure 8.43. *PI controller, disturbance rejection for a leak in tank* 1 *with a friction coefficient of* 0.65 (*test* 1).

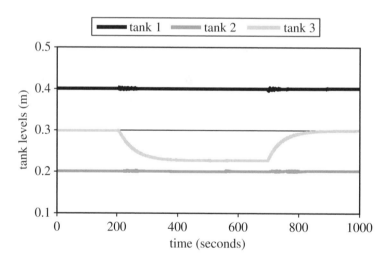

Figure 8.44. *Fuzzy disturbance rejection for a plug between tanks* 1 *and* 3 *with a friction coefficient of* 0.40 (*test* 2).

8.6   Conclusions

This chapter was intended to demonstrate the differences between fuzzy and probabilistic controllers. In addition, we attempted comparisons between these controllers and the industry standard, the PID controller. The fuzzy and probabilistic controllers appear to be very good. They provide very similar responses, at least for the problems studied here. The major differences seem to be in the philosophies behind the two controllers. Since most control engineers are paid for results and not philosophy, it is suggested that the developer use the control paradigm with which he/she is most comfortable. The control systems developed from these two paradigms can range from quite simple to quite complex. The developer has the ability to easily customize these controllers so that they provide an excellently controlled response for the system under control.

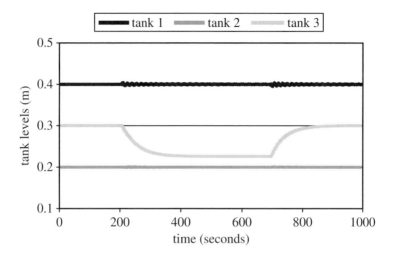

Figure 8.45. *Probabilistic disturbance rejection for a plug between tanks 1 and 3 with a friction coefficient of 0.40 (test 2).*

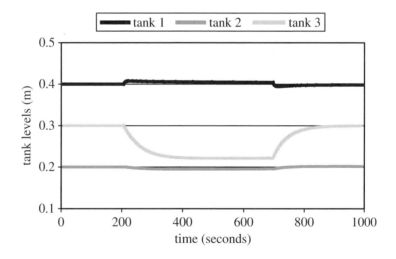

Figure 8.46. *PI controller, disturbance rejection for a plug between tanks 1 and 3 with a friction coefficient of 0.40 (test 2).*

Table 8.8. *Integral errors for each controller for disturbance rejection tests.*

Test and tank	Fuzzy error	Probabilistic error	PI error
Test 1 (leak in tank 1)			
tank 1	0.3933	0.4622	12.1279
tank 2	0.0035	0.0051	0.3293
tank 3	0.1455	0.1492	0.56128
Test 2 (plug between tanks 1 and 3)			
tank 1	0.3975	0.3057	3.3104
tank 2	0.4057	0.0565	2.5549
tank 3	35.4034	35.2975	37.0451

The PID controller is a simple and very robust controller. One can often apply this type of controller with much less knowledge of the "plant" and still get pretty good results. This does not mean zero knowledge, however, because control constants must be set for stable control. Without an automated tuning routine (Smith (1994)), tuning these controllers for optimal control can be very time consuming. The PID controller is a fine controller. It is and probably will remain the workhorse of industry. There are many options that can be used with PID controllers to make them perform better. One can use programmed control where a set of rules is used to adjust the control constants for different conditions. These rules are often fuzzy but could also be probabilistic. In the case of the two- and three-tank problems in this chapter where two controllers are used, the controllers can be decoupled as suggested in Example 5c. A good discussion on decoupling controllers can be found in Ogunnaike and Ray (1994).

Nonlinear controllers based upon a modification of the linear algebra used in Example 5c are another type of controller that shows a great deal of promise for problems like the three-tank problem. For further study of the three-tank problem using a nonlinear controller and two decoupled PI controllers in addition to the three controllers discussed here, see Parkinson (2001a, b).

References

J. D. BARRETT AND W. H. WOODALL (1997), A probabilistic alternative to fuzzy logic controllers, *IIE Trans.*, 29, pp. 459–467.

P. HENRICI (1964), *Elements of Numerical Analysis*, Wiley, New York.

B. KOSKO (1983), *Fuzzy Thinking: The New Science of Fuzzy Logic*, Hyperion, New York.

F. KREITH (1958), *Principles of Heat Transfer*, International Textbook Company, Scranton, PA.

M. LAVIOLETTE, J. W. SEAMAN, JR., J. D. BARRETT, AND W. H. WOODALL (1995), A probabilistic and statistical view of fuzzy methods, *Technometrics*, 37, pp. 249–261.

W. L. LUYBEN (1989), *Process Modeling, Simulation and Control for Chemical Engineers*, 2nd ed., McGraw–Hill, New York.

P. W. MURRILL (1988), *Application Concepts of Process Control*, Instrument Society of America, Research Triangle Park, NC.

P. W. MURRILL (1991), *Fundamentals of Process Control Theory*, 2nd ed., Instrument Society of America, Research Triangle Park, NC.

B. A. OGUNNAIKE AND W. H. RAY (1994), *Process Dynamics, Modeling, and Control*, Oxford University Press, New York.

C. L. PHILLIPS AND R. D. HARBOR (1996), *Feedback Control Systems*, 3rd ed., Prentice–Hall, Englewood Cliffs, NJ.

W. J. PARKINSON (2001A), *Fuzzy and Probabilistic Techniques Applied to the Chemical Process Industries*, Ph.D. dissertation, University of New Mexico, Albuquerque, NM.

W. J. PARKINSON (2001B), *Fuzzy and Probabilistic Techniques Applied to the Chemical Process Industries*, thesis and Technical Report LA-13845-T, Los Alamos National Laboratory, Los Alamos, NM.

F. G. SHINSKEY (1988), *Process Control Systems: Applications, Design, and Tuning*, 3rd ed., McGraw–Hill, New York.

R. E. SMITH (1994), *Automated Computer-Aided Design of Control Systems through Artificial Intelligence*, Ph.D. dissertation, University of Wisconsin, Madison, WI.

F. SZIDAROVSZKY AND A. T. BAHILL (1992), *Linear Systems Theory*, CRC Press, Boca Raton, FL.

Chapter 9

Structural Safety Analysis: A Combined Fuzzy and Probability Approach

Timothy J. Ross and Jonathan L. Lucero

Abstract. This chapter describes a safety analysis using a combined fuzzy and probability approach for multistory frame structures. The subjective information and uncertain data which must be adequately incorporated into the analysis are described. The current probabilistic approach in treating system and human uncertainties and their inadequacies is discussed. The alternative approach of using fuzzy sets to model and analyze these uncertainties is proposed and illustrated by examples. Fuzzy set models in the treatment of some uncertainties in safety assessment complement probabilistic models and can be readily incorporated into the current analysis procedure. This chapter shows that fuzzy set models can be used to treat various kinds of uncertainties and to complement probabilistic models in the safety assessment of structures.

9.1 Introduction

Professor Zadeh stated that our degree of confidence in understanding a system diminishes as the system description becomes more complex (Zadeh (1973)). This statement is certainly the case for multistory frame structures subjected to earthquake effects. The safety assessment of multistory frame structures is a complex civil engineering problem which can be approached only by assessing a number of components of the problem. Many of these components involve uncertainties. Classical statistics theories have been used to treat some of these uncertainties. Other uncertainties, especially those involving descriptive and linguistic variables, as well as those based on very scarce information, have usually not been incorporated satisfactorily into the analysis. In this chapter, some of the difficulties in using probabilistic models in the safety analysis of multistory frame structures will be described. The use of fuzzy set theory to handle these difficulties also is described.

  The typical procedure used in the safety assessment is described in the following sections. Emphasis is placed on the uncertain information and data that must be correctly incorporated into the analysis. The current approach to handling system and human uncertainties and their inadequacies is discussed. The alternative approach of using fuzzy sets to model and analyze these uncertainties is suggested and illustrated by examples. It is shown that the use of fuzzy set models to treat various uncertainties is natural because they are consistent with the degree of information available. When the information becomes more refined, it is shown that the fuzzy description progresses logically into the probability description. Fuzzy set models complement probabilistic models and can be readily incorporated into the current analysis procedure. This chapter is an upgraded and revised version of an earlier work on which the senior author collaborated (Wong et al. (1987)).

9.2   An example

Consider the following problem. The safety of a multistory frame structure to the influence of an assumed extreme environment, specified in terms of the mechanical effects of earthquake shock, is to be assessed. We presume that safety and failure are complementary states of a structure (i.e., $p_f = 1 - p_s$). The analysis is approached by formulating the problem into the following components:

1. Identify the important modes of failure.

2. Identify the ground acceleration effects and loading mechanisms responsible for these failure modes and the uncertainty in the description of the environment.

3. Determine the response function (i.e., the relationship between the loading and response parameters) and its associated uncertainties.

4. Compute the system failure probability.

These steps are represented schematically in Figure 9.1 for a multistory frame structure.

9.3   Typical uncertainties

For our multistory frame structure, a partial list of information and data required in each step of the analysis is shown in Figure 9.2. The information is further divided into three categories, deterministic, probabilistic, and fuzzy, corresponding to decreasing precision in the information. Most current methods acknowledge only the first two categories, i.e., deterministic and probabilistic.

Very imprecise information, subjective judgment, and linguistic data are treated as deterministic or probabilistic, or are ignored altogether. For example, torsion and its associated effects (due to the choice of the foundation–structure interaction model) and the damage evaluation criterion are treated deterministically or treated vaguely as a source of probabilistic uncertainty called systematic uncertainty.

Figure 9.2 describes a fairly complex process where imprecision exists even in the identification and classification of the sources of uncertainties in the frame structure analysis. Furthermore, tabulation of uncertainties such as those shown in the figure is very subjective. People of similar backgrounds and experiences could arrive at differing conclusions.

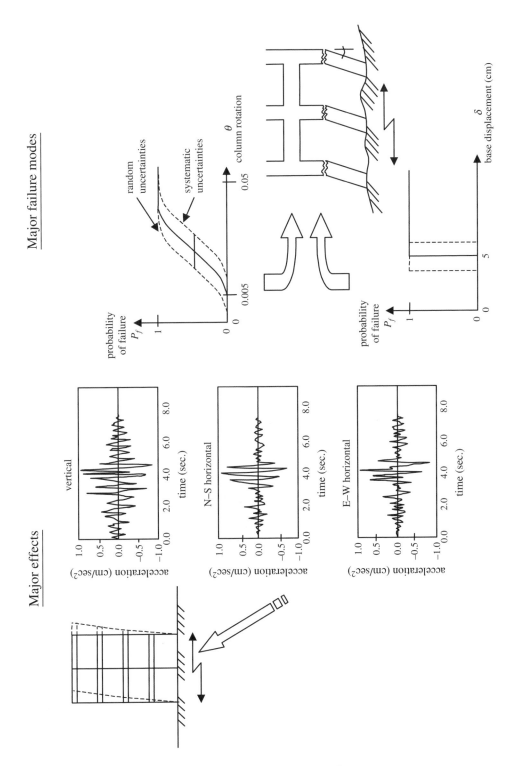

Figure 9.1. *Typical structure analysis.*

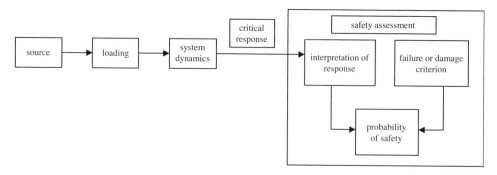

	Natural hazard	Loads	Structural dynamics	Safety assessment
Deterministic	location geology	accelerations	configuration network	failure modes
Probabilistic	azimuthal epidistance		material properties	component capacity
Fuzzy	energy	torsional effects	interaction model structural models	damage failure impact on system

Figure 9.2. *Partial list of uncertainties.*

9.4 Current treatment of uncertainties

In an analytic treatment, it is assumed that the total uncertainty, Ω, consists of the uncertainty due to inherent randomness, ε, and the uncertainty associated with the error in the prediction, Λ, such that (see Ang and Tang (1975, 1984))

$$\Omega^2 = \varepsilon^2 + \Lambda^2. \tag{9.1}$$

Hence for a response–input relationship such as

$$R = g(X_1, X_2, \ldots, X_n), \tag{9.2}$$

where R is the dependent variable and X_i, $i = 1, 2, \ldots, n$, are the independent variables, the uncertainty in R can be evaluated through first-order analysis. The mean value of R is

$$\bar{r} = g(\bar{x}_1, \bar{x}_2, \ldots, \bar{x}_n) \tag{9.3}$$

and the variance of R is

$$\sigma_R^2 = \rho_{ij} c_i c_j \sigma_{Xi} \sigma_{Xj}, \tag{9.4}$$

where ρ_{ij} is the correlation coefficient between X_i and X_j, $c_i = \frac{\partial g}{\partial X_i}$ evaluated at x_i, and $\sigma_{Xi} = \Omega_{Xi} \bar{x}_i$. Alternatively, the random and systematic uncertainties are propagated separately to give the random and systematic uncertainty in the dependent variable R as described in a safety assessment procedure due to Rowan (1976).

Monte Carlo schemes also have been devised to treat more complex safety problems based on the same principle. Collins (1976), for example, used the two-tiered sampling approach illustrated in Figure 9.3 to distinguish between random and systematic uncertainties. A limited number of inner loop samples are taken to estimate the average safety probability due to random variations. Each outer loop sample results in an estimate of the median value of the average safety probability. The outer loop estimates can then be used to assess the range of variation of the median probability and to establish the confidence levels.

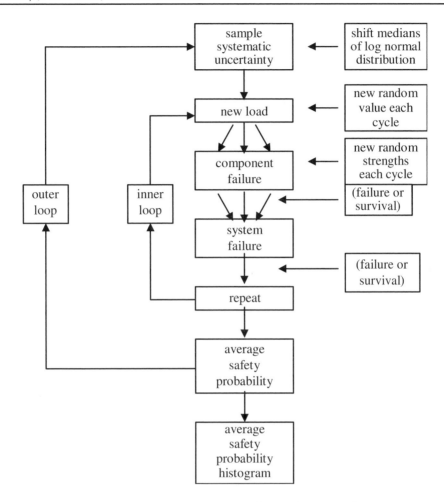

Figure 9.3. *Two-tiered Monte Carlo sampling procedure.*

Saito, Abe, and Shibata (1997) suggest an alternative approach to seismic damage detection comparable to the Monte Carlo simulation, where a reduction in calculations of nonlinear analyses of response characteristics is achieved. The "response surface method" is used to reduce the number of nonlinear analyses for risk evaluation of buildings. Lateral resistance is assumed to be expressed by the "ultimate base shear coefficient" C_y. This is a calculation based on the total areas of columns and walls in one direction of the first story. The acceleration response spectrum S_A is based on an attenuation formula. S_A depends on factors taken from a variety of natural periods T and ground conditions. Basic data of earthquakes with magnitude M are used to calculate a probability density function (PDF) for the acceleration response spectrum, S_A which is evaluated as a log-normal distribution. This study uses the three parameters $S_A(T)$, T, and C_y as random variables to evaluate seismic reliability of building groups.

9.4.1   Response surface method

$S_A(T)$, C_y, and T are transformed into 14 random variables in Gaussian space and are used as sample points. The response surface is then calculated by the least squares method after calculating the structural response at these points.

The unconditional failure probability is then calculated as

$$p_f(y) = \int p_f(y|x) f_x(x), \qquad (9.5)$$

where $p_f(y|x) = (1 - f_{Y|X}(y|x))$ is the conditional failure probability of the maximum displacement y given the value x, and $f_x(x)$ is the joint probability function.

The conclusions drawn from this study, which used statistics of building codes before and after structural code revisions in the city of Sendai, Japan, suggest results similar to the Monte Carlo simulation.

Whether the analytic or Monte Carlo method is used, the end result of such an analysis appears as in Figure 9.4. The median probability of failure curve $(1 - p_s)$ gives the probability of failure of the system for a particular loading and is often referred to as the fragility curve. The curve is a result of the incorporation of random uncertainties only. The variability of this curve when systematic uncertainties are taken into consideration is also illustrated in the figure. The left- and right-hand bounds normally correspond to the 10% and 90% confidence levels of the probability that the failure will be between these bounds in spite of imprecise data, uncertainties in the analysis model, and so forth. In some works, such as those by Wong and Richardson (1983) and Rowan (1976), modeling errors are treated as a source of systematic uncertainties. Another method by Sohn and Law (1997) identifies multiple damage locations using estimated modal parameters that account for uncertainty in measurements and modeling. This method uses a "branch-and-bound" search scheme to avoid large computational costs with a simplified approach for modeling multistory frame structures.

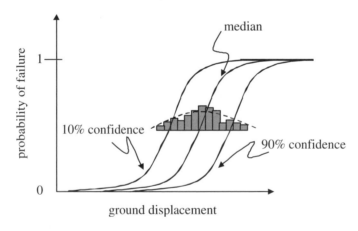

Figure 9.4. *Typical fragility curves from a survivability analysis.*

This method is model-based using modal vibration test data in its assessment of the structure. Model-based methods use analytical results compared against test data of a damaged structure and can provide quantitative results while locating damage. There are usually four steps in model-based approaches:

1. *Model construction*: Construct an analytical model.

2. *Modal testing*: Estimate modal parameters.

3. *Damage localization*: Locate the most likely damage regions using estimated modal parameters.

4. *Damage assessment and system updating*: Assess the severity of damage and update the system parameters.

For damage detection, a refinement of the analytical model is made using the estimated parameters. This is then used as the initial model for the undamaged structure. Still, differences in responses between this refined theoretical model and the actual model are unavoidable. These differences are due to two types of uncertainty in the method. The first uncertainty is from measurement errors or noise. The second is from a modeling error due to assumptions made while abstracting from a complex system.

A Bayesian probabilistic approach is applied as a heuristic means for combining prior experimental data with new test data. In order to avoid the potential high computational cost, this method uses a branch-and-bound scheme with a simplified approach to modeling multistory frames.

Comparisons of relative probabilities for damage events rather than the difficult residual force vectors (or damage vectors) are used in Sohn and Law (1997) to locate the most probable damage locations. This relative probability is a function of the posterior probability of damage based on the estimated modal data sets of the structure. This relative posterior probability is computed from the modal output, which is the difference between the estimated parameters and the theoretical parameters of the analytical model.

Theoretical formulation. The method consists of three parts: (i) Formulate the relative posterior probability for an assumed damage event, taking into account modeling errors and noise (measurement errors); (ii) apply a branch-and-bound search scheme to locate the most likely damage event without searching all possible damage events; (iii) simplify three-dimensional multistory frame analyses.

For N_{sub} substructures, obtain the stiffness matrix as

$$K(\theta) = \sum_{i=1}^{N_{\text{sub}}} \theta_i K_{si} \tag{9.6}$$

(K_s is the stiffness matrix of the substructure; θ_i, $0 \leq \theta_1 \leq 1$, is a nondimensional parameter representing the contribution of the ith substructure stiffness to the system stiffness matrix; θ_i models damage in the ith substructure), where damage (substructure) $= \theta <$ threshold value.

Damage locations and amount are determined according to the θ values; therefore, the stiffness matrix is expressed as a function of $\theta = \{\theta_i; i = 1, N_{\text{sub}}\}$.

The model data is collected and estimated from vibration tests for N_s vibration tests. The total collection of N_s modal data is represented as

$$\hat{\Psi}_{N_s}, \quad \{\hat{\Psi}(n) : n = 1, \ldots, N_s\};$$

$\hat{\Psi}(n)$ consists of both the frequencies and modal vectors estimated from the nth vibration test; i.e.,

$$\hat{\Psi}(n) = [\hat{\omega}_1^n, \ldots, \hat{\omega}_{N_m}^n, \hat{v}_1^{nT}, \ldots, \hat{v}_{N_m}^{nT}]^T \in R^{N_t} \quad P(H_j|\hat{\Psi}_{N_s}) = \frac{P(\hat{\Psi}_{N_s}|H_j)}{P(\hat{\Psi}_{N_s})} P(H_j),$$

where $\hat{\omega}_i^n$ is the ith estimated frequency vector in the nth data set and $\hat{v}_i^n$ is the ith estimated modal vector in the nth data set.

Bayes' theorem is next used to calculate posterior probability $P(H_j|\hat{\Psi}_{N_s})$ after observing a set of estimated modal parameters $\hat{\Psi}_{N_s}$,

$$P(H_j|\hat{\Psi}_{N_s}) = \frac{P(\hat{\Psi}_{N_s}|H_j)}{P(\hat{\Psi}_{N_s})} P(H_j), \qquad (9.7)$$

where H_j is the hypothesis for a damage event (containing any number of substructures as damaged) and $P(H_j)$ is the prior probability, the initial degree of belief about the hypothesis H_j. The state of the structure is defined as a binary variable: damaged or undamaged. This enables the construction of a tree to represent all possible damage cases.

The tree begins with a null hypothesis H_0 that represents no damage present. From this root, a first level of branches is constructed by adding one branch for each damage case. For a system with N_{sub} substructures, the number of first-level branches is $_{N\,sub}C_1$, where this notation is defined as $_N C_K = \frac{N!}{K!(N-K)!}$, the number of combinations of K items out of a population N.

The next level is again extended by adding another substructure as damaged. The number of second level branches is $_{N\,sub}C_2$.

Therefore, for a system with N_{sub} substructures, the damage tree has a total of

- N_{sub} levels of branches;

- $2^{N\,sub}$ branches $= {}_{N\,sub}C_0 + {}_{N\,sub}C_1 + \cdots + {}_{N\,sub}C_{N\,sub}$.

To reduce this complexity, a branch-and-bound search scheme is used.

Branch-and-bound. A branch-and-bound scheme is used to expedite the search for the most likely damage case without searching all possible combinations. The main idea is to rule out further extension of a hypothesis by using the following two main conditions:

1. Let $H_j \cup D_i$ denote an extension of hypothesis H_j by adding the ith substructure as damaged. If a posterior probability of $H_j \cup D_i$ is less than that of H_j, then further extension of $H_j \cup D_i$ is ruled out; i.e.,

   $$\text{if } P(H_j \cup D_i|\Psi_{Ns}), \quad \text{then stop extending } H_j \cup D_i.$$

2. If a posterior probability of H_j is less than P_{max}, which is the largest posterior probability among all the hypotheses examined so far, then further extension of H_j is ruled out; i.e.,

   $$\text{if } (P(H_j|\Psi_{Ns}) < P_{max}, \quad \text{then stop extending } H_j.$$

9.5   Problems with current methods

The analysis procedures described above are very appealing since they produce precise summaries of the different effects of uncertainties on the system response, e.g., in the form of Figure 9.4. Such information can be fed into the hierarchy of decision making in maintenance and replacement plans for the repair of damage to structures, for example. In actuality the separation of random and systematic uncertainties is not a simple task and simply represents yet another judgment of the engineer or analyst. The unilateral treatment of all uncertainties by probabilities implies many assumptions. For example, it assumes that the database exists, that random and systematic uncertainties are independent, and that all types of systematic uncertainties such as biases, judgments, and modeling errors are similar and can be treated in the same fashion. These assumptions are seldom justified due to very limited data, lack of knowledge, and incomplete understanding of the complex physical phenomena and structural behavior (Ross (1988), Cooper and Ross (1997)).

For example, consider one particular step in the analysis procedure shown in Figure 9.5, which depicts the choice of a model for the dynamic response of the frame structure due to ground acceleration loading. Many models can be considered, including a single-degree-of-freedom model commonly used in slab analysis, a multiple-spring–mass model, or a finite element model of the complete beam–column interaction configuration. Subjective judgment enters into such a choice, and the uncertainty associated with the choice (i.e., the computational error to be expected) is assessed from experience with past similar computations and comparative analyses. Furthermore, the evaluation of modeling uncertainties relies heavily on comparing model predictions with results from controlled tests which are seldom feasible. Ignoring these concerns for the time being (they are addressed in Wong and Richardson (1983)), suppose that the finite element modeling approach is selected. One is then faced with more modeling decisions, e.g., the choice of foundation, a concrete model, how reinforcing steel can be incorporated, the size of the finite element mesh, and so on. Another layer of judgment and decision is encountered. To proceed, let us limit the discussion to the modeling of the column of the structure. One can consider using a composite reinforced-concrete model where the reinforcement is smeared over the volume of the finite element, an explicit reinforcing steel model, or is comprised of two or three elements across the thickness of the column, and so on. Details of modeling notwithstanding, the fact is that many subjective judgments and recollections of past experience are used in the analysis. More important, these "data" cannot be summarized readily by a PDF.

Figure 9.6 illustrates the impacts on the response of the multistory frame due to the modeling options mentioned previously. The figure shows the variation in the lateral force and lateral displacement exerted by the earthquake on the frame system as a result of these modeling assumptions. This variation will be referred to for the time being as the modeling or systematic uncertainty associated with the structure model. It is apparent that such uncertainty is not amenable to a probabilistic description. Furthermore, experience shows that the explicit rebar model gives fairly good results in the slightly plastic range, that the same model is inadequate in the membrane tension mode, and that shear failure modes are completely absent from such a model. These are important "data" that should be incorporated into a realistic analysis. Such inputs are ignored in the current probabilistic formulation since they cannot be readily assimilated in the probabilistic description of uncertainties.

9.6   The fuzzy set alternative

In view of the complexities outlined previously, satisfactory treatment of uncertainties encountered in the safety assessment of multistory frame structures is still a tenuous prospect. However, the continued use of the current all-probabilistic approach also is inadequate for the reasons mentioned. It is therefore necessary to seek other descriptions for some of the uncertainties encountered in the damage detection of multistory frame structure analysis and which are not readily adapted to probabilistic models. One such alternative is a fuzzy set description which has been used in civil engineering applications, e.g., see Brown (1980), Brown and Yao (1983), Blockley (1979), Boissonnade (1984), and Ross (1988). The fuzzy set approach is attractive in multistory frame structure studies for the same reason that the probabilistic approach has been found lacking. Subjective information, engineering opinion,

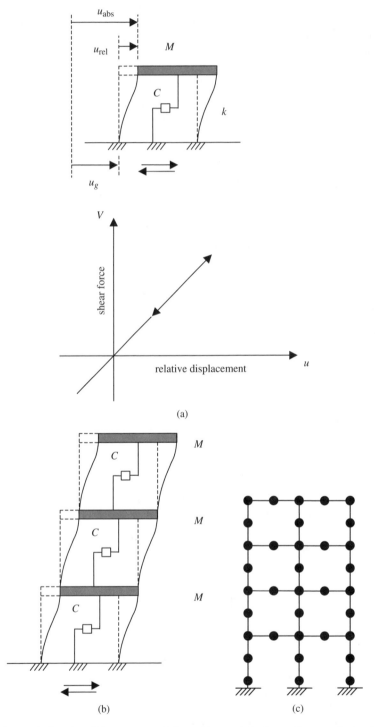

Figure 9.5. *Possible structural dynamic models for safety analysis:* (a) *single degree-of-freedom model;* (b) *multiple degree-of-freedom model;* (c) *two-dimensional finite element model.*

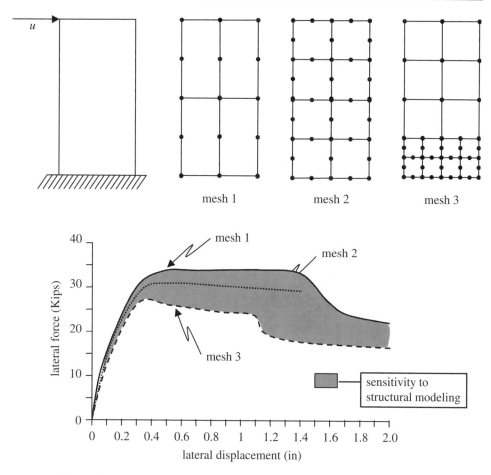

Figure 9.6. *Sensitivity of building deflection to seismic load and response to uncertainties in structural modeling (Seible and Kingsley (1991)).*

and disjoint data can be quantified and manipulated using fuzzy sets and fuzzy algebra in a manner which reflects the degree of imprecision in the original information and data. A fuzzy set approach is illustrated in the following examples; similar sets of examples were provided earlier in Wong, Ross, and Boissonnade (1987).

9.7   Examples

These examples focus on the definition of damage and the relationship between structural response and damage for the beams and columns of the reinforced-concrete multistory frame. To narrow the scope of the discussion, suppose the variable governing the failure (or safety) of the frame structure is $x = \frac{\theta}{d}$, where θ is the column rotation of the frame at the top of the first floor and d is the base displacement of the structure. This is a simplification of the problem and is used for illustration only. (The frame structure is capable of several failure modes, and each mode may be governed by more than one response parameter.) A relationship between the variable x and the damage of the structure is required to continue the analysis. Alternately, a definition of damage or failure is required in order to assess the safety of the multistory frame structure (Figure 9.1). It is helpful to consider this example in an order of progress going from all deterministic data/models to imprecise data/models.

9.7.1 Deterministic/random uncertainties

Suppose the environment (loading) and dynamic response models are deterministic. The response x also will be deterministic. Denote this response by x_{R_0}. If one imposes a failure criterion which also is deterministic, say, x_{F_0}, then the failure of the structure in the environment of interest is a binary proposition: the structure "does not fail" when $x_{R_0} < x_{F_0}$, and it "does fails" when $x_{R_0} \geq x_{F_0}$. This assessment process is illustrated in Figure 9.7 showing the "does not fail" case. The abscissa is x; the response x_R and the failure criterion x_F are represented by lines in the lower and upper half-planes, respectively, even though they are point quantities in the deterministic formulation. In particular, the lower ordinate is the PDF of the response x_R. A deterministic response x_{R_0} therefore will appear as a Dirac delta function located at x_{R_0}. Similarly, the upper ordinate is labeled "failure criterion," and the delta function at x_{F_0} corresponds to the deterministic criterion used. The reason for this portrayal of the problem will become apparent presently.

Now suppose some random uncertainties are introduced into the analytic process. They can be uncertainties in the loading environment, in material properties, or elsewhere in the analysis chain. The response x_R from such an analysis will be random and is illustrated by the PDF in Figure 9.8, with mean value x_{R_0}. If the failure criterion remains deterministic, the probability of failure of the structure is represented by the area of the probability density of the response, f_R, above x_{F_0}, and the probability of survival is given by the area below x_{F_0}. This also is indicated in Figure 9.8.

Such a formulation can be readily extended to the case where random uncertainties exist in the definition of the failure criterion, i.e., x_F also is random. The failure probability is given by the well-known convolution integral

$$p_f = \int_0^\infty \int_0^v f_F(u-v) f_R(v) du dv, \tag{9.8}$$

where $f_F(\bullet)$ is the probability density function of x_F. This case is illustrated in Figure 9.9. In this figure, the function $F_F(r)$ is the cumulative distribution function for the failure criterion, and this form of the integral is similar to (9.8). Note that (9.8) reduces to the special case of a deterministic failure criterion x_{F_0} when $f_F(x) = \delta(x - x_{F_0})$, a Dirac delta function. This approach is well known and exemplifies the predominant practice in structure analysis. The computation of the probability of failure for a particular component and failure mode is repeated for other levels of loading, and the results are summarized neatly in the form of a fragility curve, i.e., the curve relating the probability of failure and the loading level, which was discussed in previous sections (also Figures 9.7(c), 9.8(c), and 9.9(c)). It is now appropriate to show how fuzzy sets can be used to incorporate nonrandom uncertainties into this analysis framework.

9.7.2 Modeling uncertainties

It is desirable to estimate the influence of modeling uncertainties on the estimate of the deflection ratio x obtained from the analysis, which may be deterministic or random as indicated previously. Major features of the model, such as those illustrated in Figures 9.5 and 9.6, are assessed linguistically in terms of gravity characters (G_K) and consequences (E_K), following Brown (1979). Suppose engineering analyst opinions on these modeling features are solicited, resulting in values as shown in Table 9.1, in both tabular and figure form. The linguistic labels large, medium, and small are defined here, for example, as

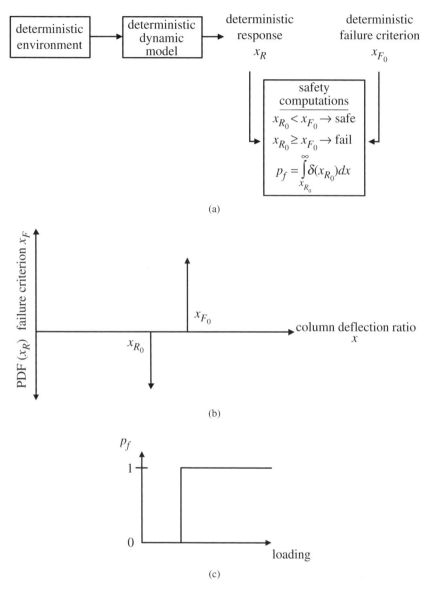

Figure 9.7. *Safety assessment with all deterministic models:* (a) *block diagram,* (b) *probabilistic description;* (c) *fragility.*

$$\text{large} = \frac{0.1}{0.6} + \frac{0.2}{0.7} + \frac{0.5}{0.8} + \frac{0.9}{0.9} + \frac{1.0}{1.0},$$

$$\text{medium} = \frac{0.2}{0.3} + \frac{0.6}{0.4} + \frac{1.0}{0.5} + \frac{0.6}{0.6} + \frac{0.2}{0.7},$$

$$\text{small} = \frac{1.0}{0.0} + \frac{0.9}{0.1} + \frac{0.5}{0.2} + \frac{0.2}{0.3} + \frac{0.1}{0.4},$$

$$\text{not very small} = \frac{0.0}{0.0} + \frac{0.19}{0.1} + \frac{0.75}{0.2} + \frac{0.96}{0.3} + \frac{0.99}{0.4},$$

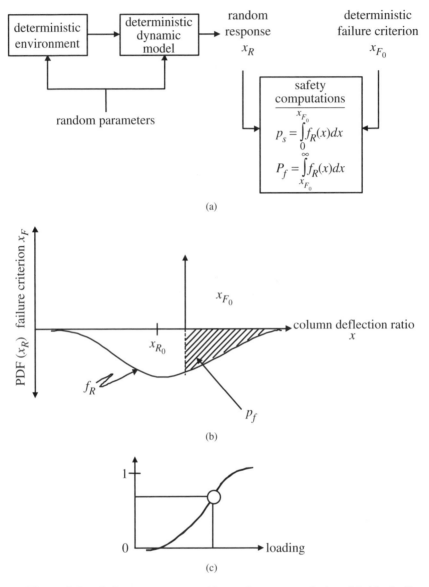

(a)

(b)

(c)

Figure 9.8. *Safety assessment with random uncertainties:* (a) *block diagram;* (b) *probabilistic description;* (c) *fragility.*

where "+" and "−" denote the set operations of union and association, respectively. Following Zadeh, the membership function of the fuzzy set "not very small," $\mu_{NVS}(x)$, is obtained from the membership function of the fuzzy set "small," $\mu_s(x)$, as $\mu_{NVS}(x) = 1 - (\mu_s(x))^2$.

The effect of the kth modeling feature, denoted by P_K, is computed as

$$P_k = G_k \times E_k = G_k \cap E_k = \iint \mu_{G_k}(g) \wedge \mu_{E_k}(e)|(g, e). \tag{9.9}$$

The total effect of all the features in modeling is then $P = \sum_k P_k = \bigcup_k G_k \times E_k$. The individual effects P_k and the total effect P are summarized in Figures 9.10 and 9.11.

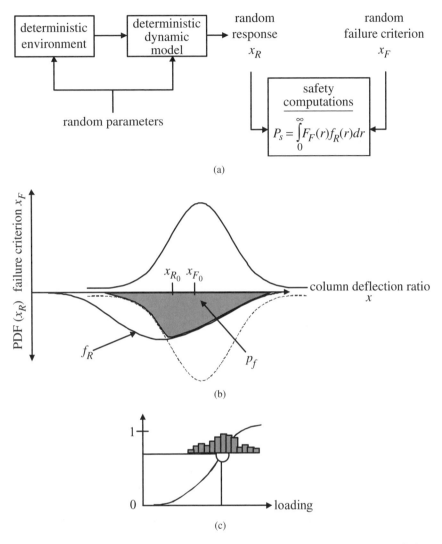

Figure 9.9. *Safety assessment with random uncertainties and random failure criterion:* (a) *block diagram;* (b) *probabilistic description;* (c) *fragility.*

Next, a fuzzy conditional relationship R is defined to link the consequences of the features to the estimation of x. This can be done through a correction factor called C, and the relation R can be expressed linguistically as R: if the consequences are large, then C is large; or else if the consequences are medium, then C is medium; or else if the consequences are small, then C is small, with C defined as (using Zadeh's notation for discrete fuzzy sets)

$$C(\text{large}) = \frac{0.1}{1.0x} + \frac{0.33}{1.1x} + \frac{0.55}{1.2x} + \frac{0.78}{1.3x} + \frac{1.0}{1.4x},$$

$$C(\text{medium}) = \frac{0.1}{1.0x} + \frac{0.16}{1.1x} + \frac{1.0}{1.2x} + \frac{0.16}{1.3x} + \frac{0.1}{1.4x},$$

$$C(\text{small}) = \frac{1.0}{1.0x} + \frac{0.78}{1.1x} + \frac{0.55}{1.2x} + \frac{0.33}{1.3x} + \frac{0.1}{1.4x}.$$

Table 9.1. *Character and consequence of modeling features, example* 1.

Feature	G (gravity)	E (consequence)
1	large	medium
2	medium	not very small
3	medium	medium

"linguistic modeling features"

Note the supports of C are equal to or greater than $1.0x$. This choice of supports implies that modeling uncertainties have deleterious effects on the assessed evaluation of structural integrity. The modifier C (large) expresses the fact that there is a strong bias towards values higher than x can take. The relation R may be expressed as $R = \bigcup_k E_k \times C_k$ and is shown in Figures 9.12 and 9.13.

The fuzzy set P represents the total effect of all the features. The fuzzy conditional relationship R links the consequences of these features to the impact on the estimation of x. It is now possible to obtain a quantitative measure of how much x will be affected. For this purpose, the fuzzy set F, defined by the composition relation $F = P \circ R$, is used to estimate the correction factor $K(x)$ on the deflection ratio x for a given character of the model uncertainty. We recalled that F is defined as

$$F = P \circ R = \iint_{P \times R} \vee_e [\mu_P(g, e) \wedge \mu_R(e, c)] | (g, c), \tag{9.10}$$

where $\vee$ and $\wedge$ represent the maximum and minimum operations, respectively. Carrying out the algebra, the fuzzy matrix F is obtained and is shown in Figure 9.14.

The fuzzy matrix F represents a relationship between the global character of the features and the impact on the estimation of x. In this example, the gravity character of the features can be taken as $G' =$ medium. Then its impact on the estimation of x is obtained through the max–min composition rule $K(x) = G' \circ F$.

It can be seen in Figure 9.14 that a strong emphasis is placed on the value $1.2x$, which means that the response is approximately 20% higher than that obtained from an analysis that does not account for modeling uncertainties.

It is easy to extend the above procedure to the case when x is a random variable with PDF $f_R(x)$. $K(x)$ is expressed in a continuous manner and the updated distribution on x

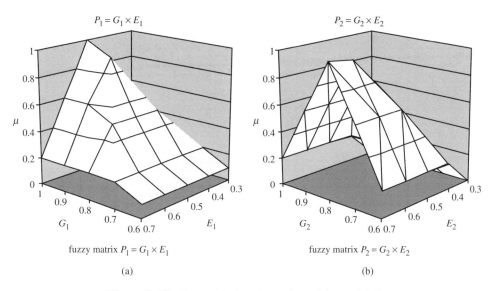

Figure 9.10. *Fuzzy relational matrices:* (a) P_1; (b) P_2.

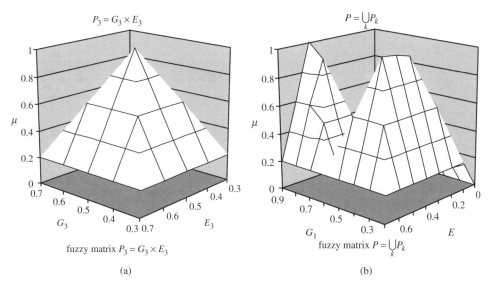

Figure 9.11. *Fuzzy relational matrices:* (a) P_3; (b) P.

can be expressed as, for example,

$$f(x) = \frac{\int_v f_R(x - v)\mu_K(v)dv}{\int_x \int_v f_R(x - v)\mu_K(v)dvdx},$$
(9.11)

where $v \in [0., 0.4]$, and μ_K denotes the membership function associated with $K(x)$.

There are other ways in which fuzzy information on features of the model can be merged with statistical information. For example, the membership function

$$K(m_x) = \frac{0.2}{m_x} + \frac{0.5}{1.1m_x} + \frac{1.0}{1.2m_x} + \frac{0.33}{1.3m_x} + \frac{0.2}{1.4m_x}$$
(9.12)

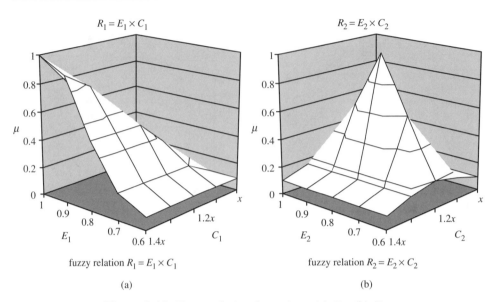

Figure 9.12. *Fuzzy relational matrices:* (a) R_1; (b) R_2.

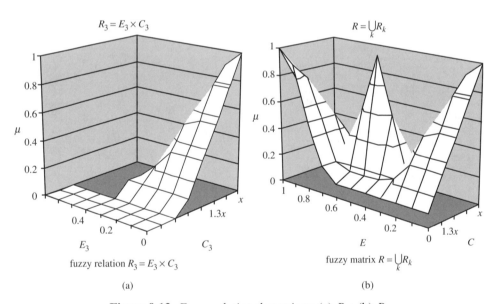

Figure 9.13. *Fuzzy relational matrices:* (a) R_3; (b) R.

can represent the possible values of m_x, the mean value of x. To incorporate this information into the probabilistic analysis, the membership functions are normalized such that $K(m_x)$ becomes a distribution for m_x. Furthermore, let $\bar{x}$ be the true deflection ratio to distinguish it from x, the deflection ratio obtained from the statistical model. The parameter K can then be considered a correction factor that accounts for imperfect information in the model. Hence

$$\bar{x} = Kx. \tag{9.13}$$

From (9.12), K has a mean value $m_K = 1.19$ and cov $\omega_K = 0.085$. The method of Ang and Tang (1975, 1984) can then be used to find the mean and cov of $\bar{x}$ using (9.13).

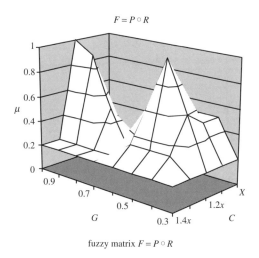

fuzzy matrix $F = P \circ R$

Figure 9.14. *Fuzzy relational matrix F.*

9.7.3   Interpretation of damage

In many fields of structural engineering, damage and its interpretation are not unambiguous. This is especially true for multistory frame structures because under the influence of a severe earthquake they are subjected to extensive plastic deformation. Furthermore, the relation between the deflection ratio x and the degree of damage of the structural component can be assessed only from limited tests on damaged structures. These tests are usually performed on small-scale structures using simulated earthquake loadings. Some representative results of such tests on multistory reinforced-concrete frame structures are shown schematically in Figure 9.15. It is seen that damage encompasses a wide spectrum of deflection states. Based on such limited data, one may interpret the damage in Figure 9.15(a) as medium, that in Figure 9.15(b) as severe, and that in Figure 9.15(c) as collapse. However, the ranges in the response parameter (column deflection in this example) corresponding to light, medium, and severe damage and collapse are not as easy to define. Furthermore, these ranges of response are expected to overlap; i.e., damage does not change abruptly from light to medium and from medium to severe upon reaching a crisp threshold. Other factors of complication are the scarcity of data and the need to extrapolate the data to realistic loading, full-size prototypes, and imperfect foundation–structure configuration.

 Returning to the example illustrated in Figures 9.7–9.9, it is plausible to try the following in an attempt to accommodate some of the inexact information on damage. Since damage (deformation) states are ill-defined, a given value of the variable x that governs one mode of failure of the structure may be assigned to several damage states. In the context of fuzzy set theory, each damage state is a fuzzy set and its elements are the values of the variable x to which are associated a membership value, $\mu_{DS}(x)$, which represents the degree of belief that variable x belongs to a given damage state j. These memberships are generally assigned through engineering judgment. A hypothetical representation of these memberships is given in the upper part of Figure 9.16. In the figure, DS_1 may correspond to light damage, DS_2 to medium damage, and so on. The ranges of x corresponding to the different damage states may overlap, as modeled by overlapping membership functions.

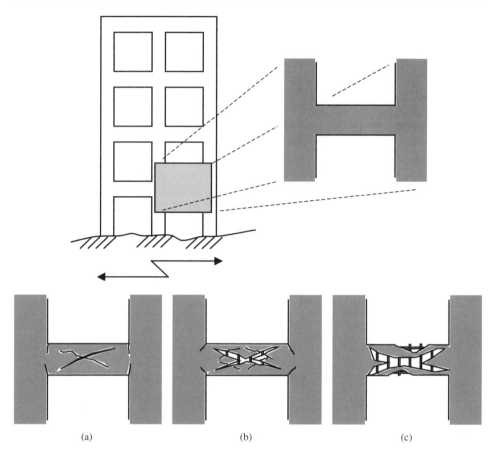

Figure 9.15. *Different levels of damage from scaled tests:* (a) *medium damage,* (b) *severe damage,* (c) *collapse (Davidovici (1993)).*

If the probability density function on the response x_R is known, as shown in the lower part of Figure 9.16, the probabilistic information on various damage states for a given critical variable such as x can then be obtained, after the appropriate normalization, as in Boissonnade (1984):

$$p(DS_i) = \int \mu_{DS_i}(v) f_R(v) dv,$$
$$\sum_i p(DS_i) = 1. \tag{9.14}$$

It should be noted that if a crisp definition of the different damage states is possible and if there are sufficient data to substantiate the variability in x_F, corresponding to a given damage state, then an all-probabilistic approach similar to that used for the failure criterion in Figure 9.9 would be possible. For example, a damage criterion can be defined for each of the three damage levels identified, i.e., light, medium, and severe. This approach is illustrated in Figure 9.17 and would result in the three fragility curves shown in Figure 9.17(c). Again, in Figure 9.17 the quantities F_l, F_m, and F_s are cumulative distribution functions. Figure 9.17 can be compared with Figure 9.16 to illustrate the similarities and differences between the all-probabilistic and probabilistic-fuzzy approaches, respectively.

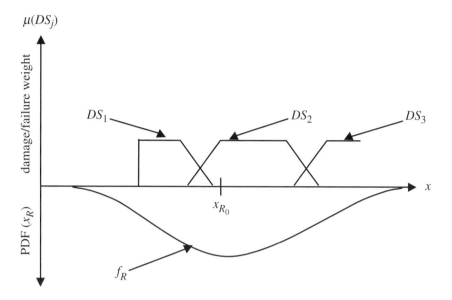

Figure 9.16. *Nonuniform and overlapping supports for three fuzzy damage states and a probabilistic response.*

To illustrate how the weighting membership function(s) in Figure 9.16 may be derived, one may proceed from available damage data that are mainly restricted to small-scale laboratory structures. Our hypothesis is to achieve a quantifiable relationship between damage to full-scale frame structures and the column deflection ratio x. This relationship does not exist a priori, however, so it must be assembled from objective elements of information that are available and from information expressed through subjective assessments. The known information on damage in this case is the relationship between damage on small-scale laboratory structures and the parameter x. The subjective information is on the relationship between small-scale damage and full-scale damage and on the linguistic interpretation of damage states.

To proceed with the hypothetical example, a fuzzy conditional relation R that expresses known deflection ratio information (denoted X) on small-scale structures to damage (say, D_s) is established (by a poll of experts; e.g., see Boissonnade (1984)) as follows: R: if X is light, then D_S is very small; or else if X is moderate, then D_S is medium; or else if X is severe, then D_S is large, where D_S stands for damage to small-scale structures and X is the deflection ratio. The fuzzy sets for X are based on engineering knowledge of the small-scale database, and the values $x = 0.05, 0.1, 0.15$, and 0.2 are chosen for clarity in illustration. One assignment of the membership functions to X(light), X(moderate), and X(severe) is for brevity; the linguistic terms for D_s, namely, very small (D_{s1}), medium (D_{s2}), and large (D_{s3}), are taken from Brown (1980) and are not repeated here. The fuzzy relation R can now be formed as $R = \bigcup_i (X_i \times D_{si})$, $i = 1, 2, 3$, and is shown in Figure 9.18.

The next step is to seek support for membership of specific deflection ratios x for full-scale structures. A possible subjective relationship between small-scale and full-scale damage for specific damage levels is shown in Table 9.2, where D_f stands for damage to full-scale structures. Note that for each damage level in Table 9.2 the linguistic terms for D_f are larger than the associated terms for D_s. This reflects the subjectivity that deformations at larger scales are larger because of reduced quality control, weaknesses in construction

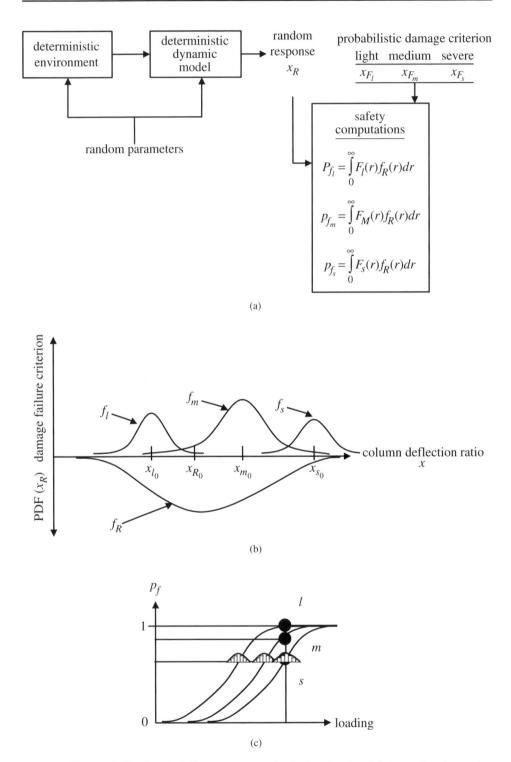

Figure 9.17. *Survivability assessment including levels of damage for the probabilistic approach:* (a) *block diagram;* (b) *probabilistic description;* (c) *fragility curves.*

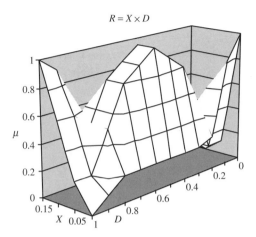

fuzzy damage relation R

Figure 9.18. *Fuzzy relational matrix R, example* 2.

Table 9.2. *Subjective information on scaling, example* 2.

Damage level	Small-scale (D_s)	Full-scale (D_f)
1	very small	small
2	medium	large
3	large	very large

materials based on material sizing and detailing, differences in the fabrication environment (e.g., field versus laboratory), and scale-dependent phenomena such as concrete microcracking per unit volume. The complete statement of subjective information on scaling is given by $P = \bigcup_i (P_i = (D_{si} \times D_{fi})), i = 1, 2, 3$ (Figure 9.19(a)). The influence of X on D_f can now be determined by the fuzzy composition

$$N = R \circ P = \int_{XXD_f} \vee_{D_s}[\mu_R(x, d_s) \wedge \mu_p(d_s, d_f)]|(x, d_f). \qquad (9.15)$$

The fuzzy set N contains the membership support for the deflection ratios ($x = 0, 0.1,$ 0.15, and 0.2) in view of the subjective information in P. The selection of a subset of N (sometimes referred to as a fuzzified kernel) for membership values of a specific full-scale damage level also is a subjective process. In other words, the relationship between X and D_f can be obtained by altering the X versus D_s relationship with the D_s versus D_f subjective information on scaling. The result is shown in Figure 9.19(b).

From Table 9.2, the membership values associated with the x values for a given damage state of the full-scale structure, D_f, can be estimated. For example, for "small" damage with support values in the range [0, 0.2], the membership values associated with x might be determined from the maximum membership within this range as

$$\text{small} = \frac{1.0}{0.05} + \frac{0.25}{0.10} + \frac{0.25}{0.15} + \frac{0.2}{0.2}.$$

Similarly, for "large" damage with support values in the range [0, 0.2], one may obtain

$$\text{large} = \frac{0.25}{0.05} + \frac{1.0}{0.10} + \frac{1.0}{0.15} + \frac{1.0}{0.2}.$$

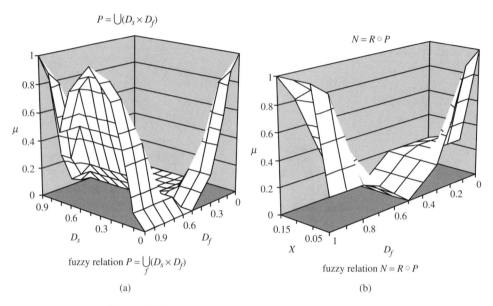

Figure 9.19. *Fuzzy relational matrices:* (a) P; (b) N.

Fuzzy algebra also can be used to obtain the membership values of x corresponding to "very large" damage; i.e.,

$$\text{very large} = \frac{0.06}{0.05} + \frac{1.0}{0.10} + \frac{1.0}{0.15} + \frac{1.0}{0.2},$$

where $\mu_A(x) = (\mu_B(x))^2$ with A corresponding to "very large" and B corresponding to "large." Damage states for full-scale structures are more diffused than small-scale structures due to the additional uncertainty in scaling.

9.8   Summary

The determination of the safety (or failure) of a multistory frame structure is a complex process. This complexity arises from not knowing completely the physical process that is taking place, the uncertainty in modeling some components of the problem, and the fact that other components of the problem, such as damage characterization, are prescribed linguistically. In particular, the current practice of treating systematic uncertainties or biases in engineering models or in damage interpretation as all-probabilistic processes was shown in earlier work to be ineffective (Ross (1988), Cooper and Ross (1997)). Current practitioners continue to struggle to recognize the different kinds of uncertainties present and which must be accommodated, e.g., time, loading, and quality control, as well as those arising from random processes.

The analytic treatment outlined in (9.1)–(9.4) tends to exemplify the simplicity currently attached to the treatment of systematic uncertainties (denoted in the equations as Λ). Random and systematic uncertainties are combined as identical entities and treated in the same analytic fashion. This is equivalent to the presumption that objective and subjective elements of a problem have similar properties. In the Monte Carlo numerical approach in the development of fragility (probability of failure) curves, the inherent assumption on systematic uncertainties is that they can be sampled from an assumed distribution. The common presumption is that these uncertainties (and especially a large number of them) can be modeled as Gaussian variables following the central limit theorem. Hence, it is possible to shortcut the "outer loop" in the Monte Carlo analysis (Figure 9.3) and specify the solution of the convolution integral shown in (9.8) using an analytical expression for the first two moments of the systematic uncertainty distribution. These assumptions are unnecessary when nonprobabilistic descriptions such as fuzzy sets are allowed.

The use of fuzzy set theory in the assessment of the probability of failure of a structure complements the current approach. The probabilistic methods deal with objective information and the fuzzy set methods deal with subjective information. In the example problems, it is shown that noncrisp and judgmental data such as those existing in the description of damage to a structure and the effects of modeling can be represented more realistically using the fuzzy set approach. Furthermore, it is shown that fuzzy descriptions can be readily incorporated into the current probabilistic analysis framework.

For the example problems, the division of subjective information into gravity and consequence elements has additional merits not present in an all-probabilistic approach in safety assessments. For example, there may be a high possibility of a multistory frame structure sustaining large, irrecoverable deformations in a base column, but the consequence to internal occupants may not be high if the structure is designed with sufficient spiral reinforcing to prevent total collapse and loss of axial capacity. Hence, in designing a structure to survive an earthquake and when subjective considerations indicate a high possibility of failure, reducing either the damage level (gravity) or the consequence on the building's function will constitute remedial action. Such an assessment can be best performed using fuzzy algebra.

To pursue the use of fuzzy set theory, more research is needed in constructing membership functions for various degrees of damage, types of damage, and different failure modes. This requires a much more intense effort in data analysis and reduction than is illustrated herein. In a wider scope, more research is needed in the area of defining systematic uncertainties in safety assessment and how they should be modeled in the analysis. For simplicity, the discussion in this paper was limited to one component and one damage mode.

References

A. H.-S. ANG AND W. H. TANG (1975), *Probability Concepts in Engineering Planning and Design, Vol. I: Basic Principles*, Wiley, New York.

A. H.-S. ANG AND W. H. TANG (1984), *Probability Concepts in Engineering Planning and Design, Vol. II: Decision, Risk, and Reliability*, Wiley, New York.

D. I. BLOCKLEY (1979), The role of fuzzy sets in civil engineering, *Fuzzy Sets Systems*, 2, pp. 267–278.

A. C. BOISSONNADE (1984), *Earthquake Damage and Insurance Risk*, Ph.D. dissertation, Stanford University, Stanford, CA.

A. C. BOISSONNADE (1984), Seismic damage forecasting of structures, in *Proceedings of the 8th World Conference on Earthquake Engineering*, Prentice–Hall, Englewood Cliffs, NJ, pp. 361–368.

C. B. BROWN (1979), A fuzzy safety measure, *ASCE J. Engrg. Mech.*, 105, pp. 855–872.

C. B. BROWN (1980), The merging of fuzzy and crisp information, *ASCE J. Engrg. Mech.*, 106, pp. 123–133.

C. B. BROWN AND J. T. P. YAO (1983), Fuzzy sets and structural engineering, *ASCE J. Structural Engrg.*, 109, pp. 1211–1225.

J. COLLINS (1976), Hardened system vulnerability analysis, *Shock Vibration Bull.*, pp. 87–98.

J. A. COOPER AND T. J. ROSS (1997), Improved safety analysis through enhanced mathematical structures, in *Proceedings of the* 1997 *International Conference on Systems, Man, and Cybernetics*, IEEE, Piscataway, NJ, pp. 1656–1661.

V. E. DAVIDOVICI (1993), Strengthening existing buildings, in *Proceedings of the* 17*th Regional European Seminar on Earthquake Engineering*, A. A. Balkema, ed., Rotterdam, The Netherlands.

T. J. ROSS (1988), Approximate reasoning in structural damage assessment, in *Expert Systems in Construction and Structural Engineering*, H. Adeli, ed., Chapman and Hall, London, pp. 161–192.

W. H. ROWAN (1976), Recent advances in failure analysis by statistical techniques, *Shock Vibration Bull.*, pp. 19–32.

T. SAITO, S. ABE, AND A. SHIBATA (1997), Seismic damage analysis of reinforced concrete buildings based on statistics of structural lateral resistance, *Structural Safety*, 19, pp. 141–151.

F. SEIBLE AND G. R. KINGSLEY (1991), Modeling of concrete and masonry structures subjected to seismic loading, in *Experimental and Numerical Methods in Earthquake Engineering*, Kluwer Academic Publishers, Dordrecht, The Netherlands, pp. 281–318.

H. SOHN AND K. H. LAW (1997), A Bayesian probabilistic approach for structure damage detection, *Earthquake Engrg. Structural Dynam.*, 26, pp. 1259–1281.

F. S. WONG AND E. RICHARDSON (1983), Design of underground shelters including soil-structure interaction effects, in *Proceeding of the Symposium on the Interaction of Non-Nuclear Munitions with Structures*, U.S. Air Force Academy, Colorado Springs, CO.

F. S. WONG, T. J. ROSS, AND A. C. BOISSONNADE (1987), Fuzzy sets and survivability analysis of protective structures, in *The Analysis of Fuzzy Information*, Vol. 3, James Bezdek, ed., CRC Press, Boca Raton, FL, pp. 29–53.

L. ZADEH (1973), Outline of a new approach to the analyses of complex systems and decision processes, *IEEE Trans. Systems Man Cybernet.*, 3, pp. 28–44.

Chapter 10

Aircraft Integrity and Reliability

Carlos Ferregut, Roberto A. Osegueda, Yohans Mendoza, Vladik Kreinovich, and Timothy J. Ross

Abstract. In his unpublished recent paper "Probability theory needs an infusion of fuzzy logic to enhance its ability to deal with real-world problems" (see also Zadeh (2002)), L. A. Zadeh explains that probability theory needs an infusion of fuzzy logic to enhance its ability to deal with real-world problems. In this chapter, we give an example of a real-world problem for which such an infusion is indeed successful: the problems of aircraft integrity and reliability.

10.1   Case study: Aircraft structural integrity: Formulation of the problem

10.1.1   Aerospace testing: Why

One of the most important characteristics of the aircraft is its weight: every pound shaved off means a pound added to its carrying ability. As a result, planes are made as light as possible, with their "skin" as thin as possible. However, the thinner its layers, the more vulnerable is the resulting structure to stresses and faults, and consequently a flight is a very stressful experience. Therefore, even minor faults in the aircraft's structure, if undetected, can lead to disaster. To avoid possible catastrophic consequences, we must thoroughly check the structural integrity of the aircraft before the flight.

10.1.2   Aerospace testing: How

Some faults, like cracks, holes, etc., are external and can therefore be detected during the visual inspection. However, to detect internal cracks, holes, and other internal faults, we must somehow scan the inside of the thin plate that forms the skin of the plane. This skin is not transparent when exposed to light or to other electromagnetic radiation; very energetic radiation, e.g., X-rays or gamma rays, can go through the metal but is difficult to use on such a huge object as a modern aircraft.

The one thing that easily penetrates the skin is vibration. Therefore, we can use sound, ultrasound, etc. to detect the faults. Usually, a wave easily glosses over an obstacle whose size is smaller than its wavelength. Therefore, since we want to detect the smallest possible faults, we must choose sound waves with the smallest possible wavelength, i.e., the largest possible frequency. This frequency usually is higher than the frequencies that humans can hear, so it corresponds to ultrasound.

Ultrasonic scans are indeed one of the main nondestructive evaluation (NDE) tools; see, e.g, (Chimenti (1997), Clough and Penzien (1986), Grabec and Sachse (1997), Mal and Singh (1991), and Viktorov (1967)).

10.1.3 Aerospace integrity testing is very time-consuming and expensive

One possibility is to conduct a point-by-point ultrasound testing, the so-called S-scan. This testing detects the exact locations and shapes of all the faults. Its main drawback, however, is that since we need to cover every point, the testing process is very time-consuming (and therefore very expensive).

A faster test sends waves through the material so that, with each measurement, we will be able to test not just a single point but the entire line between the transmitter and the receiver. To make this procedure work, we need signals at special frequencies called Lamb waves. There are other testing techniques, all which aim to determine whether there are faults and, if so, what are the location and size of each fault. All these methods require significant computation time. How can we speed up the corresponding data processing?

10.2 Solving the problem

10.2.1 Our main idea

The amount of data coming from the ultrasonic test is huge, and processing this data takes a lot of time. It is therefore desirable to uncover some structure in the data and to use this structure to speed up the processing of this data.

The first natural idea is to divide the tested structure into pieces and consider these pieces as different clusters. However, physically the tested piece is a solid body, so the observed vibrations of different points are highly correlated and cannot be easily divided into clusters.

Instead of making a division in the original space, we propose to make a division, crudely speaking, in the frequency domain, i.e., a division into separate vibration modes.

For each vibration mode, we can estimate the energy density at each point; if this measured energy density is higher than in the original (undisturbed) state, this is a good indication that a fault may be located at this point. The larger the increase in energy density, the larger the probability of a fault. After we relate the probabilities to different modes, we must combine them into an overall probability of having a fault at this particular point.

10.2.2 Steps necessary for implementing the main idea

Our idea leads to the following steps:

- First, we must be able to transform the information about the excess energy of each mode at different points into the probability of having a fault at the corresponding point.

- Second, for each point, we must combine the probabilities coming from different modes into a single probability of a fault.

To make the first step more accurate, we should take into consideration not only the value of the excess energy but also the value of the original mode-related energy. For example, if the original energy at some point is 0, then this means that the vibrations corresponding to this mode have zero amplitude at this point; in other words, this mode does not affect our point at all and therefore cannot give us any information about the faults at this point. Similarly, points with small mode energy can give little information about the presence of the fault. Therefore, when computing the probability of a fault at different points based on different modes, we must take into consideration not only the excess energy but also the original mode energy at this point.

To apply our idea, we must know

- the function that transforms the value of the excess energy (and of the original mode energy) into the probability of a fault and

- the combination function which transforms probabilities coming from different modes into a single probability.

10.2.3   We do not have sufficient statistical data, so we must use expert estimates

Ideally, we should get all these probability functions from the experiments; however, in real life, we do not have enough statistics to get reliable estimates for probabilities; we have to complement the statistics with expert estimates. In other words, we must use intelligent methods for nondestructive testing as described, e.g., in Ferregut, Osegueda, and Nunez (1997).

10.2.4   Soft computing

One of the most natural formalisms for describing expert estimates is fuzzy theory, which together with neural networks, genetic algorithms, simulated annealing, etc. form a combined intelligent methodology called soft computing. Therefore, in this chapter we will be using methods of soft computing (including fuzzy set theory).

10.2.5   The choices of transformation and combination functions are very important

The quality of fault detection essentially depends on the choice of these methods:

- For some choices of a transformation function and a combination method, we get a very good fault detection.

- For others, however, the quality of detection is much worse.

It is therefore desirable to find the optimal transformation and combination functions.

10.2.6   How can we solve the corresponding optimization problem?

Solving this optimization problem is very difficult for two reasons:

- First, due to the presence of expert uncertainty, it is difficult to formulate this problem as a precise mathematical optimization problem.

- Second, even when we succeed in formalizing this problem, it is usually a complicated nonlinear optimization problem which is extremely difficult to solve using traditional optimization techniques.

In our previous work (see, e.g., Nguyen and Kreinovich (1997)), we developed a general methodology for finding the optimal uncertainty representation. In this chapter, we show how this general methodology can be used to find the optimal uncertainty representations for this particular problem—namely, the following:

- The problem of assigning probability to excess energy is solved similarly to the problem of finding the best *simulated annealing* technique.

- The problem of finding the best approximation to the probability of detection (POD) curve (describing the dependence of probability of detection on the mode energy; see Barbier and Blondet (1993), Gros (1997), Hovey and Berens (1988)) is solved similarly to the problem of finding the best activation function in *neural networks*.

- A solution to the probability combination problem; it turns out that under certain reasonable conditions, probability combination methods can be described by so-called Frank's *t*-norms; this problem is solved using *fuzzy* techniques.

Some of our previous results have been published in conference proceedings (see Kosheleva, Longpre, and Osegueda (1999); Krishna, Kreinovich, and Osegueda (1999); Osegueda et al. (1999a); Ross et al. (1999); Yam, Osegueda, and Kreinovich (1999)).

10.3   How to determine probabilities from observed values of excess energy: Optimal way (use of simulated annealing)

10.3.1   An expression for probabilities

For every point x, we estimate the value of the excess energy $J(x)$ at this point. We want to transform these values into the probabilities $p(x)$ of different points containing the fault.

The larger the value of $J(x)$, the more probable it is that the point x contains a fault, i.e., the larger the value $p(x)$. Therefore, a natural first guess would be to take $p(x) = f(J(x))$ for some increasing function $f(z)$.

However, this simple first guess does not work: In many applications, there is usually only one fault. As a result, the total probability $\sum_x p(x)$ that a fault is located somewhere should be equal to 1. If we simply take $p(x) = f(J(x))$, then this condition is not satisfied. In order to satisfy this condition, we must *normalize* the values $f(J(x))$; i.e., consider the probabilities

$$p(x) = \frac{f(J(x))}{\sum_y f(J(y))}. \tag{10.1}$$

Which function $f(z)$ should we choose? This question is very important, because numerical experiments show that different choices lead to drastically different efficiency of the resulting method; so to increase detection rate, we would like to choose the best possible function $f(z)$.

10.3.2 Best in what sense?

What do we mean by "the best"? It is not so difficult to come up with different criteria for choosing a function $f(z)$:

- We may want to choose the function $f(z)$ for which the resulting fault location error is, on average, the smallest possible $P(f) \to$ min (i.e., for which the *quality of the answer* is, on average, the best).

- We may also want to choose the function $f(z)$ for which the *average computation time* $C(f)$ is the smallest (average in the sense of some reasonable probability distribution on the set of all problems).

At first glance, the situation seems hopeless: we cannot estimate these numerical criteria even for a single function $f(z)$, so it may look as though we cannot undertake the even more ambitious task of finding the *optimal* function $f(z)$. The situation is not as hopeless as it may seem because there is a symmetry-based formalism (actively used in the foundations of fuzzy, neural, and genetic computations; see, e.g., Nguyen and Kreinovich (1997)) that also will enable us to find the optimal function $f(z)$ for our situation. (Our application will be mathematically similar to the optimal choice of a nonlinear scaling function in genetic algorithms; see Kreinovich, Quintana, and Fuentes (1993) and Nguyen and Kreinovich (1997).)

Before we make a formal definition, let us make two comments.

The first comment is that our goal is to find probabilities. Probabilities are always nonnegative numbers, so the function $f(z)$ must also take only nonnegative values.

The second comment is that all we want from the function $f(z)$ are the probabilities. These probabilities are computed according to formula (10.1). From expression (10.1), one easily can see that if we multiply all the values of this function $f(z)$ by an arbitrary constant C, i.e., if we consider a new function $\tilde{f}(z) = C \cdot f(z)$, then this new function will lead (after the normalization involved in (10.1)) to exactly the same values of the probabilities. Thus whether we choose $f(z)$ or $\tilde{f}(z) = C \cdot f(z)$ does not matter. Therefore, what we are really choosing is not a *single* function $f(z)$ but a *family* of functions $\{C \cdot f(z)\}$ (characterized by a parameter $C > 0$).

In the following text, we will denote families of functions by capital letters such as F, F', G, etc.

10.3.3 An optimality criterion can be nonnumeric

Traditionally, optimality criteria are *numerical*; i.e., to every family F we assign some value $J(F)$ expressing its quality and we choose a family for which this value is minimal (i.e., when $J(F) \leq J(G)$ for every other alternative G). However, it is not necessary to restrict ourselves to such numerical criteria only.

For example, if we have several different families F that have the same average location error $P(F)$, we can choose from them the one that has the minimal computational time $C(F)$. In this case, the actual criterion that we use to compare two families is not numeric, but more complicated: A family F_1 is better than the family F_2 if and only if either $P(F_1) < P(F_2)$ or $P(F_1) = P(F_2)$ and $C(F_1) < C(F_2)$.

The only thing that a criterion *must* do is allow us, for every pair of families (F_1, F_2), to make one of the following conclusions:

- The first family is better with respect to this criterion; we will denote it by $F_1 \succ F_2$ or $F_2 \prec F_1$.

- The second family is better $(F_2 \succ F_1)$ with respect to the given criterion.

- The two families have the same quality with respect to this criterion; we will denote it by $F_1 \sim F_2$.

- This criterion does not allow us to compare the two families.

Of course, it is necessary to demand that these choices be consistent. For example, if $F_1 \succ F_2$ and $F_2 \succ F_3$, then $F_1 \succ F_3$.

10.3.4   The optimality criterion must be final

A natural demand is that this criterion must choose a *unique* optimal family (i.e., a family that is better, with respect to this criterion, than any other family). The reason for this demand is very simple.

If a criterion *does not choose* any family at all, then it is of no use.

If *several* different families are the best according to this criterion, then we still have the problem of choosing the best among them. Therefore, we need some additional criterion for that choice, as in the above example: If several families $F_1, F_2, \ldots$ turn out to have the same average location error $(P(F_1) = P(F_2) = \cdots)$, then we can choose among them a family with minimal computation time $(C(F_i) \to \min)$.

So what we actually do in this case is abandon that criterion for which there were several "best" families and we consider a new "composite" criterion instead: F_1 is better than F_2 according to this new criterion if either F_1 was better according to the old criterion, or if both F_1 and F_2 had the same quality according to the old criterion and F_1 is better than F_2 according to the additional criterion.

In other words, if a criterion does not allow us to choose a unique best family, it means that this criterion is not final; we will have to modify it until we come to a final criterion that will have that property.

10.3.5   The criterion must not change if we change the measuring unit for energy

The exact mathematical form of a function $f(z)$ depends on the exact choice of units for measuring the excess energy $z = J(x)$. If we replace this unit by a new unit that is λ times larger, then the same physical value that was previously described by a numerical value $J(x)$ will now be described in the new unit by a new numerical value $\tilde{J}(x) = \frac{J(x)}{\lambda}$.

How will the expression for $f(z)$ change if we use the new unit? In terms of $\tilde{J}(x)$, we have $J(x) = \lambda \cdot \tilde{J}(x)$. Thus if we change the measuring unit for energy, the same probabilities $p(x) \sim f(J(x))$ that were originally represented by a function $f(z)$ will be described in the new unit as $p(x) \sim f(\lambda \cdot \tilde{J}(x))$, i.e., as $p(x) \sim \tilde{f}(\tilde{J}(x))$, where $\tilde{f}(z) = f(\lambda \cdot z)$.

There is no reason why one choice of unit should be preferable to the other. Therefore, it is reasonable to assume that the relative quality of different families should not change if we simply change the unit; i.e., if a family F is better than a family G, then the transformed family $\tilde{F}$ also should be better than the family $\tilde{G}$.

We are now ready to present the formal definitions.

10.3.6 Definitions and the main result

Definition 10.1. Let $f(z)$ be a differentiable strictly increasing function from real numbers to nonnegative real numbers. By a family that corresponds to this function $f(z)$, we mean a family of all functions of the type $\tilde{f}(z) = C \cdot f(z)$, where $C > 0$ is an arbitrary positive real number. (Two families are considered equal if they coincide, i.e., consist of the same functions.)

In the following text, we will denote the set of all possible families by Φ.

Definition 10.2. By an optimality criterion, we mean a consistent pair $(\prec, \sim)$ of relations on the set Φ of all alternatives that satisfies the following conditions for every $F, G, H \in \Phi$:

1. If $F \prec G$ and $G \prec H$, then $F \prec H$.

2. $F \sim F$.

3. If $F \sim G$, then $G \sim F$.

4. If $F \sim G$ and $G \sim H$, then $F \sim H$.

5. If $F \prec G$ and $G \sim H$, then $F \prec H$.

6. If $F \sim G$ and $G \prec H$, then $F \prec H$.

7. If $F \prec G$, then $G \not\prec F$ and $F \not\succ G$.

Comment.   The intended meaning of these relations is as follows:

- $F \prec G$ means that, with respect to a given criterion, G is better than F.

- $F \sim G$ means that, with respect to a given criterion, F and G are of the same quality.

Under this interpretation, conditions 1–7 have simple intuitive meaning; e.g., 1 means that if G is better than F and H is better than G, then H is better than F.

Definition 10.3.

- We say that an alternative F is optimal (or best) with respect to a criterion $(\prec, \sim)$ if for every other alternative G either $F \succ G$ or $F \sim G$.

- We say that a criterion is final if there exists an optimal alternative, and this optimal alternative is unique.

Definition 10.4. Let $\lambda > 0$ be a positive real number.

- By a λ-rescaling of a function $f(x)$, we mean a function $\tilde{f}(x) = f(\lambda \cdot x)$.

- By a λ-rescaling $R_\lambda(F)$ of a family of functions F, we mean the family consisting of λ-rescaling of all functions from F.

Definition 10.5. We say that an optimality criterion on Φ is unit-invariant if for every two families F and G, and for every number $\lambda > 0$, the following two conditions are true:

1. If F is better than G in the sense of this criterion (i.e., $F \succ G$), then $R_\lambda(F) \succ R_\lambda(G)$.

2. If F is equivalent to G in the sense of this criterion (i.e., $F \sim G$), then $R_\lambda(F) \sim R_\lambda(G)$.

Theorem 10.1. *If a family F is optimal in the sense of some optimality criterion that is final and unit-invariant, then every function $f(z)$ from this family F has the form $C \cdot z^\alpha$ for some real numbers C and α.*

Comment.   Experiments show that for nondestructive testing, the best choice is $\alpha \approx 1$.

10.4   How to determine the probability of detection: Optimal way (use of neural networks)

In the previous section, we assumed that the probability of having a fault at a certain point x depends only on the value of the excess energy $J(x)$ at this point x. This assumption, however, is not always true. For example, if at some point x the energy $E(x)$ of the original (no-fault) vibration mode is equal to 0, then this means that this point does not participate in this vibration mode at all; therefore, from observing this vibration mode, we cannot deduce whether or not there is a fault at this point. Similarly, if the strain energy $E(x)$ is small, this means that this point x is barely moving and hardly participating in the vibration. Therefore, this vibration mode is barely affected by the presence or absence of the fault at this point.

With this in mind, we can say that the probability $p(x)$ described in the previous section is not the "absolute" probability of a fault at a point x based on this mode but rather a *conditional* probability that there is fault $P(\text{fault}|C)$—under the condition C that the analysis of this mode can detect a fault at a point x. The actual probability $P(\text{fault})$ of detecting a fault from the measurements related to this mode can be therefore computed as a product $P(\text{fault}) = P(\text{fault}|C) \cdot P(C)$.

We already know how to compute $P(\text{fault}|C)$. Therefore, to compute the desired probability $P(\text{fault})$, we must find the probability $P(C)$ that the analysis of this mode can detect a fault at a given point x. We have already mentioned that this detection probability depends on the energy $E(x)$ of the original vibration mode at the point x: $P(\text{fault}) = p(E(x))$: if $E = 0$, then $p(E) = 0$; if E is small, then $p(E)$ is small; if E is large enough, then $p(E)$ is close to 1. Let us describe this POD dependence $p(E)$.

Our application will be mathematically similar to the optimal choice of an activation function in neural networks (Nguyen and Kreinovich (1997)).

10.4.1   The POD function must be smooth and monotonic

If we change the energy E slightly, the probability $p(E)$ of detecting the fault should not change drastically. Thus we expect the dependency $p(E)$ to be *smooth* (differentiable).

For a POD function, the probability of detection should be equal to 0 when the point is not affected by the vibration ($E = 0$) and should be equal to 1 when the point is highly affected ($E \to \infty$). The larger the energy E, the more probable it is that we will be able to find the fault; thus the dependence $p(E)$ should be *monotonic*.

10.4.2   We must choose a family of functions, not a single function

For practical applications, we need the function $p(E)$ to determine the probability that if a point with an energy E is presented to a certain NDE technique, then the corresponding fault will be detected. In order to determine this function empirically, we must have statistics of samples for which the fault was discovered; from these statistics, we can determine the desired probability.

This probability, however, depends on how we select the samples presented to the NDE techniques. For example, most structures are inspected visually before using a more complicated NDE technology. Some aerospace structures are easier to inspect visually, so we can detect more faults visually, and only harder-to-detect faults are presented to the NDE technique; as a result of this preselection, for such structures the success probability $p(E)$ is lower than in other cases. Other structures are more difficult to inspect visually; for these structures, all the faults (including easy-to-detect ones) are presented to the NDE techniques, and the success probabilities $p(E)$ will be higher.

In view of this preselection, for one and the same NDE technique, we may have different POD functions depending on which structures we apply it to. Therefore, instead of looking for a *single* function $p(E)$, we should look for a *family* of POD functions which correspond to different preselections.

How are different functions from this family related to each other? Preselection means, in effect, that we are moving from the original unconditional detection probability to the conditional probability, under the condition that this particular sample has been preselected. In statistics, the transformation from an unconditional probability $P_0(H_i)$ of a certain hypothesis H_i to its conditional probability $P(H_i|S)$ (under the condition S that a sample was preselected) is described by the Bayes formula

$$P(H_i|S) = \frac{P(S|H_i) \cdot P_0(H_i)}{\sum_j P(S|H_j) \cdot P_0(H_j)}.$$

In mathematical terms, the transformation from $p(E) = P_0(H_i)$ to $\tilde{p}(E) = P(H_i|S)$ is *fractionally linear*, i.e., has the form $p(E) \to \tilde{p}(E) = \varphi(p(E))$, where

$$\varphi(y) = \frac{k \cdot y + l}{m \cdot y + n}$$

for some real numbers k, l, m, and n. Therefore, instead of looking for a single function $p(E)$, we should look for a family of functions $\{\varphi(p(E))\}$, where $p(E)$ is a fixed function and $\varphi(y)$ are different fractionally linear transformations. In the following text, when we say "a family of functions," we will mean a family of this very type.

Similarly to the previous section, we are looking for a family which is optimal with respect to some final optimality criterion, and it is reasonable to require that this criterion should not change if we change the measuring unit for energy. Thus we arrive at the following definitions.

10.4.3   Definition and the main result

Definition 10.6.

- By a probability function, we mean a smooth monotonic function $p(E)$ defined for all $E \geq 0$ for which $p(0) = 0$ and $p(E) \to 1$ as $E \to \infty$.

- By a family of functions, we mean the set of functions is obtained from a probability function $p(E)$ by applying fractionally linear transformations.

Theorem 10.2. *If a family F is optimal in the sense of some optimality criterion that is final and unit-invariant, then every function p from the family F is equal to*

$$p(E) = \frac{A \cdot E^\beta}{A \cdot E^\beta + 1} \tag{10.2}$$

for some A and $\beta > 0$.

10.5 How to combine probabilities (use of fuzzy techniques)

Based on the formulas given in the previous two sections, we can determine, for each point x and for each mode i, the probability $p_i(x)$ that, based on this mode, there is a fault at this point x. Since there are several modes, we must therefore *combine*, for each point x, these probabilities $p_i(x)$ into a single probability $p(x)$ that there is a fault in the structure at this point. In this section, we show how different techniques can help us to find the best combination.

10.5.1 Traditional probabilistic approach: Maximum entropy

For each point x, after we obtain the probabilities $p_i(x)$ $(1 \le i \le n)$ that this point has a fault in it, we must combine these n probabilities into a single probability $p(x)$ that there is a fault. A fault exists if it is detected by one of the modes, i.e., if it is detected by the first mode, *or* detected by the second mode, etc. In other words, if $p_i(x)$ is a probability of the event $D_i \overset{\text{def}}{=}$ "a fault at point x is detected by ith mode," then $p(x)$ is the probability of the disjunction $D_1 \vee \cdots \vee D_n$.

In general, if we know only the probabilities of events D_i, then it is not possible to uniquely determine the probability of the disjunction; to select a unique probability, we use a *maximum entropy* approach. The idea of this approach is as follows: to find the probabilities of all possible logical combinations of the events D_i, it is sufficient to determine the probabilities of all 2^n events W of the type $D_1^{\varepsilon_1} \& \cdots \& D_n^{\varepsilon_n}$, where $\varepsilon_i \in \{-, +\}$, D_i^+ means D_i, and D_i^- means its negation $-D_i$. Therefore, we must determine the probabilities $p(W)$ so that $\sum p(W) = 1$, and for every i, the sum of $p(W)$ for all events for which D_i is true is equal to $p_i(x)$. There are many distributions of this type; we select the one for which the entropy $-\sum p(W) \cdot \log(p(W))$ takes the largest possible value.

For this distribution, the desired probability of the disjunction is equal to $p(x) = 1 - \prod_{i=1}^n (1 - p_i(x))$ (see, e.g., Kreinovich, Nguyen, and Walker (1996)).

10.5.2 Traditional approach is not always sufficient

From the statistical viewpoint, the maximum entropy (MaxEnt) formula corresponds to the case in which all modes are statistically independent. In reality, the detection errors in different modes can occur from the same cause, and therefore these are not necessarily independent.

If we knew the correlation between these errors, then we could use traditional statistical methods to combine the probabilities $p_i(x)$. In reality, however, we typically do not have sufficient information about the correlation between the components. Therefore, we need to find a new method of fusing probabilities $p_1(x), \ldots, p_n(x)$ corresponding to different modes.

10.5.3 Main idea: Describe general combination operations

We must choose a method for combining probabilities. In mathematical terms, we must describe, for every n, a function that takes n numbers $p_1, \ldots, p_n \in [0, 1]$ as inputs and returns the "fused" (combined) probability $f_n(p_1, \ldots, p_n)$.

This description is further complicated by the fact that the division into modes is rather subjective; e.g., for three close modes, we can

- either divide the vibration into these three modes 1, 2, and 3;

- or divide the vibration into two: mode 3 and a "macromode" $\{1, 2\}$ combining modes 1 and 2;

- or divide it into 1 and $\{2, 3\}$.

Depending on the division, we get different expressions for the resulting probability:

- If we divide the vibration into three modes, the resulting probability is $p = f_3(p_1, p_2, p_3)$.

- If we divide the vibration into two macromodes $\{1, 2\}$ and 3, then for the first macromode, we get $p_{12} = f_2(p_1, p_2)$, and thus $p = f_2(p_{12}, p_3) = f_2(f_2(p_1, p_2), p_3)$.

- Similarly, if we divide the vibration into two macromodes 1 and $\{2, 3\}$, then for the second macromode, we get $p_{23} = f_2(p_2, p_3)$, and thus $p = f_2(p_1, p_{23}) = f_2(p_1, f_2(p_2, p_3))$.

The resulting probability should not depend on the (subjective) subdivision into modes. As a result, we should get $f_3(p_1, p_2, p_3) = f_2(f_2(p_1, p_2), p_3) = f_2(p_1, f_2(p_2, p_3))$. In other words, we can make two conclusions:

- First, the combination function for arbitrary $n > 2$ can be expressed in terms of a combination function corresponding to $n = 2$ as $f_n(p_1, \ldots, p_n) = f_2(p_1, f_2(p_2, \ldots, f_2(p_{n-1}, p_n) \cdots))$.

- Second, the function f_2 which describes the combination of two probabilities should be *associative* (i.e., $f_2(a, f_2(b, c)) = f_2(f_2(a, b), c)$ for all $a, b, c \in [0, 1]$).

Associativity is a reasonably strong property, but associativity alone is not sufficient to determine the operation $f_2(a, b)$ because there are many different associative combination operations. We hope there is another property that we can use. To describe this property, let us recall that for fault detection, the function $f_2(p_1, p_2)$ has the following meaning:

- p_1 is the probability $P(D_1)$ of the event D_1 defined as "the first mode detected a fault";

- p_2 is the probability $P(D_2)$ of the event D_2 defined as "the second mode detected a fault"; and

- $f_2(p_1, p_2)$ is the probability $P(D_1 \vee D_2)$ that one of the two modes detected a fault.

Thus far, our main goal was to detect a fault; this problem is difficult because faults are small and therefore show up only on one of the modes. Different faults present different dangers to the aerospace structure: a small fault can be potentially dangerous, but if the faults are small enough, they may not require grounding the aircraft. On the other hand, large faults are definitely dangerous. Therefore, in addition to detecting all the faults, we would like to know which of them are large (if any). A large fault is probably causing strong changes in all vibration modes. Thus we can expect a large fault if we have detected a fault in all the modes. In other words, in this problem, we are interested in the value $P(D_1 \& D_2)$. From the probability theory, we know that $P(D_1 \& D_2) = P(D_1) + P(D_2) - P(D_1 \vee D_2)$, i.e., $P(D_1 \& D_2) = p_1 + p_2 - f_2(p_1 \vee p_2)$. We can describe this expression as $P(D_1 \vee D_2) = g_2(p_1, p_2)$, where $g_2(a, b) = a + b - f_2(a, b)$.

Similar to the case of the "or" combination of different components, we can describe the probability $P(D_1 \& D_2 \& D_3)$ (that all three modes detect a fault) in two different ways:

- either as $P((D_1 \& D_2) \& D_3) = g_2(g_2(p_1, p_2), p_3)$,

- or as $P(D_1 \& (D_2 \& D_3)) = g_2(g_2(p_1, p_2), p_3)$.

The expression for $P(D_1 \& D_2 \& D_3)$ should not depend on how we compute it, and therefore we should have $g_2(g_2(p_1, p_2), p_3) = g_2(p_1, g_2(p_2, p_3))$. In other words, not only should the function $f_2(a, b)$ be associative, but also the function $g_2(a, b) = a + b - f_2(a, b)$ should be associative.

10.5.4   The notions of *t*-norms and *t*-conorms

In the above text, we described everything in terms of combining probabilities. However, from the mathematical viewpoint, the resulting requirements were exactly the requirements traditionally used for combining membership values in the fuzzy approach (see, e.g., Klir and Yuan (1995) and Nguyen and Walker (1999)):

- The function $f_2(p_1, p_2)$ that describes the degree of truth of the statement $P(D_1 \vee D_2)$, provided that we know the degrees of truth for statement D_1 and D_2, is a *t-conorm*.

- The function $g_2(p_1, p_2)$ that describes the degree of truth of the statement $P(D_1 \& D_2)$, provided that we know the degrees of truth for statement D_1 and D_2, is a *t-norm*.

Main result: Frank's *t*-norms

Functions $f(a, b)$ for which both the function itself and the expression $g(a, b) = a + b - f(a, b)$ are associative have been classified (Frank (1979)). These functions (called *Frank's t-conorms*) are described by a formula

$$f_2(a, b) = 1 - \log_s \left[1 + \frac{(s^{1-a} - 1) \cdot (s^{1-b} - 1)}{s - 1} \right],$$

and the corresponding *t*-norms are

$$g_2(a, b) = \log_s \left[1 + \frac{(s^a - 1) \cdot (s^b - 1)}{s - 1} \right]$$

for some constant s. As a particular case of this general formula, we get the expression $1 - (1 - a) \cdot (1 - b)$ (for $s \to 1$).

Thus to combine the probabilities $p_i(x)$ originating from different modes, we should use Frank's *t*-conorms.

Comment. Fuzzy techniques have been successfully used, in particular, in nondestructive testing (for a recent survey, see, e.g., Ferregut, Osegueda, and Nunez (1997)) and in damage assessment in general (see, e.g., Terano, Asai, and Sugeno (1987) and Ulieru and Isermann (1993)).

The fact that, for probabilistic data, we get similar formulas makes us hope that this algebraic approach will be able to combine probabilistic and fuzzy data.

10.6   Preliminary results

As a case study, we applied the new method to the problem of the nondestructive evaluation of the structural integrity of the U.S. Space Shuttle's vertical stabilizer. To prove the applicability of our method, we applied these techniques to the measurement results for components with known fault locations.

The value s was determined experimentally so as to achieve the best performance; it turns out that the best value is $s \approx 1$ (corresponding to the independent modes). For this value s, our method detected all the faults in $\approx 70\%$ of the cases, a much larger proportion than with any previously known techniques (for details, see Andre (1999), Osegueda et al. (1999), Pereyra et al. (1999), and Stubbs, Broom, and Osegueda (1998)).

For other values of s, we got an even better detection, but at the expense of false alarms. We are currently trying different data fusion techniques (as described, e.g., in Gros (1997)) to further improve the method's performance.

10.7   Alternative approach to fusing probabilities: Fuzzy rules

10.7.1   Main problems with the above approach

There are two main problems with the above approach:

- First, because we used several different (and reasonably complicated) formalisms, the resulting computational models are rather time-consuming and are not very intuitive.

- Second, although we got better fault detection than was gotten by all previously known methods, there is still some room for improvement.

10.7.2   The use of fuzzy rules

The main problem we face is that of complexity of the computational models we use. Complex models are justified in such areas as fundamental physics, where simpler first approximation models are not adequate. However, in our case, the computational models are chosen not because simpler models have been tried, but because these complex models were the only ones which fit our data and which were consistent with the expert knowledge.

The very fact that a large part of our knowledge comes from expert estimates, which have a high level of uncertainty, makes us believe that within this uncertainty we can find simpler computational models which will work equally well. How can we find such models?

A similar situation, when unnecessarily complex models were produced by existing techniques, was what motivated fuzzy logic. Namely, L. Zadeh proposed using new simplified models based on the direct formalization of expert knowledge instead of traditional analytical models.

In view of the success of fuzzy techniques, it is reasonable to use a similar approach in fault detection as well. Let us first describe the corresponding rules.

10.7.3   Expert rules for fault detection

As a result of the measurements, for each location we get five different values of the excess energy $E_1, \ldots, E_5$ which correspond to five different modes. An expert can look at these values and tell whether we have a definite fault here, or a fault with a certain degree of certainty, or definitely no fault at all.

Before we formulate the expert rules, we should note that for each node, the *absolute* values of excess energy are not that useful because, e.g., a slight increase or decrease in the original activation can increase or decrease all the values of the excess energy, while the fault locations remain the same. Therefore, it is more reasonable to look at the *relative* values of the excess energy. Namely, for each mode i, we compute the mean square average σ_i of all the values and then divide all values of the excess energy by this mean square value to get the corresponding relative value of the excess energy $x_i = \frac{J_i}{\sigma}$.

In accordance with standard fuzzy logic methodology, we would like to describe some of these values as "small positive" (SP), some as "large positive" (LP), etc. To formalize these notions, we must describe the corresponding membership functions $\mu_{SP}(x)$ and $\mu_{LP}(x)$.

Some intuition about the values x_i comes from the simplified situation in which the values of excess energy J_i are random, following a normal distribution with 0 average. In this simplified situation, the mean square value σ_i is (practically) equal to the standard deviation of this distribution. For normal distributions, deviations which exceed $2\sigma_i$ are rare and are therefore usually considered to be definitely large; on the other hand, deviations smaller than the average σ_i are, naturally, definitely small. Deviations $J_i \geq 2\sigma_i$ correspond to the values $x_i = \frac{J_i}{\sigma_i} \geq 2$, and deviations $J_i \leq \sigma_i$ correspond to $x_i = \frac{J_i}{\sigma_i} \leq 1$. Therefore, we can conclude that values $x_i \geq 2$ are definitely large and positive values $x_i \leq 1$ are definitely small.

Thus for the fuzzy notion "small," we know that

- values from 0 to 1 are definitely small; i.e., $\mu_{SP}(x_i) = 1$ for these values;

- values 2 and larger are definitely not small; i.e., $\mu_{SP}(x_i) = 0$ for these values.

These formulas determine the value of the membership function for all positive values of x_i, except for the values from 1 to 2. In accordance with standard fuzzy techniques, we use the simplest (linear) interpolation to define $\mu_{SP}(x_i)$ for values from this interval; i.e., we take $\mu_{SP}(x_i) = 2 - x_i$ for $x_i \in [1, 2]$.

Similarly, we define the membership function for "large" as follows: $\mu_{LP}(x_i) = 0$ for $x_i \in [0, 1]$; $\mu_{LP}(x_i) = x_i - 1$ for $x_i \in [1, 2]$; and $\mu_{LP}(x_i) = 1$ for $x_i \geq 2$.

Similarly, we describe the membership functions corresponding to "small negative" (SN) and "large negative" (LN): in precise terms, for $x_i < 0$ we set $\mu_{SN}(x_i) = \mu_{SP}(|x_i|)$ and $\mu_{LN}(x_i) = \mu_{LP}(|x_i|)$.

This takes care of fuzzy terms used in the condition of expert rules. To present our conclusion, we determine that experts use five different levels of certainty from level 1 to level 5 (absolute certainty). We can identify these levels with numbers from 0.2 to 1.

Now we are ready to describe the rules:

1. If the "total" excess energy $x_1 + \cdots + x_5$ attains its largest possible value, or is close to the largest possible value (by ≤ 0.06), then we definitely have a fault at this location (this conclusion corresponds to level 5).

2. If all five modes show increase, then we have a level 4 certainty that there is a fault at this location.

3. If four modes show increase, and one mode shows small or large decrease, then level 4 holds.

4. If three modes show increase and two show small decrease, then level 4 holds.

5. If three modes show increase, and we have either one small and one large decrease or two large decreases, then level 3 holds.

6. If two modes show large increase and three modes show small decrease, then level 3 holds.

7. If two modes show large increase, one or two modes show large decrease, and the rest show decrease, then level 2 holds.

8. If one mode shows large increase, one mode shows small increase, and three modes show small decrease, then level 2 holds.

9. In all other cases, level 1 holds.

10.7.4   The problem with this rule base and how we solve it

The techniques of fuzzy modeling and fuzzy control enable us to translate rule bases (like the one above) into an algorithm which transforms the inputs $x_1, \ldots, x_n$ into a (defuzzified) value of the output y. In principle, we can apply these techniques to our rule base, but the problem is that we will need too many rules. Indeed, standard rules are based on conditions such as "if x_1 is $A_1, \ldots$ and x_n is A_n, then y is B." In our case, we have five input variables, each of which can take four different fuzzy values (LN, SN, SP, and LP). Therefore, to describe all possible combinations of inputs, we must use $4^5 = 1024$ rules. This is feasible but definitely not the simplification we seek.

To decrease the number of the resulting rules, we can use the fact that all the rules do not distinguish between different modes. Therefore, if we permute the values x_i (e.g., swap the values x_1 and x_2), the expert's conclusion will not change. Hence instead of considering all possible combinations of x_i, we can first apply some permutation to decrease the number of possible combinations. One such permutation is *sorting* the values of x_i, i.e., reordering these values in decreasing order. Let us show that if we apply the rules to the reordered values, then we can indeed drastically decrease the number of resulting fuzzy rules.

Let $y_1 \geq y_2 \geq \cdots \geq y_5$ denote the values $x_1, \ldots, x_5$ reordered in decreasing order. Let us show how, e.g., rules 2, 3, and 4 from the above rule base can be reformulated in terms of these new values y_i:

Rule 2. To say that all five values x_i are positive is the same as saying that the smallest of these values is positive, so the condition of rule 2 can be reformulated as $y_5 > 0$.

Rule 3. When four modes are positive and the fifth is negative, it means that $y_4 > 0$ and $y_5 < 0$.

Notice that since rules 2 and 3 have the same conclusion, they can be combined into a single rule with a new (even simpler) condition $y_4 > 0$. (Indeed, we have either $y_5 > 0$ or $y_5 \leq 0$; if $y_4 > 0$ and $y_5 > 0$, then the conclusion is true because of rule 2; if $y_4 > 0$ and $y_5 < 0$, then the conclusion is true because of rule 3.)

Rule 4. Similarly, this condition can be reformulated as $y_3 > 0$, y_4 is SN, and y_5 is SN. As a result, we get the following new (simplified) rule base:

1. If the "total" excess energy $y_1 + \cdots + y_5$ attains its largest possible value, or is close to the largest possible value (by ≤ 0.06), then level 5 holds.

2. If $y_4 > 0$, then level 4 holds.

3. If $y_3 > 0$, y_4 is SN, and y_5 is SN, then level 4 holds.

4. If $y_3 > 0$, $y_4 < 0$, and y_5 is LN, then level 3 holds.

5. If y_2 is LP, $y_3 < 0$, and y_5 is SN, then level 3 holds.

6. If y_2 is LP, $y_3 < 0$, and y_5 is LN, then level 2 holds.

7. If y_1 is LP, y_2 is SP, $y_3 < 0$, y_4 is SN, and y_5 is LN, then level 2 holds.

8. In all other cases, level 1 holds.

To transform these fuzzy rules into a precise algorithm, we must select a fuzzy "and" operation (t-norm) and a fuzzy "or" operation (t-conorm), e.g., $\min(a, b)$ and $\max(a, b)$, and a defuzzification; in this paper, we use centroid defuzzification (see Chapter 2).

For each rule (except the last one), we can compute the degree of satisfaction for each of the conditions. The rule is applicable if its first condition holds *and* the second condition holds, etc. Therefore, to find the degree with which the rule is applicable, we apply the chosen "and" operation to the degrees with which different conditions of this rule hold.

For each level > 1, we have two rules leading to this level. The corresponding degree of certainty is achieved if either the first *or* the second of these rules is applicable. Therefore, to find a degree to which this level is justified, we must apply the chosen "or" operation to the degrees to which these two rules are applicable.

As a result, we get the degrees $d(l)$ with which we can justify levels $l = 2 \div 5$. Since the last rule (about level 1) says that this rule is applicable when no other rule applies, we can compute $d(1)$ as $1 - d(2) - \cdots - d(5)$. Now, centroid defuzzification leads to the resulting certainty $1 \cdot d(1) + 2 \cdot d(2) + \cdots + 5 \cdot d(5)$. This is the value that the system outputs as the degree of certainty (on a 1 to 5 scale) that there is a fault at a given location.

10.7.5   Experimental results

We have applied the resulting fuzzy model to simple beams with known fault locations. The results are as follows:

When there is *only one fault*, this fault can be determined as the location where the degree of certainty attains its largest value, 5. This criterion leads to a *perfect fault localization*, with no false positives and no false negatives.

When there are *several faults*, all the faults correspond to locations with degree 4 or larger. This criterion is not perfect; it *avoids the most dangerous errors of false negatives* (i.e., all the faults are detected), but it is false positive; i.e., sometimes faults are wrongly indicated in the areas where there are none.

To improve the fuzzy algorithm, we take into consideration that the vibration corresponding to each mode has points in which the amplitude of this vibration is 0. The corresponding locations are not affected by this mode, and therefore the corresponding excess energy values cannot tell anything about the presence or absence of a fault. Therefore, it makes sense to consider only those values x_i for which the corresponding mode energy is at least, say, 10% of its maximum. If we thus restrict the values x_i, then the number of false positives decreases.

Similar results follow for two-dimensional cases.

We tried different t-norms and t-conorms. Thus far, we have not found a statistically significant difference between the results obtained by using different t-norms and t-conorms; therefore, we recommend using the simplest possible operations $\min(a, b)$ and $\max(a, b)$.

10.8   Applications to aircraft reliability

A similar approach works not only for fault detection but also for another important problem, reliability.

10.8.1   Reliability: General problem

A typical system (e.g., an airplane) consists of several heterogeneous components. For a system to function normally, it is important that all these components function well.

For example, for an airplane to function normally, it is important that its structural integrity is intact, that its engines are running normally, that its communication system is functioning, and that the controlling software is performing well.

The reliability of each component is normally analyzed by different engineering disciplines that use slightly different techniques. As a result of this analysis, we get the probabilities f_i of each component's failure (or, equivalently, the probability $p_i = 1 - f_i$ that the ith component functions correctly). To estimate the reliability of the entire system, we must combine the probabilities of each component functioning correctly into a single probability p that the whole system functions correctly.

10.8.2   Traditional approach to reliability

The simplest case typically covered by statistical textbooks is when all n components are independent. In this case, for reliability, the probability p of the system's correct functioning is equal to the product of the *correctness* probabilities (the n components):

$$p = p_1 \cdots p_n.$$

Another case with a known solution is when all failures are caused by one and the same cause (the case of full correlation). In this case, the failure probability is determined by its weakest link, so $f = \max(f_1, \ldots, f_n)$ and $p = \min(p_1, \ldots, p_n)$.

10.8.3   Traditional approach is not always sufficient: A problem

Most real-world situations lie in between these two extremes described above (see, e.g., Petroski (1994) and Ross (1998)), i.e.,

- components are not completely independent (e.g., a structural fault also can damage sensors, and thus computational ability suffers), or

- components are not fully correlated.

Typically, however, we do not have sufficient information about the correlation between the components.

10.8.4   Proposed approach to fusing probabilities: Main idea

We must choose a method for combining probabilities. In mathematical terms, we must describe, for every n, a function that takes n numbers $p_1, \ldots, p_n \in [0, 1]$ as inputs and returns the "fused" (combined) probability $f_n(p_1, \ldots, p_n)$.

Similar to modes of vibration of an aircraft, the division into components is rather subjective; e.g., for three components, we can

- either divide the system into these three subsystems 1, 2, and 3,

- or divide the system into two "macrocomponents" $\{1, 2\}$ and 3,

- or divide it into 1 and $\{2, 3\}$.

Depending on the division, we get different expressions for the resulting probability:

- If we divide the system into three components, the resulting probability is $p = f_3(p_1, p_2, p_3)$.

- If we divide the system into two macrocomponents $\{1, 2\}$ and 3, then for the first macrocomponent we get $p_{12} = f_2(p_1, p_2)$, and thus for a system as a whole, $p = f_2(p_{12}, p_3) = f_2(f_2(p_1, p_2), p_3)$.

- Similarly, if we divide the system into two macrocomponents 1 and $\{2, 3\}$, then for the second macrocomponent we get $p_{23} = f_2(p_2, p_3)$, and thus for a system as a whole, $p = f_2(p_1, p_{23}) = f_2(p_1, f_2(p_2, p_3))$.

The resulting probability should not depend on the (subjective) subdivision into components. As a result, we should get $f_3(p_1, p_2, p_3) = f_2(f_2(p_1, p_2), p_3) = f_2(p_1, f_2(p_2, p_3))$. In other words, we can make two conclusions:

- First, the combination function for arbitrary $n > 2$ can be expressed in terms of a combination function corresponding to $n = 2$, as $f_n(p_1, \ldots, p_n) = f_2(p_1, f_2(p_2, \ldots, f_2(p_{n-1}, p_n) \cdots))$.

- Second, the function f_2 that describes the combination of two probabilities should be *associative* (i.e., $f_2(a, f_2(b, c)) = f_2(f_2(a, b), c)$ for all $a, b, c \in [0, 1]$).

For reliability, the function $g_2(p_1, p_2)$ has the following meanings:

- p_1 is the probability $P(C_1)$ of the event C_1 defined as "the first component is functioning correctly";

- p_2 is the probability $P(C_2)$ of the event C_2 defined as "the second component is functioning correctly"; and

- $f_2(p_1, p_2)$ is the probability $P(C_1 \& C_2)$ that both components are functioning correctly.

In some reliability problems, several components serve as backups for one another; in such situations, the system as a whole functions correctly if at least one of the components functions correctly. In other words, in such problems, we are interested in the value $P(C_1 \vee C_2)$. From probability theory, we know that $P(C_1 \vee C_2) = P(C_1) + P(C_2) - P(C_1 \& C_2)$; i.e., $P(C_1 \vee C_2) = p_1 + p_2 - f_2(p_1, p_2)$. We can describe this expression as $P(C_1 \vee C_2) = g_2(p_1, p_2)$, where $g_2(a, b) = a + b - f_2(a, b)$.

Similar to the case of the "and" combination of different components, we can describe the probability $P(C_1 \vee C_2 \vee C_3)$ (that at least one of three components functions correctly) in two different ways:

- either as $P((C_1 \vee C_2) \vee C_3) = g_2(g_2(p_1, p_2), p_3)$,

- or as $P(C_1 \vee (C_2 \vee C_3)) = g_2(g_2(p_1, p_2), p_3)$.

The expression for $P(C_1 \vee C_2 \vee C_3)$ should not depend on how we compute it, and therefore we should have $g_2(g_2(p_1, p_2), p_3) = g_2(p_1, g_2(p_2, p_3))$. In other words, not only the function $f_2(a, b)$ but also the function $g_2(a, b) = a + b - f_2(a, b)$ should be associative.

10.8.5   Resulting solution

We already know that functions $f_2(a, b)$ for which both the function itself and the expression $a + b - f(a, b)$ are associative have been classified and are known as Frank's t-norms:

$$f_2(a, b) = \log_s \left[1 + \frac{(s^a - 1) \cdot (s^b - 1)}{s - 1} \right]$$

for some constant s. As a particular case of this general formula, we get the above two expressions $a \cdot b$ (for $s \to 1$) and $\min(a, b)$ (for $s \to 0$).

10.9   Closing thoughts

We have presented a technique to extend standard reliability theory to accommodate nonnumeric forms of information, such as expert judgment. Such an extension would be useful in other fields as well. For example, the main problem of mammography is to detect small faults (small clots, cracks, etc.) in the mammary gland, which may indicate a tumor. This problem is very similar to the problem of aerospace testing: in both cases, we must detect possible faults.

Appendix: Proofs

Proof of Theorem 10.1

This proof is based on the following lemma.

Lemma 10.1. *If an optimality criterion is final and unit-invariant, then the optimal family F_{opt} is also unit-invariant, i.e., $R_\lambda(F_{opt}) = F_{opt}$ for every number λ.*

Proof. Since the optimality criterion is final, there exists a unique family F_{opt} that is optimal with respect to this criterion; i.e., for every other F, either $F_{opt} \succ F$ or $F_{opt} \sim F$.

To prove that $F_{opt} = R_\lambda(F_{opt})$, we will first show that the rescaled family $R_\lambda(F_{opt})$ is also optimal, i.e., that for every family F, either $R_\lambda(F_{opt}) \succ F$ or $R_\lambda(F_{opt}) \sim F$.

If we prove this optimality, then the desired equality will follow from the fact that our optimality criterion is final, and thus there is only one optimal family (and since the families F_{opt} and $R_\lambda(F_{opt})$ are both optimal, they must be the same family).

Let us show that $R_\lambda(F_{opt})$ is indeed optimal. How can we, e.g., prove that $R_\lambda(F_{opt}) \succ F$? Since the optimality criterion is unit-invariant, the desired relation is equivalent to $F_{opt} \succ R_{\lambda^{-1}}(F)$. Similarly, the relation $R_\lambda(F_{opt}) \sim F$ is equivalent to $F_{opt} \sim R_{\lambda^{-1}}(F)$.

These two equivalences allow us to complete the proof of the lemma. Indeed, since F_{opt} is optimal, we have one of two possibilities: either $F_{opt} \succ R_{\lambda^{-1}}(F)$ or $F_{opt} \sim R_{\lambda^{-1}}(F)$. In the first case, we have $R_\lambda(F_{opt}) \succ F$; in the second, we have $R_\lambda(F_{opt}) \sim F$.

Thus whatever family F we take, we always have either $R_\lambda(F_{opt}) \succ F$ or $R_\lambda(F_{opt}) \sim F$. Hence $R_\lambda(F_{opt})$ is indeed optimal, and thence $R_\lambda(F_{opt}) = F_{opt}$. The lemma is proved. $\Box$

Let us now prove the theorem. Since the criterion is final, there exists an optimal family $F_{opt} = \{C \cdot f(z)\}$. Due to the lemma, the optimal family is unit-invariant.

From unit-invariance, it follows that for every λ there exists a real number $A(\lambda)$ for which $f(\lambda \cdot z) = A(\lambda) \cdot f(z)$. Since the function $f(z)$ is differentiable, we can conclude that the ratio $A(\lambda) = \frac{f(\lambda \cdot z)}{f(z)}$ is differentiable as well. Thus we can differentiate both sides of the above equation with respect to λ, and substitute $\lambda = 1$. As a result, we get the following differential equation for the unknown function $f(z)$:

$$z \cdot \frac{df}{dz} = \alpha \cdot f,$$

where by α we denoted the value of the derivative $\frac{dA}{d\lambda}$ taken at $\lambda = 1$. Moving terms dz and z to the right-hand side and all the term containing f to the left-hand side, we conclude that

$$\frac{df}{f} = \alpha \cdot \frac{dz}{z}.$$

Integrating both sides of this equation, we conclude that $\ln(f) = \alpha \cdot \ln(z) + C$ for some constant C and therefore that $f(z) = \text{const} \cdot z^\alpha$. The theorem is proved. $\Box$

Proof of Theorem 10.2

Since the optimality criterion is final, there exists an optimal family F_{opt}. Similarly to the proof of Theorem 10.1, we prove that this optimal family is unit-invariant; i.e., $R_\lambda(F_{opt}) = F_{opt}$ for all real numbers $\lambda > 0$.

Therefore, if a function $p(E)$ belongs to the optimal family F_{opt}, then for every $\lambda > 0$, the rescaled function $p(\lambda \cdot E)$ of multiplying E to this function f belongs to F_{opt}; i.e., due to the definition of a family, there exist values $k(\lambda)$, etc. for which

$$p(\lambda \cdot E) = \frac{k(\lambda) \cdot p(E) + l(\lambda)}{m(\lambda) \cdot p(E) + n(\lambda)}. \qquad (10.3)$$

The solution to this functional equation is, in essence, described in Aczel (1966). For completeness, let us describe the proof in detail.

For $\lambda = 1$, we have $k = n = 1$ and $l = m = 0$, so since p is smooth (hence continuous), for $\lambda \approx 1$, we have $n(\lambda) \neq 0$; hence we can divide both the numerator and the denominator of (10.3) by $n(\lambda)$ and thus get a similar formula with $n(\lambda) = 1$. If we multiply both sides of the resulting equation by the denominator, we get the following formula:

$$m(\lambda) \cdot p(E) \cdot p(\lambda \cdot E) + p(E) = k(\lambda) \cdot p(E) + l(\lambda).$$

If we fix λ and take three different values of E, we get three linear equations for determining three unknowns $k(\lambda)$, $l(\lambda)$, and $m(\lambda)$, from which we can determine these unknowns using Cramer's rule. Cramer's rule expresses every unknown as a fraction of two determinants, and these determinants polynomially depend on the coefficients. The coefficients either do not depend on λ at all (like $p(E)$) or depend smoothly ($p(\lambda \cdot E)$ smoothly depends on λ because $p(E)$ is a smooth function). Therefore, these polynomials also are smooth functions of λ, as are their ratios $k(\lambda)$, $l(\lambda)$, and $m(\lambda)$.

Now that we know that all the functions in (10.3) are differentiable, we can differentiate both sides with respect to λ and set $\lambda = 1$. As a result, we get the following differential equation:

$$E \cdot \frac{dp}{dE} = C_0 + C_1 \cdot p + C_2 \cdot p^2$$

for some constants C_i. To solve this equation, we can separate the variables, i.e., move all the terms related to E to one side and all the terms related to p to the other side, and get the differential equation

$$\frac{dp}{C_0 + C_1 \cdot p + C_2 \cdot p^2} = \frac{dE}{E}. \tag{10.4}$$

Let us first show that $C_2 \neq 0$. Indeed, if $C_2 = 0$ and $C_1 = 0$, then $\frac{p}{C_0} = \ln(E) + \text{const}$, which contradicts our assumption that $p(0) = 0$. If $C_2 = 0$ and $C_1 \neq 0$, then we get $C_1^{-1} \cdot \ln(C_1 \cdot p + C_0) = \ln(E) + \text{const}$; hence $C_1 \cdot p + C_0 = A \cdot E^\alpha$, which for $\alpha < 0$ contradicts the assumption that $p(0) = 0$ and for $\alpha > 0$ contradicts the assumption that $p(E) \to 1$ as $E \to \infty$. Thus the case $C_2 = 0$ is impossible, and $C_2 \neq 0$. For $C_2 \neq 0$, in general, the left-hand side of (10.4) can be represented as a linear combination of elementary fractions $(p + z_1)^{-1}$ and $(p + z_2)^{-1}$ (where z_i are—possibly complex—roots of a quadratic polynomial $C_1 + C_1 \cdot p + C_2 \cdot p^2$):

$$\frac{1}{C_0 + C_1 \cdot p + C_2 \cdot p^2} = c \cdot \left(\frac{1}{p + z_1} - \frac{1}{p + z_2} \right).$$

(The case of a double root can be handled in a similar manner.) Thus integrating (10.4), we conclude that

$$c \ln \left(\frac{p + z_1}{p + z_2} \right) = \ln(E) + \text{const}$$

and

$$\frac{p + z_1}{p + z_2} = P \cdot E^\beta$$

for some A and β. Therefore, the expression $A \cdot E^\beta$ can be obtained from $p(E)$ by a fractional linear transformation; hence by applying the inverse transformation (and it is known that the inverse to a fractionally linear transformation is also fractionally linear), we conclude that

$$p(E) = \frac{A \cdot E^\beta + B}{C \cdot E^\beta + D}$$

for some numbers A, B, C, and D. One can easily check that we get a monotonic everywhere-defined function $p(E)$ only for real values A–D and β.

If $\beta < 0$, then we can multiply both numerator and denominator by $E^{-\beta}$ and get a similar expression with $\beta > 0$. Thus without losing generality, we can assume that $\beta > 0$. Now, the condition $p(0) = 0$ leads to $\frac{B}{D}$ and hence to $B = 0$. The condition also leads to $A = C$, i.e., to

$$p(E) = \frac{A \cdot E^\beta}{C \cdot E^\beta + D}.$$

Since $p(E)$ is not identically equal to 1, we have $D \neq 0$. Therefore, we can divide both the numerator and the denominator by D and get the desired expression (10.2). The theorem is proved. $\square$

References

J. ACZEL (1966), *Lectures on Functional Equations and Their Applications*, Academic Press, New York, London.

G. ANDRE (1999), *Comparison of Vibrational Damage Detection Methods in an Aerospace Vertical Stabilizer Structure*, Master's thesis, Civil Engineering Department, University of Texas at El Paso, El Paso, TX.

P. BARBIER AND P. BLONDET (1993), *Using NDT Techniques in the Maintenance of Aeronautical Products*, Report 93-11587/1/GAL, Aerospatiale, Paris.

D. E. CHIMENTI (1997), Guided waves in plates and their use in materials characterization, *Appl. Mech. Rev.*, 50, pp. 247–287.

R. W. CLOUGH AND J. PENZIEN (1986), *Dynamics of Structures*, McGraw–Hill, New York.

C. FERREGUT, R. A. OSEGUEDA, AND A. NUNEZ, EDS. (1997), *Proceedings of the International Workshop on Intelligent NDE Sciences for Aging and Futuristic Aircraft*, El Paso, TX.

M. J. FRANK (1979), On the simultaneous associativity of $F(x, y)$ and $x + y - F(x, y)$, *Aequationes Math.*, 19, pp. 194–226.

I. GRABEC AND W. SACHSE (1997), *Synergetics of Measurement, Prediction, and Control*, Springer-Verlag, Berlin, Heidelberg.

X. E. GROS (1997), *NDT Data Fusion*, Wiley, London.

P. W. HOVEY AND A. P. BERENS (1988), Statistical evaluation of NDE reliability in the aerospace industry, in *Review of Progress in QNDE*, Plenum, New York, pp. 1761–1768.

G. KLIR AND B. YUAN (1995), *Fuzzy Sets and Fuzzy Logic: Theory and Applications*, Prentice–Hall, Upper Saddle River, NJ.

O. KOSHELEVA, L. LONGPRE, AND R. A. OSEGUEDA (1999), Detecting known non-smooth structures in images: Fuzzy and probabilistic methods, with applications to medical imaging, non-destructive testing, and detecting text on web pages, in *Proceedings of the 8th International Fuzzy Systems Association World Congress (IFSA'99)*, Taipei, Taiwan, pp. 269–273.

V. Kreinovich, H. T. Nguyen, and E. A. Walker (1996), Maximum entropy (MaxEnt) method in expert systems and intelligent control: New possibilities and limitations, in *Maximum Entropy and Bayesian Methods*, K. M. Hanson and R. N. Silver, eds., Kluwer, Dordrecht, The Netherlands, pp. 93–100.

V. Kreinovich, C. Quintana, and O. Fuentes (1993), Genetic algorithms: What fitness scaling is optimal?, *Cybernet. Systems*, 24, pp. 9–26.

M. Krishna, V. Kreinovich, and R. A. Osegueda (1999), Fuzzy logic in non-destructive testing of aerospace structures, in *Proceedings of the 1999 IEEE Midwest Symposium on Circuits and Systems*, IEEE, Piscataway, NJ.

A. K. Mal and S. J. Singh (1991), *Deformation of Elastic Bodies*, Prentice–Hall, Englewood Cliffs, NJ.

H. T. Nguyen and V. Kreinovich (1997), *Applications of Continuous Mathematics to Computer Science*, Kluwer, Dordrecht, The Netherlands.

H. T. Nguyen and E. A. Walker (1999), *A First Course in Fuzzy Logic*, CRC Press, Boca Raton, FL.

R. A. Osegueda, Y. Mendoza, O. Kosheleva, and V. Kreinovich (1999a), Multiresolution methods in non-destructive testing of aerospace structures and in medicine, in *Proceedings of the 14th IEEE International Symposium on Intelligent Control/Intelligent Systems and Semiotics* (*ISIC/ISAS'99*), IEEE, Piscataway, NJ, pp. 208–212.

R. A. Osegueda, A. Revilla, L. Pereyra, and O. Moguel (1999b), Fusion of modal strain energy differences for localization of damage, in *Nondestructive Evaluation of Aging Aircraft, Airports, and Aerospace Hardware* III: *Proceedings of SPIE*, A. K. Mal, ed., Paper 3586-28, SPIE, Bellingham, WA.

L. R. Pereyra, R. A. Osegueda, C. Carrasco, and C. Ferregut (1999), Damage detection in a stiffened plate using modal strain energy differences, in *Nondestructive Evaluation of Aging Aircraft, Airports, and Aerospace Hardware* III: *Proceedings of SPIE*, A. K. Mal, ed., Paper 3586-29, SPIE, Bellingham, WA.

H. Petroski (1994), *Design Paradigms: Case Histories of Error and Judgment in Engineering*, Cambridge University Press, Cambridge, UK.

T. J. Ross (1998), Case studies in civil engineering: Fuzzy logic applications, in *Current Trends and Developments in Fuzzy Logic: Proceedings of the 1st International Workshop*, B. Papadopoulos and A. Syropoulos, eds., Thessaloniki, Greece, pp. 211–236.

T. J. Ross, C. Ferregut, R. A. Osegueda, and V. Kreinovich (1999), System reliability: A case when fuzzy logic enhances probability theory's ability to deal with real-world problems, in *Proceedings of the 18th International Conference of the North American Fuzzy Information Society* (*NAFIPS'99*), IEEE, Piscataway, NJ, pp. 81–84.

N. S. Stubbs, T. Broom, and R. A. Osegueda (1998), Non-destructive construction error detection in large space structures, in *AIAA ADM Issues of the International Space Station*, AIAA, Williamsburg, VA, pp. 47–55.

T. Terano, K. Asai, and M. Sugeno (1987), *Fuzzy Systems Theory and Its Applications*, Academic Press, San Diego, CA.

M. ULIERU AND R. ISERMANN (1993), Design of a fuzzy-logic based diagnostic model for technical processes, *Fuzzy Sets Systems*, 52, pp. 249–272.

I. A. VIKTOROV (1967), *Rayleigh and Lamb Waves: Physical Theory and Applications*, Plenum, New York.

Y. YAM, R. A. OSEGUEDA, AND V. KREINOVICH (1999), Towards faster, smoother, and more compact fuzzy approximation, with an application to non-destructive evaluation of Space Shuttle's structural integrity, in *Proceedings of the 18th International Conference of the North American Fuzzy Information Society* (*NAFIPS'99*), IEEE, Piscataway, NJ, pp. 243–247.

L. A. ZADEH (2002), Towards a perception-based theory of probabilistic reasoning with imprecise probabilities, *J. Statist. Planning Inference*, 105, pp. 233–264.

Chapter 11

Auto Reliability Project

Jane M. Booker and Thomas R. Bement

Abstract.   Reliability is defined as the probability that an item (system, subsystem, or component) performs according to specifications. When test data are sparse (such as with a new concept design), expert judgment can be used to estimate reliability. Using expert judgment information in conjunction with data and historical information brings into focus ways of handling various kinds of uncertainties. In this chapter, we examine the use of a traditional probabilistic approach, and also the use of fuzzy membership functions for reliability estimation under uncertainty.

11.1   Description of the reliability problem

Over the years, many advancing techniques have surfaced for estimating the reliability[1] or performance of a system. These techniques rely on extensive test data from prototypes or produced systems. In the traditional military context, a product would be developed, or a similar product that already existed would be modified, and then the product would be tested. The typical long-term test was designed to statistically demonstrate a reliability requirement at a specified confidence.

However, many practical considerations have arisen such as whether the testing involves additional time, cost, and resources. There also is the possibility that the end of development will overlap with the start of production. Problems with the system encountered from testing on a tight production schedule can cause a delay in production and/or result in insufficient time to fix problems—the latter being how faulty products get into customers' hands, resulting in recalls and lack of customer confidence.

There is a definite need to understand the reliability of a new product during its development program, even during its initial concept phase. Such a need can be met by estimating reliability through combining all available information at any lifetime phase. Information sources can include the traditional test data but also may include what is known about similar

---

[1]Reliability is defined as the probability that the item (system, subsystem, or component) performs according to specifications. *Performance* is a more general term than reliability that could be measured in quantities such as miles/gallon or failures per unit time.

systems, what can be learned from simulations or computer models of physical phenomenon, and what can be elicited from the experts, whose knowledge and expertise are a valuable resource. Combining all this information produces an estimate for the performance or reliability of a system. This estimate can be revised and updated as new information becomes available, providing tracking of the system throughout its lifetime. These ideas form a methodology called PREDICT[2] (Meyer et al. (1999)) and are the basis of the examples and discussion here.

The reliability of the system or product includes the product design and the manufacturing process. The reliability at any given point in time or at any given step in the overall development and lifetime of the product is hereby referred to as *reliability characterization*. Reliability characterization refers to both the functional calculation of the reliability (point estimate value) and the uncertainty (usually represented by a distribution function) that accompanies that reliability value.

Reliability characterizations are calculated from formulas or models for each component, process, failure mode, etc. and then propagated through the many parts and processes which comprise the system. The way the characterizations are combined or propagated through the system depends on how the individual parts and processes are interrelated in the system. To specify these relationships, it is helpful to depict the system with a logical structure, such as a reliability block diagram. Figure 11.1 is a block diagram for a simple series of 10 higher-level parts called subsystems—each subsystem is composed of a number of underlying components or processes and corresponding failure modes for each. Diagrams such as these are specified for all levels of the system, and reliabilities can be calculated for individual items (boxes) or the different levels (component, assembly, subsystem, and system). Nuclear reactor systems can have thousands of basic components, resulting in dozens of levels. Automotive systems typically contain 50–80 components, each having 1–9 failure modes/mechanisms, and 10–12 processes.

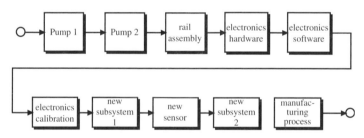

Figure 11.1. *Reliability block diagram for subsystem level.*

A mathematical equation based on Boolean algebra for combining the 10 subsystems in Figure 11.1 is the product of the individual subsystem reliabilities. Because reliability characterization involves uncertainties attached to reliability estimates, the reliability characterization for each subsystem in the figure is a distribution function (such as a probability distribution function (PDF)). These distributions are then multiplied together using Monte Carlo simulation to form the total system reliability. In turn, each subsystem depicted is a product of all the lower level items beneath it, assuming these also are connected in series.[3]

During the various time steps of the product's development and lifetime, new information (from design or manufacturing changes, vendor corrective actions, prototype tests,

---

[2]PREDICT stands for performance and reliability evaluation with diverse information combination and tracking.
[3]Boolean expressions for items in parallel (redundant items) or dependent items are more complex (Henley and Kumomoto (1981)).

production and usage) becomes available. The reliability, R_i, is calculated at each time step, i, along with a new uncertainty distribution, $f(R_i)$, and monitored according to Figure 11.2 relative to the *reliability requirement* or target. For example, a target might be that the system must have 98% or greater reliability. In the automotive industry, such a requirement may be dictated by emissions requirements set by government standards, or it may be based on customer need in a competitive environment. Certain requirements may change throughout the system's lifetime. For example, a requirement for a product to enter the production phase may be stricter than one for entering the prototyping stage. Therefore, monitoring reliability characterization is a decision tool for deciding which new designs are viable for future production and use.

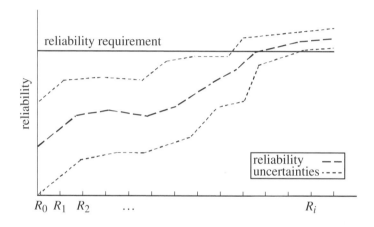

Figure 11.2. *Reliability growth plot.*

Calculating the value of R_i and $f(R_i)$ (the uncertainty distribution for R_i) requires combining information from various sources, e.g., test data, product data, engineering judgment, specifications/requirements, data from similar components, and data at different levels (component, system, etc.). Reevaluating R_i and $f(R_i)$ in light of new information requires methods for incorporating new information with existing information, e.g., adding new test data or accommodating design changes. Weighting schemes, rules, expert judgment, and Bayes' theorem are all useful for this purpose.

With each evaluation of R_i and $f(R_i)$, gaps in the current state of knowledge become apparent, providing the basis for a strategy for deciding where to devote future testing and analysis resources. If a gap from poor information results in a large uncertainty in one important area (e.g., unknown performance of a component), then time and/or effort should be devoted to understanding why and where additional information can be useful in improving the reliability and/or reducing the uncertainty. Understanding the effects of changes in the design, uncertainties, and test results can be gained from recalculating the reliability characterization without actually making such changes. These so-called "what if" cases provide decision makers with a mechanism for experimenting with anticipated or proposed changes or tests. For example, decision makers may ask "What happens to R_i and $f(R_i)$ if component A is replaced by component B or if 10 successful new tests are performed on subsystem C?" "What if we changed vendors/supplies of component A?" "What if we replaced a plastic component with metal?" Test planning and resource allocation can then be done with greater chances of success.

11.2   Implementing the probability approach

11.2.1   Logic and reliability models

As noted in section 11.1, the first step in a reliability or performance analysis includes defining what is meant by performance or reliability for a system and establishing a structure which specifies how all the parts of the system interrelate. With a probability approach, that definition includes the construction of a probability model for estimating the reliability or performance. This, in turn, requires the formation of a logic model, which specifies how all the pieces of the system interrelate. If all the components, subsystems, etc. are in series (meaning that all subsystems must function for the entire system to function), then the model of such a system is simply the product of the reliabilities of the parts/processes. Reliability for the system at time i is

$$R_i(\text{system}) = \prod_{j=1}^{k} R_i(x_j) \quad \text{for } j = 1, k \text{ parts/processes } P. \tag{11.1}$$

The next step involves specifying the way reliability is calculated. In many manufacturing applications, this involves the concept of the hazard function which defines portions of the product's lifetime such as infant mortality, useful life, and wearout (Kerscher (1989)). "Infant mortality" is mainly due to the latent defects generated during the manufacturing process. The "useful life" portion is primarily due to latent design defects that manifest themselves over the life of the product. "Wearout" of parts is due to failure modes associated with operating the product beyond its useful life. Figure 11.3 illustrates these life phases in the so-called *bathtub* curve. Here the hazard function is plotted as a function of time. Good engineering practice has long held that wearout failure modes should be identified during the development process and that those failure modes that cannot be removed from the design should at least be designed to occur beyond the useful life of the product. Wearout failure modes are assumed to occur beyond useful life and therefore are not considered.

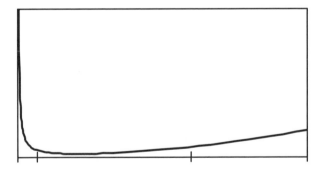

Figure 11.3. *Hazard function versus time indicating the infant mortality phase (rapidly decreasing), useful life (flat), and wearout (gradually increasing).*

The reliability model must be physically appropriate and mathematically correct for the system. Of equal importance, the model and its usage must be culturally acceptable to the organization using it. Neither of these requirements is a small challenge. With regard to the first requirement, the two-parameter Weibull probability density function may be used to model the infant mortality and useful lifetime (Kerscher (1989)). With regard to the second requirement, this model of the reliability perspective is well suited to the implicit understanding of the design and manufacturing community, given their awareness of the "bathtub" curve, and therefore may be culturally acceptable.

The two-parameter Weibull expression for reliability as a function of time, t, is

$$R(t) = \exp(-(\lambda t)^\beta). \tag{11.2}$$

The next challenge is to specify what the two parameters are for β (the slope) and λ (the failure rate) for each component/process. Because β influences how the failure rate changes with time (degrades for $\beta < 0$, stays constant for $\beta = 0$, and improves for $\beta > 0$), potential estimates for β can be found from the previous performance of similar products, expert judgment, or some combination of information relating to this phenomenon. The initial estimate of the failure rate of the distribution may be made from whatever information is available relative to previous designs.

Typically, at the outset of a new product development program actual test data are very limited due to the absence of hardware. Considerable information about the potential reliability is available, however, in the form of expert judgment possessed by members of the product development teams. For instance, these experts may be accustomed to thinking about the reliability of the product at a particular time of the product's life, such as 12 months. Eliciting these judgments, the subpopulation due to latent design defects may be modeled, and the two parameters of the Weibull model may be estimated.

A similar approach may be taken with respect to the latent defects generated by the manufacturing process. Again, a two-parameter Weibull distribution may be used as a model. The physical situation here, however, can be different enough to argue in favor of an alternative approach to estimating the parameters. When mistakes are made in the manufacturing process, they certainly can occur in matters of degree (e.g., varying solder bath temperature). More typically, however, mistakes tend to be very significant, such as putting parts together upside down or leaving parts out of an assembly. These latent manufacturing defects tend to manifest themselves relatively quickly, if not immediately. It may therefore be suggested that because the bulk of these defects will emerge in a definable time period, an estimate could be made of the fraction of the defects that would manifest themselves almost immediately, or at the beginning of the product's useful life. Specifically, if the fraction of defects could be estimated, as well as the endpoint at which the whole subpopulation would manifest itself, in addition to the fraction of the subpopulation that would manifest itself almost immediately, then sufficient information would exist to estimate the two parameters directly. Simply put, this would reduce the determination of the two Weibull parameters to understanding the reliability at two different points in time, specifically at or near time 0 and at the time at which all the elements failed. This approach to estimating the reliability performance of manufactured products fits well within the culture of automotive manufacturing facilities because of the nature of their business and manufacturing operations.

Having identified how the parameters can be specified, individual Weibull distributions can be identified for the latent design and manufacturing defects. They are then combined to produce the distribution representative of the whole component or subsystem and then ultimately combined to form the distribution representative of the entire system. Estimates of reliability are then calculated at key points in time for predicting long-term performance such as for warranty periods and government-mandated requirements.

11.2.2   Expert elicitation

To obtain an initial overall reliability estimate, R_0, of the entire system in the absence of test data, estimates of component and subsystem reliabilities (with uncertainties) are elicited from teams of subject matter experts. The experts are selected by their managers and peers as being knowledgeable of specific subsystems, processes, or components. Separate elicitations can be conducted for teams working on product design and those working on the manufacturing process.

It is recommended that the entire elicitation procedure be documented in a *script*, which is made available to the experts. This can be conveniently done as an official memo describing the goals of the elicitation, specifying the purpose of the product development effort, providing motivation for experts to participate, estimating the time and effort required for the experts, and supplying contact personnel information.

It may also be possible to design elicitation *worksheets*—documents containing instructions for the questionnaire use, descriptions of the parts/processes, and questions about the parts/processes performance in terms of the specified performance measures. To account for uncertainties, elicited estimates should always be accompanied by uncertainty ranges of values. These could be in the form of asking for three cases: the most likely, the best, and the worst. Assumptions, conditions, and requirements related to performance should also be asked of the experts to obtain proper estimates and uncertainties and to provide documentation for updating and tracking. Separate worksheets should be constructed for different definitions of reliability/performance. For example, if performance is defined in terms of failure rates for the product design, but in terms of successes per total units for processes, then slightly different questions may be necessary, resulting in different worksheet versions.

The worksheets (so-called *interview instruments*) must be pilot tested on an individual expert before the expert teams meet. Proper wording of the elicitation questions is crucial to avoid confusion and misinterpretation by the teams of experts later on.

It may be necessary to schedule more than one meeting for each team of experts. For example, the first meeting may be an introduction to the elicitation process. The roles of the facilitator, team leaders, and experts are defined in the script developed earlier. Experts can be handed the worksheets and other supporting material for background. Team leaders can review the logic model diagram of the system structure with the experts for any adjustments at this time. A second meeting would be the actual elicitation for the initial reliability prediction. The leader of each team then completes the questions on the worksheet, indicating the consensus of the team. Any final adjustments to the logic model should be made at this meeting.

As part of the elicitation, and indicated on the worksheet, experts are asked to specify all known or potential failure mechanisms, or failure modes, for each component or subsystem. Failure modes are failures in the components themselves, such as a valve wearing out, mistakes made during the manufacture of components, or assembly of multiple components into a subsystem. For updating and documenting purposes, the percent or proportion contribution of each failure mode also should be specified by the experts.

For processes, it may be necessary to specify the percent or proportion of items that slip through the quality control procedures (called *spills*). Other estimated quantities are elicited that are necessary for characterizing spills (e.g., frequency and duration of the spills). Also, the experts are asked how these spills relate to the specified failure modes. As noted earlier, this information can be used to estimate the appropriate parameters of the Weibull distributions.

The experts do not have to be asked to estimate reliabilities, per se, but should provide their estimates about component, subsystem, and system performance in terms familiar to them. (This approach and its benefits are described in further detail in Meyer and Booker (2001).) For example, the experts in the design process might think in terms of *incidence per thousand vehicles* (IPTV), while those in the manufacturing process might use *parts per million* (PPM). As part of their estimates, the experts are asked to give a very brief explanation of their reasoning. In addition, the experts provide ranges on their estimates, which were used to represent the uncertainty and to ultimately formulate $f(R_0)$.

The results from the elicitations are presented to all the participating experts for their review and reconciliation across the entire system. The results include reliabilities with uncertainties for components, subsystems, and the system at various times (e.g., 12 months, 36 months, 100,000 miles). The design and process information from the elicitations is combined for the calculation of R_0. From the resulting uncertainty distribution, $f(R_0)$, measures can be extracted that are meaningful to the experts, such as the median reliability of the system at 12 with a 90% lower uncertainty limit.

Subsequent information, including new test data, is reflected in subsequent values of R_i and $f(R_i)$ as described next. In this way reliability may be monitored over time (reliability growth) and plans formulated accordingly.

11.2.3   Updating methods

As discussed previously, the system under study needs to be defined and characterized with a logic model, such as a reliability block diagram that, in turn, determines how reliabilities are calculated at various levels. New information becomes available at various stages in the system lifetime. This information could be in the form of design changes, supplier changes, process changes, prototype test data, production data, engineering judgment, etc.

Pooling information/data from different sources or of different types (e.g., tests, process capability studies, engineering judgment) is usually done with methods that combine the uncertainty distribution functions associated with the various information sources. Bayes' theorem offers one mechanism for such a combination. As seen in Chapter 4, Bayesian pooling combines information with the following structure: The existing information (data) forms the likelihood distribution. Parameter(s) of the likelihood distribution is (are) not considered a fixed quantity (quantities) but instead, has (have) probability distribution(s), which form(s) the prior. The prior is combined with the likelihood using Bayes' theorem to form the posterior distribution. Bayesian combination is often referred to as an updating process, where new information is combined with existing information. As noted in Chapter 4, if the prior and likelihood distributions overlap (i.e., reinforce each other), then Bayesian combination will produce a posterior whose variance is narrower than if the two were combined via other methods, such as a linear combination. This is an advantage of using Bayes' theorem.

Simulation methods often are used to combine or propagate uncertainties (represented as distribution functions) through the logic model and the corresponding reliability model. Unless conjugate priors (see Chapter 4) are used, simulation is needed to calculate the posterior in the Bayesian combination or any other weighting scheme. Monte Carlo simulation is used to propagate reliability characterizations through the various levels of the diagram, with the accuracy being dependent on the number of simulations (e.g., 10,000) used to sample from the distributions.

It is not necessary to develop prior information for subsystems above the most basic level, which usually is the component level. Priors for higher levels are formed by combining the reliability characterizations from the lower levels. However, if there is prior information on these higher levels (e.g., subsystems), the reliability characterization from that information can be combined with the distribution formed from lower levels using methods in Martz and Almond (1997), Martz and Waller (1990), and Martz, Waller, and Fickas (1988). More important, test data and other new information can be added to the existing reliability characterization at any level and/or block (e.g., system, subsystem, or component).

This data/information may be applicable to the entire block or to only a single failure mode within the block. When new data or information becomes available at a higher level (e.g., subsystem) for a reliability characterization at step i, it is necessary to *back propagate* the effects of this new information to the lower levels (e.g., component). The reason for this is because at some future step, $i + j$, updating may be required at the lower level and the effect propagated up the diagram. The statistical issues involved with this back propagation are difficult (Martz and Almond (1997)). A process for back propagating the effects of new information/data from higher levels to lower levels is presented in Martz, Waller, and Fickas (1988) for series systems and in Martz and Waller (1990) for series/parallel systems.

In general, the initial reliability characterization R_0 is developed from expert judgment and is referred to as the native prior distribution for the system. Data/information may be developed regarding each element (e.g., system, subsystem, component), and this would be used to form likelihood distributions for Bayesian updating. All of the distribution information in the items at the various levels must be combined through the logic flow diagram to produce a final estimate of the reliability and its uncertainty at the top, or system, level. Three different combination methods are used:

1. For each prior distribution that needs combining with a data-based or likelihood distribution, Bayes' theorem is used and a posterior distribution results.

2. Posterior distributions within a given level are combined according to the logic model to form the induced prior distribution of the next higher level.

3. Induced prior and native prior distributions at the higher levels are combined within the same item using a method in Martz, Waller, and Fickas (1988) to form the combined prior (for that item), which is then merged with the data (for that item) via method 1. This approach is continued up through the diagram until a posterior distribution is developed at the system level.

As more data become available and incorporated into the reliability characterization through the Bayesian updating process, this data will tend to dominate the effects of the initial estimates developed through expert judgment. In other words, R_i formulated from many test results will look less and less like R_0 from expert estimates.

After the experts review the results of each R_i, they determine the ensuing courses of action for system development. These included decisions about which parts or processes to change, what tests to plan, what prototypes to build, what vendors to use, etc. Before any (expensive) actions are taken (e.g., building prototypes), "what if" cases are calculated to predict the effects on estimated reliability of such proposed changes or tests. The experts can run several "what if" cases, guiding their decisions on design changes, prototypes, and planning for tests. These cases can involve changes in the logic model, changes in expert estimates, effects of proposed test data results, or changes in the terms of the reliability equations. Further breakdown of components into the failure modes affecting them may be required to properly map design changes and proposed test data into the logic model.

11.2.4 Calculating reliabilities

While the methodology does not require $f(R_i)$ to conform to any particular distributional form or family, a useful approximation that sometimes may be helpful for planning purposes can be organized around the beta and binomial distributions:

$$\text{beta}(a, b) = \frac{\Gamma(a + b)}{\Gamma(a)\Gamma(b)} x^{(a-1)} (1 - x)^{(b-1)}, \tag{11.3}$$

$$\text{binomial}(n, p) = \frac{n!}{x!(n-p)!} p^x (1-p)^{n-x}. \qquad (11.4)$$

As illustrated in Chapter 4, the beta distribution is the conjugate prior distribution for the binomial parameter p (Martz and Waller (1990)) and can in some cases be used to approximate the empirical distribution (resulting from the simulation) of the R_i. The beta is often well suited for representing possible values for p because it ranges between 0 and 1, and it is an extremely flexible distribution with many possible shapes (e.g., symmetric, asymmetric, unimodal, uniform, U-shaped, or J-shaped). Its usefulness derives from the fact that the two parameters a and b of the beta are sometimes referred to as the *pseudosuccesses* and *pseudofailures*. This calls to mind the image of a *pseudotest*, where $a + b$ equals the number of pseudotests.

A useful planning application involves situations where new test data is—or will be—of the form of x number of successes out of n number of trials. Such data is binomially distributed. In a Bayesian reliability problem, if a beta distribution with parameters a and b is considered to be the prior distribution for R_0, then the posterior distribution for R_1 also will be a beta, with parameters $a + x$ and $b + n - x$. Thus using the beta formulation may be useful in characterizing the possible value of additional tests. Because both the posterior distribution and the prior distribution are of the beta family, this process could be iterated indefinitely.

For example, the beta distribution shown in Figure 11.4 was fit to a component reliability distribution, R_0, derived solely from the estimates of the subject matter experts. In this case, a beta approximation yielded $a = 81.9$ pseudosuccesses and $b = 1.01$ pseudofailures. New information, in the form of a test of 40 of these components resulting in zero failures, was introduced and a new component reliability distribution for R_1 was determined. An approximate beta fit to this new distribution is shown in Figure 11.5. The beta parameters of this new distribution are $a = 121.9$ and $b = 1.01$. A good fit to the resulting distribution is

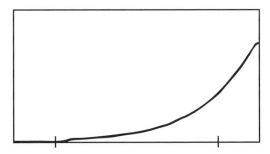

Figure 11.4. *Beta distribution, $a = 81.9$, $b = 1.01$.*

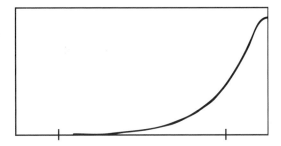

Figure 11.5. *Beta distribution, $a = 121.9$, $b = 1.01$.*

therefore made by simply adding the n and x test results of the binomial data distribution to the a and b parameters of the beta prior distribution.

It should be noted that often a beta distribution cannot do an acceptable job of approximating a distribution of R_1 and remain consistent with the intentions of the experts who provided the input. This is especially true when experts state a very high level of reliability with a small but significant chance that the reliability may be quite low. For example, suppose that the most likely reliability is estimated to be 0.997, the best reliability probability does not exceed 0.9995 but, because of some uncertainty (e.g., an untested scenario), there is a small chance that it could be as low as 0.70. Also, it often becomes more difficult and inaccurate to attempt to fit the beta to distributions formed from composite information at higher levels of the model or after several updates have introduced significant unrelated data. When the beta does not provide an acceptable approximation, one must rely on using parameters such as the mean and percentiles to characterize the reliability uncertainty distribution.

These examples illustrate cases where new test information or data are introduced to update a reliability, R_1, to the form R_{i+1}. The continuous monitoring of R_1 and $f(R_i)$ is possible as new information or changes become available. Not all changes may be beneficial, as reliability can decrease and/or the uncertainty can increase at any given change step i. However, by judiciously planning new tests or changes for the purposes of reducing uncertainty and/or improving reliability, the overall trend will indicate such desired results.

Figure 11.2 is the resulting reliability growth plot of the pilot program being discussed. Note that the median reliability and 95% and 5% probability limits of reliability are plotted against product stage/development time. The individual data points correspond to the initial reliability characterization R_0 and the events associated with the updates R_i. This plot captures the results of the design teams' efforts to improve reliability, but the power of the approach is the roadmap developed which will lead to higher reliability and the ability to characterize all the efforts made to get there.

11.2.5   Results

The reliability equations permitted calculations of reliability for the overall system (and its parts—components, subsystems, and processes) at specified times of interest during the product's lifetime. Figure 11.2 shows this for the system during all lifetime stages. In addition, for the automotive industry certain times were important such as the end of the warranty period and government-mandated emission requirements at 100,000 miles. We calculate the system, subsystem, component, and process reliabilities at these important times, creating multiple figures such as those in Figure 11.2. Recall that the reliability predictions for these times are calculated using the reliability model. Examples of growth figures include

- the system reliability at 12 months (as in Figure 11.2),

- the system reliability at 36 months,

- the system reliability at 100,000 miles,

- reliabilities for each subsystem at 12 months,

- reliabilities for each subsystem at 100,000 miles,

- reliabilities for each individual process at 12 months,

- reliabilities for each individual process at 100,000 miles.

Reliability tracking and monitoring graphs of a system's various levels and times are used by the experts for test planning and resource allocation throughout the lifetime of the system.

11.2.6   Documentation

At the end of a system's lifetime, all roadmap graphs, the logic and probability models, the expert judgment used, and the analyses done (e.g., "what if" cases) provide a well-documented record—a knowledge base—of the lifetime development and performance of the system. This documentation can be used by others (especially new employees) in the future, provides a learning tool, and contributes to corporate memory.

Software packages exist for knowledge management, but it is also relatively easy to customize a graphical user interface with languages such as Java and HTML (Meyer et al. (1999)).

11.3   Auto performance using fuzzy approach

11.3.1   Aspects of the fuzzy and probability approaches

The basic description of the reliability problem in section 11.2.1 remains unchanged by the choice of methods used to estimate the performance. By this we mean that performance can be defined in a probability space (e.g., reliability) or in a fuzzy logic space. Either way, the initial steps of defining performance and defining the structure of the system, such as with a reliability block diagram, remain the same as in sections 11.2.1 and 11.2.2. Many of the other methods for combining diverse sources of information and updating also can operate in either space.

The choice between fuzzy and probability should be made based on how the experts in the application think about performance—whether in terms of probability theory or fuzzy set theory. The use of proper elicitation methods is imperative in either case (Meyer and Booker (2001)). It also is possible to move from one space to another, creating a fuzzy/probability hybrid approach. Conditions for this hybrid approach include

- if the choice between probability and fuzzy is not obvious, because definitions of performance and/or uncertainties can be defined in either space; or

- if some definitions are better suited to one space while other definitions are suited to the other space.

For example, suppose the experts are accustomed to a probabilistic definition of reliability but have a fuzzy interpretation of the uncertainties associated with the conditions affecting the performance of the system. Or suppose an expert can easily specify probabilities for the system in its nominal environment, where knowledge is extensive, but can only specify qualitative statements and rules when the system is in an unusual environment, where knowledge is lacking. Such cases require mapping the results of a fuzzy analysis (mapping condition into performance) into a probability result for the reliability, as seen below.

In traditional probabilistic risk assessments (PRAs), experts often require training in probability theory before their estimates and uncertainties can be elicited (United States Nuclear Regulatory Commission (1989)). Fuzzy logic methods, using an approach similar to that applied in the use of fuzzy control systems, have the advantage of permitting experts to assess in natural language terms (e.g., "high," "good," etc.) parameters affecting performance of components/systems. Recognizing that there is a cost associated with obtaining and using more precise information, fuzzy methods allow the experts to use the level of precision with which they understand the underlying process.

11.3.2   A fuzzy/probability hybrid approach

Fuzzy control system techniques are used to develop models of systems, often including their expert operators, for enhanced control of processes and systems. These techniques can be especially useful in applications involving highly nonlinear systems. The control system maps observed "plant" output parameter values into required control actions, or plant inputs. In a fuzzy control system, these observed plant outputs are transformed into degrees of membership in fuzzy plant-output sets. "If then" rules transform these degrees of membership into fuzzy control-action (or plant-input) sets. The precise control action is determined via *defuzzification* such as selecting the *centroid* of the fuzzy control-action set (Jamshidi, Vadiee, and Ross (1993)). The replacement of the fuzzy control system approach with a probabilistic controller approach is described in Laviolette et al. (1995).

   This same process can be applied to the development of uncertainty distributions in applications such as probabilistic risk assessment, probabilistic safety assessment, and reliability analysis. In a safety application, for instance, the plant-output parameters used by the control system may become component condition and accident-scenario parameters, and the control action may become the predicted component response to the accident conditions. Our particular interest is in cases where the relationship between the condition of the system and its performance is not well understood, especially for some sets of possible operating conditions, and where developing a better understanding is very difficult and/or expensive. For example, suppose a system was developed and tested for a specific range of operating parameters, and we now desire to characterize its reliability in untested or minimally tested parameter ranges. In applications, the manner in which uncertainty distributions are obtained ranges from being largely data driven to being highly subjective, depending upon where test results are available (Smith et al. (1997)).

   The following generic example is based on the wear example given in Chapter 14. In the automotive example in Figure 11.2, expert judgment information forms a significant (if not the sole) source of data for reliability analysis. Consider an oversimplified version of that system with one component, which can influence performance of the system. The component is subject to wear, potentially degrading performance. In fuzzy control system terms, the component condition is analogous to plant output, and performance of the system is analogous to plant input. For a given wear level, performance degradation will be variable, and the range of possible performance levels is analogous to the control-action set.

   Membership functions (for either condition or performance) are useful for describing unusual or nonnominal conditions or performance behaviors. For example, on rare occasions extreme values of temperature can result in large degradations of performance. In traditional PRAs, it is common to have analogous statements such as "on rare occasions, this failure mode occurs and the performance is severely affected." Specifying the membership functions quantifies statements such as this along with the uncertainties associated with them. Membership functions also are effective for quantifying the wearout failures occurring beyond the useful life of the product. In the probability model in section 11.2.1, these failures were not considered because they were assumed to be beyond the useful life of the product—an assumption that may be unreasonable.

   Figure 11.6 shows membership functions for three hypothetical component-condition

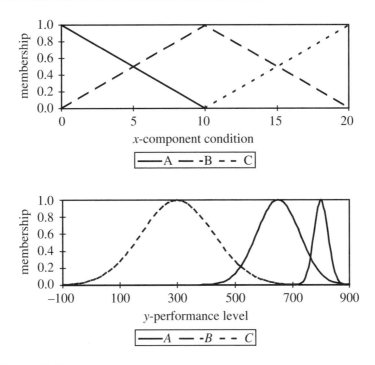

Figure 11.6. *Component-condition and performance-level sets for three member-ship functions.*

sets and three performance-level sets. The notation N (mean, standard deviation) is used for the performance-level functions that are normal distributions without the scale factor so that they range from 0 to 1 in the traditional membership function scaling. Three "if then" rules define the condition/performance relationship:

- If condition is A, then performance is A.

- If condition is B, then performance is B.

- If condition is C, then performance is C.

For the component condition, $x = 4.0$, x has membership of 0.6 in A and 0.4 in B. Using the rules, the defined component-condition membership values are mapped to performance-level weights. Performance-level set A, $N(800, 25)$, characterizes the range of performance values with a weight of 0.6. The membership function for performance-level set B, $N(650, 75)$, characterizes the range of performance values with a weight of 0.4. In fuzzy control system methods, the membership functions for performance-level sets A, $N(800, 25)$, and B, $N(650, 75)$ are combined based on the weights 0.6 and 0.4. This combined membership function can be bimodal and can be used to form the basis of an uncertainty distribution for characterizing performance for a given condition level.

It should be noted that this method of combining membership functions is not the only available method. In this particular example with scaled normal performance membership functions, an alternative method would be to take a linear combination of the two normals. For unscaled normal distributions, the linear combination is another normal, specifically

$$0.6 * N(800, 25) + 0.4 * N(650, 75) = N(740, 33.5), \tag{11.5}$$

which is a unimodal symmetric distribution, unlike the bimodal resulting from the combination above. Engineers may be more comfortable with the unimodal shape rather than with explaining the "valley" or minimal likelihood density that appears in the center of the bimodal distribution—the region of most common interest.

While the component-condition membership functions shown are equally spaced, experts can specify the number and density of these functions according to their knowledge and experience. For example, for certain values of x (say, 0–7) or y (say, 900–700), experts may place several membership functions because much is known about the conditions and performance in these regions. However, the functions could become quite sparse and less dense for those values of x and y where uncertainty is great.

Departing from standard fuzzy control systems methods, we now normalize the combined performance membership function so that it integrates to 1.0. This transformation converts the fuzzy-based combined membership function into a more convenient distribution function. The resulting function, $f(y|x)$, is the performance uncertainty distribution corresponding to the situation where component condition is equal to x, such as $x = 4$. Figure 11.7 is the cumulative function form of the uncertainty distribution, $F(y|x)$. If there is a performance requirement such that the performance must exceed some threshold, T, in order for the system to operate successfully, then the reliability of the system for the situation where component condition is equal to x can be expressed as $R(x) = 1 - F(T|x)$. As illustrated in Figure 11.7, a threshold of $T = 550$ corresponds to a reliability of $R(4.0) = 0.925$.

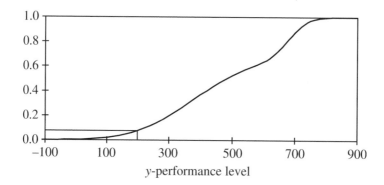

Figure 11.7. *Performance uncertainty cumulative function.*

Clearly, uncertainty or variability in the level of wear for the component induces uncertainty in $R(x)$. Suppose that the uncertainty in the wearout, at given values of x, is characterized by some distribution representing this uncertainty. The distribution of $F(T|x)$ and hence of $R(x)$ can then be obtained. The results of repeatedly sampling values from that distribution and calculating F would produce an "envelope" of cumulative functions. The approximate distribution of R can be obtained from such a numerical simulation process along with its characteristics such as a median, fifth, and 95th percentile values, as in Figure 11.8.

11.3.3 A fuzzy automotive example

For a more direct example of an automotive application (Kerscher et al. (2002)), suppose that the focus is on a single component imbedded in the auto system in Figure 11.1. New information becomes available from the engineering judgment of the supplier who provides this component. The supplier's continued development of the component brings to light

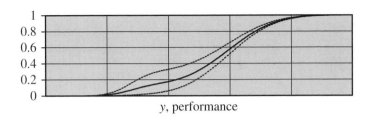

y, performance

Figure 11.8. *Median, fifth, and 95th percentiles of CDF envelope generated by sampling from the component wearout distribution.*

some issues not considered in the team's judgments used to formulate the prior, yet the prior information is still relevant for inclusion in the analysis. The new information also represents the supplier's different way of modeling the component's reliability and involves uncertainties that are less quantitative in nature. The supplier does not wish to think in terms of the Weibull model in (11.1) but instead formulates "if then" rules that directly map his feeling of the design into reliability. For example, "if" anticipated problems are very small, "then" reliability is very good. Table 11.1 illustrates such a rule base. The uncertainties associated with using verbal descriptions can be quantified using methods from fuzzy logic based on fuzzy set theory (Zadeh (1965)). Membership functions in Figures 11.9 and 11.10 provide the quantification mechanism for both linguistic terms and reliability descriptors at 36 months.

Table 11.1. *"If then" rules for mapping knowledge to reliability.*

Membership function	If anticipated problems are...	Then reliability is...	Membership function
a	very small	excellent	A
b	okay	nominal	B
c	moderate	poor	C
d	large	unacceptable	D
e	extremely large	ridiculous	E

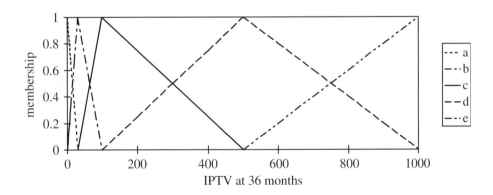

Figure 11.9. *Anticipated problems membership functions.*

Specifically, the triangular membership functions in Figure 11.9 are for five levels (a–e) of verbal descriptions corresponding to the linguistic terms shown in Table 11.1 (very small,

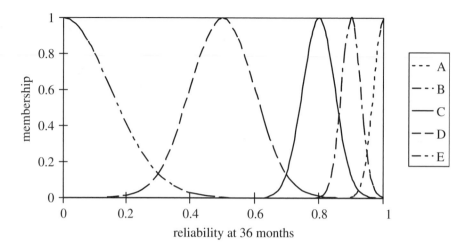

Figure 11.10. *Reliability membership functions.*

okay, moderate, large, and extremely large). Similarly, the normal-looking membership functions in Figure 11.10 indicate the five levels (A–E) for reliability corresponding to the additional linguistic terms in Table 11.1 (excellent, nominal, poor, unacceptable, ridiculous). The reliability membership functions are normal distributions without a scale factor such that they range from 0 to 1 for membership values. Five rules (Table 11.1) define mapping of the supplier's feeling of the design into reliability: if anticipated problems are very small (a), then reliability is excellent (A), and so on. It should be noted that the "if then" rules and membership functions may apply to many different components or be specific to one component/process.

The supplier tells us that the range of anticipated problems with this component at 36 months varies from almost none to very small, such that in the worst case they have membership of 0.67 in a and 0.33 in b. Using the rules, the defined anticipated problems membership values are mapped into reliability level A with a weight of 0.67 and reliability level B with a weight of 0.33. In fuzzy control system methods, the membership functions for reliability levels A and B are combined based on the weights 0.67 and 0.33. This combined membership function is reduced to a single value (in this case the median) in a process called "defuzzification." The resulting reliability in this case is 0.98. Reliabilities for the lower limit and most likely values of anticipated problems were similarly calculated to form the basis of an uncertainty distribution for characterizing reliability for this component.

To update this component in light of the new information from the supplier, the question becomes how to combine the supplier's fuzzy generated reliability uncertainty distribution with the PDF formulated from the team's prior information. It can be shown that Bayes' theorem is useful in a theoretical sense for solving this problem because membership functions can be treated as likelihoods (see Chapter 5, section 5.6). Therefore, the new information from the supplier's rules and fuzzy membership functions can be used to form a likelihood distribution function for reliability that may be combined with the reliability distribution function provided by the team's prior information.

The supplier's fuzzy generated reliability uncertainty distribution was used as the likelihood for the Bayesian updating of the prior. The improvement in reliability of the component after this update is shown by the darker line in Figure 11.11. The gray line is the reliability uncertainty before the fuzzy information is added. It is interesting to note that this fuzzy-based updating is similar to that of a probability-based Bayesian update for a proposal of "what if" we ran 40 successful tests. In either updating case, the uncertainty after the update is reduced, while the medians remain essentially unchanged. The similarity between the results of the "what if" update and the fuzzy update is noteworthy because it illustrates the fact that information of any type (test, judgment, computer simulation) can be valuable in achieving the goals of reducing uncertainty and improving reliability.

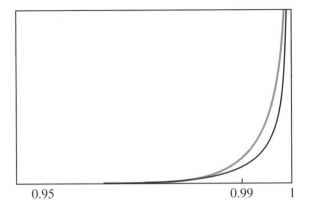

Figure 11.11. *Reliability membership functions.*

Although the use of fuzzy logic may be quite helpful in applications where only qualitative information is available, direct updates using quantitative data are always possible, as described in Kerscher et al. (2001). In addition, useful approximations exist that are helpful for planning purposes involving potential component tests and manufacturing process capability studies. These also are described in detail in Kerscher et al. (2001).

Characterizing with large uncertainty the initial reliability of a new product under development, and then working to reduce that uncertainty, has been found to be a culturally acceptable way to address the reliability issue. The PREDICT methodology (Meyer et al. (1999)) is useful in facilitating this approach. The methodology is flexible enough to integrate information at various stages of product development, including the asking of "what if" questions. Also, as has been shown, updating information that is fuzzy in nature is also possible. If application of this methodology further assists a project team in successfully including the reliability issue earlier in its day-to-day activities involving performance, cost, and timeliness, it will prove to be a powerful additional tool in the development of a high reliability product.

11.4   Comments on approaches for system performance/reliability

The most important issue in deciding on a probabilistic, fuzzy-based, or hybrid approach is how the experts define reliability or performance. That must be expressed in terms that are familiar to and comfortable for that manufacturing culture. If experts think in terms of the bathtub Weibull reliability model (probability-based), then they are better able to estimate the parameters necessary for that model using a probability approach. If they think in terms of mapping condition of the parts and processes in terms of performance with the use of heuristic rules, then the fuzzy approach is easier for the experts to use. If unusual environments push the knowledge of the experts to their limits, if they have difficulty in quantifying uncertainties, or if the relationship between condition and performance is complex, unknown, nonlinear, etc., then the fuzzy or fuzzy/probability hybrid approach can provide ways of quantifying uncertainties and handling these complexities. If several of the above issues occur together, then a hybrid approach is useful.

We have had success in implementing the hybrid approach in an application where experts faced such mixed issues. Details of that example cannot be presented for proprietary reasons, but the methodology described in section 11.3.2 was implemented.

Regardless of the space (probability or fuzzy) or transformations from one space to another (hybrid) used by the experts in their thinking, estimating and specifying the uncertainties, performance, conditions, and the system structure must be properly implemented through formal elicitation techniques, as in Meyer and Booker (2001). That is the only way to minimize cognitive, motivational, and elicitation biases to ensure the highest possible quality of information elicited. Formally elicited expert judgment can be used for analysis and combined with other sources of information, such as test data.

References

E. J. HENLEY AND H. KUMOMOTO (1981), *Reliability Engineering and Risk Assessment*, Prentice–Hall, Englewood Cliffs, NJ.

M. JAMSHIDI, N. VADIEE, AND R. ROSS (1993), *Fuzzy Logic and Control Software and Hardware Applications*, PTR/Prentice–Hall, Englewood Cliffs, NJ.

W. J. KERSCHER III (1989), A reliability model for total field incidents, in *Proceedings of the Annual Reliability and Maintainability Symposium*, IEEE, Piscataway, NJ, pp. 22–28.

W. J. KERSCHER III, J. M. BOOKER, AND M. A. MEYER (2001), PREDICT: A case study, in *Proceedings of ESREL 2001, SARS and SRA-Europe Annual Conference, Torino, Italy, September* 16–20, 2001, A. A. Balkema, ed., Rotterdam, The Netherlands.

W. J. KERSCHER III, J. M. BOOKER, M. A. MEYER, AND R. A. SMITH (2002), PREDICT: A case study using fuzzy logic, in *Proceedings of the Annual Reliability and Maintainability Symposium*, Seattle, WA.

M. LAVIOLETTE, J. SEAMAN JR., J. BARRETT, AND W. A. WOODALL (1995), Probabilistic and statistical view of fuzzy methods, *Technometrics*, 37, pp. 249–281.

H. F. MARTZ AND R. G. ALMOND (1997), Using higher-level failure data in fault tree quantification, *Reliability Engrg. System Safety*, 56, pp. 29–42.

H. F. MARTZ AND R. A. WALLER (1990), Bayesian reliability analysis of complex series/parallel systems of binomial subsystems and components, *Technometrics*, 32, pp. 407–416.

H. F. MARTZ, R. A. WALLER, AND E. T. FICKAS (1988), Bayesian reliability analysis of series systems of binomial subsystems and components, *Technometrics*, 30, pp. 143–154.

M. A. MEYER AND J. M. BOOKER (2001), *Eliciting and Analyzing Expert Judgment: A Practical Guide*, ASA–SIAM Series on Statistics and Applied Probability, SIAM, Philadelphia, ASA, Alexandria, VA.

M. A. MEYER, J. M. BOOKER, T. R. BEMENT, AND W. KERSCHER III (1999), *PREDICT: A New Approach to Product Development*, Report LA-UR-00-543 and 1999 R&D 100 Joint Entry, Los Alamos National Laboratory, Los Alamos, NM and Delphi Automotive Systems, Troy, MI.

R. E. SMITH, T. R. BEMENT, W. J. PARKINSON, F. N. MORTENSEN, S. A. BECKER, AND M. A. MEYER (1997), The use of fuzzy control systems techniques to develop uncertainty distributions, in *Proceedings of the Joint Statistical Meetings*, American Statistical Association, Alexandria, VA.

UNITED STATES NUCLEAR REGULATORY COMMISSION (1989), *Severe Accident Risks: An Assessment for Five US Nuclear Power Plants*, Report NUREG-1150, Washington, DC.

L. ZADEH (1965), Fuzzy sets, *Inform. and Control*, 8, pp. 338–353.

Chapter 12

Control Charts for Statistical Process Control

W. Jerry Parkinson and Timothy J. Ross

Abstract. Statistical process control is similar to engineering process control, described in Chapter 8, in that both types of control force the process to produce a good product. One difference between them, however, lies in the amount of effort used to determine if the process is under control before making a change or searching for a cause of lack of control. Control charts are the tools that are used to determine whether or not a process is in control. This chapter summarizes some of the rudimentary principles and ideas behind the development and use of control charts. We then present fuzzy control charts and give examples of their use.

12.1   Introduction

In this chapter, we present a brief introduction to control charts used in statistical process control (SPC) and its analogue, "fuzzy" SPC. A complete discussion of SPC requires an entire textbook, and there are many good textbooks on this subject. See, for example, Mamzic (1995), Wheeler and Chambers (1992), and Box and Luceno (1997). The entire breadth of this interesting field is beyond the scope of this chapter. The goals of this chapter, therefore, are to introduce some of the fundamental techniques used in SPC to readers not previously introduced to the subject, and also to demonstrate useful applications of fuzzy logic to SPC.

One question that we need to answer is, "How is SPC different from process control discussed in Chapter 8?" In Chapter 8, we looked briefly at classical process control, fuzzy process control, and probabilistic control. All three techniques were concerned with processes that can be controlled online. They relied on a feedback measurement used in an algorithm or a recipe that supplied an immediate correction to be applied to the plant under control. These methods are often called regulatory control or engineering process control (EPC) techniques. The SPC techniques, or control charts, discussed here, are not control techniques in the same sense: they are tools used to determine if a change or correction should be applied to a process or plant. The information supplied by the control chart may be helpful in determining not only what corrections have to be made but also their magnitude. On the other hand, this information might not help at all with the correction part of the control problem.

SPC can be used for continuous processes like quality control in chemical plants or petroleum refineries. Continuous processes are those in which the EPC techniques described in Chapter 8 are most often used. SPC is currently most often used with discrete processes, such as in manufacturing or machine shop operations, or with batch processes in batch chemical reactors or batch distillation. SPC analysis relies heavily on a tool called a control chart, discussed in the next two sections. Control charts were invented by Shewart (1980, 1986) in the United States in the late 1930s. They were used extensively by Deming (as described by Mamzic (1995) and Wheeler and Chambers (1992)) in his SPC work during the 1960s in Japan. Deming's SPC work helped to bring significant improvements to the Japanese manufacturing industry. Once these improvements were noted, SPC and control charts gained new popularity in the United States.

The basic idea behind the control chart is that there are two causes of variability (or errors from an EPC point of view). The first one is called a common-cause event. All processes have this. It is a random variability in the process and/or in the sensing or measurement of the variable of interest. It is something that is always present that cannot really be removed or controlled. The other cause of variability is often called a special-cause event. This is a disturbance that can be understood physically and corrected. The control chart is a tool that is used to determine the difference between these two causes and to thereby determine if and when corrective action should be taken. Control charts do not actually determine *what* corrective action should be taken, but often they can be used as a tool to help determine *if and when* corrective action should be taken.

12.2   Background

Improving productivity and quality of products is a major objective of statistical process control (SPC). The term "statistical" is used in the name because the procedure involves the use of numerical data and probability theory in an attempt to derive useful information about the process under observation. Although manufacturing constitutes the most popular "process" requiring the help of SPC, it can be used to improve a wide variety of processes. Your own office or business can probably benefit from the use of SPC techniques. They can be used whenever a process produces a product in which some attribute of that product is plagued with undesired variations.

We assume that the process is at steady state or not undergoing any systematic change. There are, however, variations that are beyond our control. We do not necessarily like them, but we have to live with them. This is the common-cause event mentioned above. This variability may be due to a wide variety of factors that are beyond our control, such as the weather. This would include changes in temperature and/or barometric pressure. A common cause for these variations in the chemical industry is variation in the quality of the feedstock. This problem could be caused by poor quality control on the part of the supplier or because the feedstock is a natural material such as crude oil or coal. The list of possibilities for common-cause events is almost endless. The trick is to use SPC and a control chart to determine if the variability observed over a period of time is indeed a common-cause or special-cause event. If the variability is common-cause, then continue the process as before. If it is determined that the variability is a special-cause event, exercise the control portion of the SPC and stop the process and determine the cause, if possible, and then remedy the situation before continuing the process.

There are two basic types of control charts:

- those dealing with measured variables like temperature, pressure, part diameter, etc.;

- those dealing with attributes, such as "go or no-go," flaws, etc.

Our first example will deal with the measured variable type control chart.

12.2.1   Example 1: Measured variable control chart

Suppose that you and a friend are building a big log cabin. You have never done this before, but you are pretty smart and you know that you can do it. You buy a big axe (you want this to be a challenge), and you head for the forest. You look at your plans and see that there are a great number of logs that must be cut to a length of exactly 20 feet or 240 inches. The logs are pretty thick, so you figure that you can have a tolerance of about ± 2.5 inches and still use the logs. Your friend measures and marks the trees and logs and you cut them. At regular intervals you select sets or subgroups of five logs from those that you are cutting. You measure these cut logs and take the average length and range (maximum length minus the minimum length) of each set. Over a period of time, you repeat this procedure 20 times. Since this is a very big cabin, you still have a lot of logs to cut. Therefore, you want to determine if any systematic errors or a special-cause event has crept into your process. Table 12.1 shows the results of these calculations, the average lengths (X) and ranges (R) to the nearest tenth inch, for each of the 20 five-log sets. Figure 12.1 is the Shewhart X bar chart for this 20-set run and Figure 12.2 is the R chart. The combination is commonly known as the X bar–R control chart. In many of the figures that follow, the upper and lower control limits are depicted by the same line style. There should be no confusion since the upper limit will always be above the average line and the lower limit will always be below it.

Table 12.1. *Values of the sample means (X bar) and ranges (R) of sets of five logs (first try)*.

Set number	X bar, set average (inches)	R, set range (inches)
1	233.9	30.9
2	242.9	17.5
3	241.3	22.6
4	238.6	37.2
5	238.3	26.7
6	238.1	16.3
7	234.9	21.6
8	236.3	22.8
9	238.4	13.6
10	237.9	15.5
11	238.3	28.4
12	233.7	32.9
13	240.4	35.9
14	234.1	28.7
15	234.2	14.9
16	236.1	12.4
17	249.6	29.5
18	241.3	14.3
19	235.2	17.2
20	236.4	14.9

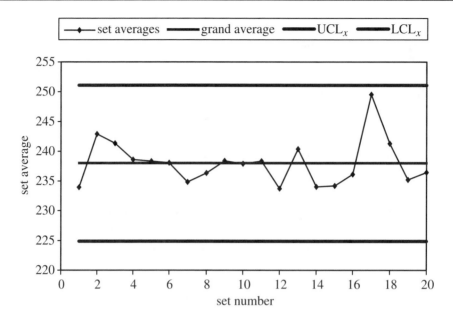

Figure 12.1. *Shewhart X bar chart for the log cabin problem (first try).*

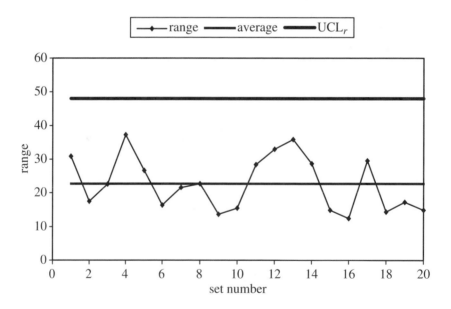

Figure 12.2. *The R chart for the log cabin problem (first try).*

There are several things that need to be explained about these charts. First, in Figure 12.1, the set averages, the grand average (the average of the set averages), and the upper and lower control limits (UCL$_x$ andLCL$_x$, respectively) are plotted for each set for a given run length. In this case the run length is 20 sets. Second, in Figure 12.2, the set ranges, the range average, and the upper range control limit, UCL$_r$, are plotted for each set for the run. (When the set size is 6 or smaller, no lower range control limit can be computed.) A discussion of how to compute the control limits and what they really stand for will follow shortly. For now, we see that both the set averages and the set ranges fall within the control limits and there are no unusual patterns in either plot. This means that the sample averages are consistent and the data scatter or dispersion from set to set is consistent, which in turn means that the variability in the process is probably from a common-cause event rather than a special-cause event. Probably the most significant thing that we notice from observing these charts is that the grand average is about 238 inches, barely within our tolerance limits. The average range is about 22.7 inches, far beyond our tolerance range of 5 inches (± 2.5 inches) that will work for our cabin. It looks like we are going to have to revamp our entire process because we have to reduce our common-cause variation. We have to get the common-cause variation within our tolerance limits. Wheeler and Chambers (1992) describe a method for computing what they call natural process limits that can be used to see if our tolerance limits are reasonable. We will discuss this computation shortly, but first let us look at Figure 12.3.

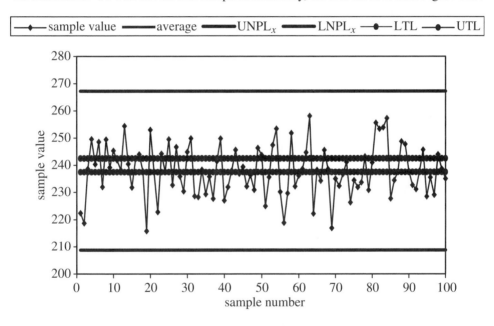

Figure 12.3. *Raw data for the cabin problem,* 100 *logs (first try).*

Figure 12.3 is a plot of the sample measurement, the average of the 100 (20 sets of 5 logs) samples (same as the grand average), the upper and lower tolerance limits (UTL and LTL, respectively), and the upper and lower natural process limits (UNPL$_x$ and LNPL$_x$, respectively) versus the sample number. Table 12.2 lists some of the important parameters from the data plotted in Figure 12.3 as well as the three-sigma limits about the mean value. The mean, or grand average, is almost equal to the lower tolerance limit, so it is difficult to see on this plot. It is worth noting that the natural process limits are very similar to the three-sigma limits ($\pm$ three standard deviations on either side of the mean). It is very important to notice that the tolerance limits are significantly inside the natural limits. This is a strong indication that not many of the products, in this case cut logs, will meet the specifications dictated by the tolerance limits. It is a signal that one should probably change the tolerance limits or change the process. In our case, we are going to change the process to reduce the common-cause variation. We are going to buy a high performance circular saw that will cut logs. However, before we continue, we will discuss the computation of the control limits

Table 12.2. *Important parameters for data in Figure* 12.3.

Parameter	Value (inches)
mean (grand average)	238.0
upper tolerance limit (UTL)	242.5
lower tolerance limit (LTL)	237.5
upper natural process limit (UNPL$_x$)	267.3
lower natural process limit (LNPL$_x$)	208.7
upper three-sigma limit	265.8
lower three-sigma limit	210.2

and the natural process limits.

The upper and lower control limits are computed using the following formulas and the factors in Table 12.3:

$$\text{UCL}_{\bar{x}} = \overline{\overline{X}} + A_2 \bar{R}, \tag{12.1}$$

$$\text{LCL}_{\bar{x}} = \overline{\overline{X}} - A_2 \bar{R}_1, \tag{12.2}$$

$$\text{UCL}_R = D_4 \bar{R}, \tag{12.3}$$

$$\text{LCL}_R = D_3 \bar{R}, \tag{12.4}$$

where $\overline{\overline{X}}$ is the grand average and $\bar{R}$ is the range average.

Table 12.3. *Factors for determining the control limits for X bar–R charts.*

n	A_2	D_3	D_4
2	1.880	—	3.268
3	1.023	—	2.574
4	0.729	—	2.282
5	0.577	—	2.114
6	0.483	—	2.004
7	0.419	0.076	1.924
8	0.373	0.136	1.864
9	0.337	0.184	1.816
10	0.308	0.223	1.777
11	0.285	0.256	1.744
12	0.266	0.283	1.717
13	0.249	0.307	1.693
14	0.235	0.328	1.672
15	0.223	0.347	1.653

The "n" value shown in both Tables 12.3 and 12.4 is the number of measurements in the set or subgroup. In our example, n is equal to 5. The upper and lower natural process limits are computed by using formulas (12.5) and (12.6) and the factors in Table 12.4:

$$\text{UNPL}_X = \overline{\overline{X}} + \frac{3\bar{R}}{d_2}, \tag{12.5}$$

$$\text{LNPL}_X = \overline{\overline{X}} - \frac{3\bar{R}}{d_2}. \tag{12.6}$$

Table 12.4. *Factors for determining natural process limits.*

n	d_2
2	1.128
3	1.693
4	2.059
5	2.326
6	2.534
7	2.704
8	2.847
9	2.970
10	3.078
11	3.173
12	3.258
13	3.336
14	3.407
15	3.472

One item worth mentioning is that all the control limits are closely related to the mean $\pm 3\sigma$ or three standard deviations from the mean. Shewhart developed the X bar–R technique as an easy approximation of the three-sigma limits. Keep in mind that there is nothing "holy" about the three-sigma limits. They are just a convenient limit that a truly random process will only very seldom go beyond. The idea behind the control chart is to determine if there is a pattern in the data that is nonrandom or noncommon-cause. Both of these limits provide a good bound that will only very seldom give "false alarms" or signal an out-of-control alarm that is only a random error. Some workers prefer to use X bar–s charts that actually use standard deviations instead of set ranges. Another technique that is appropriate to use with measurement data is the moving range chart. This is a good technique to use when only limited data are available. One would use individual measurements instead of averages, and the moving range would become the differences between the individual values. For more information on these techniques, see Mamzic (1995), Western Electric Company, Inc. (1958), and Committee E-11 (1995).

12.2.2   Example 2: Special-cause events

Now let us redo our cabin with our new saw. We are going to introduce a special-cause error this time. We will make the mistake of cutting a log and then using that log as a template for cutting the next log. We will then use the newly cut log as a template for the next log and so on. Table 12.5 shows the data from our first 20-set run with our new process. Figures 12.4 and 12.5 are the X bar and R charts, respectively, for this run. Figure 12.6 is the individual or raw data plot.

The things that should be noticed from this run are the following:

- In Figure 12.4, the X bar chart, near the end of the run the sample averages are starting to move above the upper control limit, signaling a special-cause event. This is because the length of the logs is starting to grow because we are using a different template every time. This time the control chart signaled that we were out of control before we exceeded our tolerance limits. (The tolerance limits are the endpoints on the Y-axis in Figure 12.4.)

Table 12.5. *Values of the sample means (X bar) and ranges (R) of sets of five logs (second try).*

Set number	X bar, set average (inches)	R, set range (inches)
1	239.7	1.1
2	240.1	1.6
3	240.1	3.4
4	239.5	1.5
5	239.6	2.5
6	240.4	1.0
7	239.9	1.5
8	239.8	1.6
9	240.1	2.1
10	239.6	1.4
11	239.7	1.2
12	240.3	1.1
13	239.8	0.7
14	240.2	1.4
15	240.2	1.1
16	239.9	0.7
17	239.8	2.4
18	240.4	1.0
19	241.2	1.9
20	241.2	1.9

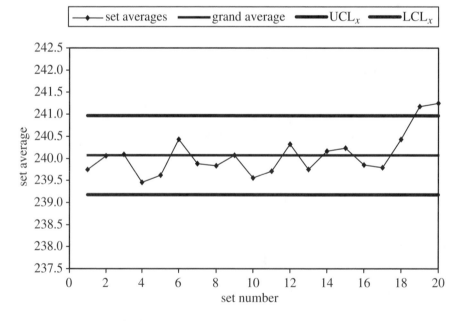

Figure 12.4. *X bar chart for the log cabin problem (second try).*

- In Figure 12.5, the *R* chart, the range goes above the upper control limit for set number 3. If we look at Figure 12.6, the individual data points, we can see that one of the longest logs and one of the shortest logs are in the same set. This can probably be classified as a false alarm because it only happens once. However, one should be aware of large differences within a set; it can signify problems, possibly a sampling error.

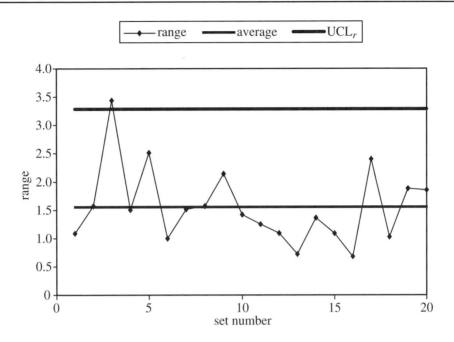

Figure 12.5. *The R chart for the log cabin problem (second try).*

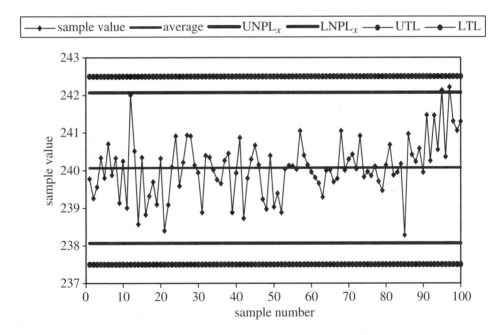

Figure 12.6. *Raw (individual) data for the cabin problem,* 100 *logs (second try).*

• Finally, Figure 12.6 shows us that our tolerance limits are outside of our natural process
  limits. This means that it is possible to meet our goals with a minimum amount of
  discarded logs.

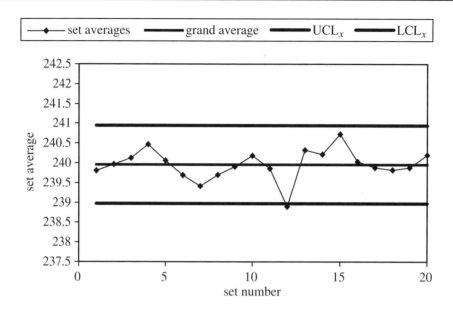

Figure 12.7. *X bar chart for the log cabin problem (third try).*

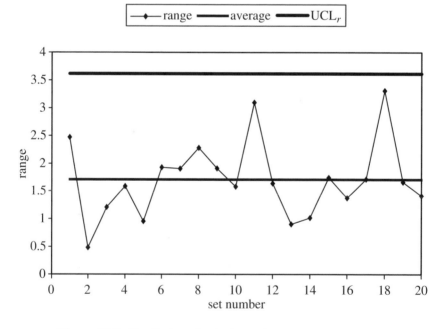

Figure 12.8. *The R chart for the log cabin problem (third try).*

12.2.3   Example 3: Traditional SPC

We are going to try one more time to build this cabin. This time we will use the same log for a template every time we make a cut. Table 12.6 shows the data from our first 20-set run with the final process. Figures 12.7 and 12.8 are the X bar and R charts, respectively, for this run. Figure 12.9 is the individual or raw data plot.

Table 12.6. *Values of the sample means (X bar) and ranges (R) of sets of five logs (third try).*

Set number	X bar, set average (inches)	R, set range (inches)
1	239.8	2.5
2	240.0	0.5
3	240.1	1.2
4	240.5	1.6
5	240.1	1.0
6	239.7	1.9
7	239.4	1.9
8	239.7	2.3
9	239.9	1.9
10	240.2	1.6
11	239.9	3.1
12	238.9	1.6
13	240.3	0.9
14	240.2	1.0
15	240.7	1.7
16	240.0	1.4
17	239.9	1.7
18	239.8	3.3
19	239.9	1.7
20	240.2	1.4

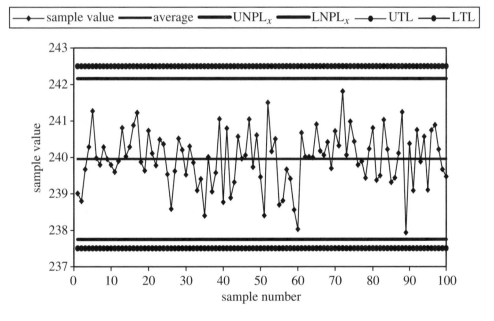

Figure 12.9. *Raw (individual) data for the cabin problem, 100 logs (third try).*

The things that should be noticed from this run are the following:

- In Figure 12.7, the X bar chart, the X bar for set number 12 is below the lower control limit, signaling a special-cause event. This is another false alarm. This run was made using a pseudorandom number generator that produces random numbers with normal statistics (Williams (1991))—this time with no interference from the authors. If this were a real process, the operators would have to look hard at this point to see what happened here. If we look at Figure 12.8, the R chart, we can see that the dispersion or range is very close to the average value—no cause for alarm there. Figure 12.9, the individuals chart (samples 56–60), shows that this may be the only set in the run where all the samples were below the average line. This may be a cause for concern if it were to happen on a regular basis. Again, the control chart provided a signal before we exceeded our tolerance limits. (The tolerance limits are the endpoints on the Y-axis in Figure 12.7.)

- Figure 12.8, the R chart, shows that all the ranges are below the upper control limit, even for set number 12. This information can be helpful when we are trying to find a cause for a system going out of control or in determining if the point really provided a false alarm.

- Again, Figure 12.9 shows us that our tolerance limits are outside our natural process limits. It will be possible to meet our goals with a minimum amount of discarded logs.

We will touch on three more points before we go onto the fuzzy work:

1. If we build our control charts and we find points out of control, should we throw them out and recompute our averages and control limits? Authors disagree on this (Mamzic (1995), Wheeler and Chambers (1992)). We did not throw away out-of-control data and recompute for any of the work shown in this chapter, and it did not make a difference in our overall outcomes. There are times when it will make a difference. Our suggestion is to keep it in mind and recompute your limits if you feel that it will truly help you solve the problem at hand. It may just depend upon whether you are building roads or watches.

2. Properly choosing a set or subgroup size can influence the usefulness of your control chart. This subject is beyond the scope of this chapter, but a good discussion can be found in Wheeler and Chambers (1992). The proper set size, or even time intervals, between sets probably can be determined by trying hard to understand the problems that might be the special-cause events and by trial-and-error. Sometimes, as in the case studies that we present next, one simply does not have a choice.

3. Other run data patterns, in addition to points beyond the upper and lower control limits, can signal process problems. Here is a list of such patterns from Mamzic (1995):

   - six successive points in one direction;

   - nine successive points on one side of the mean;

   - a recurring recycling pattern;

   - 14 successive alternating points;

   - two out of three successive points beyond $\pm 2\sigma$; one can use the upper or lower control limit as an estimate of 3σ;

   - four out of five adjacent points between 1σ and 3σ;

   - eight successive points outside 1σ on both sides of the center;

   - 15 successive points crowding the mean inside 1σ.

Use these rules at your discretion. Obviously, they might be able to tell the user something about the process operation, but overuse could lead to too many false alarms.

This section should give the uninitiated a good first glance at SPC and how to deal with measurement data. The next section is intended to do the same with fuzzy logic and measurement data.

12.3   Fuzzy techniques for measurement data: A case study

In this section, fuzzy set theory has been applied to an exposure control problem encountered in the machine manufacture of beryllium parts. This is a real problem facing the operation of a new beryllium manufacturing facility recently completed for the Department of Energy (DOE) at the Los Alamos National Laboratory. The major driving force for using fuzzy techniques rather than classical SPC in this case is that beryllium exposure is very task dependent and the manufacturing plant studied is quite atypical. Although there are purely statistical techniques available for solving this type of problem, they are more complicated than those shown in section 12.2 and are not as intuitive. If the SPC techniques described in section 12.2 are used without accounting for the task dependencies, too many false alarms will be produced. The beryllium plant produces parts on a daily basis, but every day is different. Some days many parts are produced and some days only a few. Sometimes the parts are large and sometimes the parts are small. Some machine cuts are rough and some are fine. These factors and others make it hard to define a typical day. The problem of concern in this study is worker exposure to beryllium. Even though the plant is new and very modern and the exposure levels are expected to be well below acceptable levels, the DOE has demanded that the levels for this plant be well below acceptable levels. The control charts used to monitor this process are expected to answer two questions:

1. Is the process out of control? Does management need to instigate special controls such as requiring workers to use respirators?

2. Are new, previously untested, controls making a difference?

The standard control charts, based on consistent plant operating conditions, do not adequately answer these questions. A statistical technique based on correlation and regression does work, but is not very intuitive. This approach will be briefly discussed in section 12.4. The approach described in this section is based on a fuzzy modification to the Shewhart X bar–R chart. This approach yields excellent results and is easily understood by plant operators. The work presented here is an extension of earlier work developed by Parkinson et al. (2000).

12.3.1   Background

The new facility at Los Alamos is intended to supply beryllium parts to the DOE complex and in addition act as a research facility to study better and safer techniques for producing beryllium parts. Exposure to beryllium particulate matter, especially very small particles, has long been a concern to the beryllium industry. The industrial exposure limit is set at $2\mu g/m^3$ per worker per eight-hour shift. The DOE has set limits of $0.2\mu g/m^3$ or 10 times lower than the industrial standard for this facility. In addition, they have requested continual quality improvement. In other words, in a short period of time, they intend to set even lower limits. Several controls have been implemented to ensure that the current low level can be met. However, there is a real management concern that the process remain under control and that any further process improvements are truly that—improvements. Since the facility is a research lab with manufacturing capabilities, the workload and type of work done each day can vary dramatically. This causes the average beryllium exposure to vary widely from day to day. This in turn makes it very difficult to determine the degree of control or the degree of improvement with the standard statistical control chart of the type discussed in section 12.2. For this reason, we have implemented a fuzzy control chart to improve our perception of the process.

The plant has four workers and seven machines. Each worker wears a device that measures the amount of beryllium inhaled during his or her shift. The devices are analyzed in the laboratory and the results are reported the next day after the exposure has occurred. An X bar–R chart, similar to those shown in section 12.2, can be constructed with these data and can presumably address the concerns of control and quality improvement. Although such a chart can be useful, because of the widely fluctuating daily circumstances these tests for controllability are not very meaningful.

There are four variables that have a large influence on the daily beryllium exposure. They are the number of parts machined, the size of the part, the number of machine setups performed, and the type of machine cut (rough, medium, or fine). In the fuzzy model, a semantic description of these four variables and the beryllium exposure are combined to produce a semantic description of the type of day that each worker has had. The day type is then averaged and a distribution is found. These values are then used to produce fuzzy Shewhart-type X bar and R charts. These charts take into account the daily variability. They provide more realistic control limits than the charts described in section 12.2 and make it easy to correctly determine whether or not a process change or correction has made a realistic improvement to the overall process.

12.3.2   The fuzzy system

The fuzzy system consists of five input variables or universes of discourse and one output variable. Each input universe has two membership functions and the output universe has five membership functions. The input and the output are connected by 32 rules. The five input variables are as follows:

1. Number of parts: with a range of 0 to 10 and membership functions

   (a) few,

   (b) many.

2. Size of parts: with a range of 0 to 135 and membership functions

   (a) small,

   (b) large.

3. Number of setups: with a range of 0 to 130 and membership functions

   (a) few,

   (b) many.

4. Type of cut: with a range of 1 to 5 and membership functions

   (a) fine,

   (b) rough.

5. Beryllium exposure: with a range of 0 to 0.4 and membership functions

   (a) low,

   (b) high.

The output variable is as follows:

1. The type of day: with range from 0 to 1 and membership functions

   (a) good,

   (b) fair,

   (c) OK,

   (d) bad,

   (e) terrible.

Figure 12.10 represents the input membership functions used for this problem. Figure 12.11 represents the output membership functions.

   The rules are based on some simple ideas. For example, if all four mitigating input variables indicate that the beryllium exposure should be low, and it is low, then the "type of day" is OK. Likewise, if all four indicate that the exposure should be high, and it is high, then the type of day is also OK. If all four indicate that the exposure should be low, and it is high, then the type of day is terrible. If all four indicate that the exposure should be high, and it is low, then the type of day is good. Fair and bad days fall in between the OK days and the good and terrible extremes. The 32 rules are given in Table 12.7.

   The form of the rules is as follows:

   If (number of parts) is... and if (size of parts) is... and if (number of setups) is... and if (type of cut) is... and if (beryllium exposure) is..., then (the type of day) is....

The size of parts is determined as the number of parts multiplied by the average diameter of each part, measured in centimeters. The type of cut is determined by

$$\text{roughness} = \sum_i^N \sum_j^M \frac{\text{sop}_i}{\text{RF}_{i,j}} \left[\frac{1.0}{j+1} \right], \qquad (12.7)$$

$$\text{type of cut} = \ln \left[\frac{1000}{\text{roughness}} \right], \qquad (12.8)$$

where N is the number of parts for the given day for the given worker, M is the number of setups for a given day for a given worker, sop_i is the size of part i, and $\text{RF}_{i,j}$ is the roughness factor for part i at setup j.

   A fine cut has a roughness factor (RF) of 1, a medium cut has an RF equal to 3, and a rough cut has an RF equal to 5. The calculation for type of cut is a bit complicated but it provides a daily number between 1 and 5 (fine to rough) for each worker, which is meaningful. An example of the use of the fuzzy technique will follow a discussion of the plant simulation.

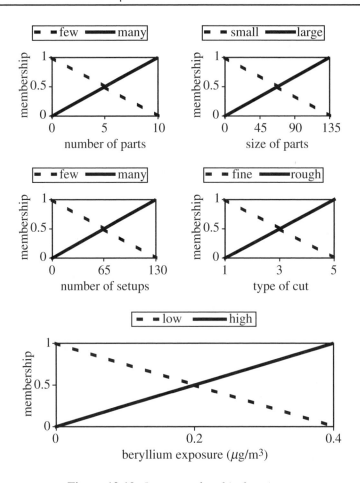

Figure 12.10. *Input membership functions.*

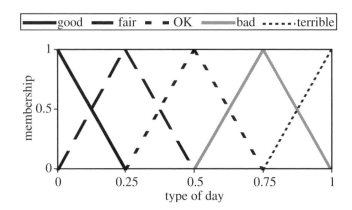

Figure 12.11. *Output membership functions.*

12.3.3   Plant simulation

The Los Alamos beryllium facility has been completed but has not yet been put into production. For this reason a computer program was written in order to provide a simulation of

Table 12.7. *Rules for beryllium exposure: "type of day."*

Rule number	Number of parts	Size of parts	Number of setups	Type of cut	Beryllium exposure	Type of day
1	few	small	few	fine	low	fair
2	few	small	few	fine	high	terrible
3	few	small	few	rough	low	ok
4	few	small	few	rough	high	terrible
5	few	small	many	fine	low	fair
6	few	small	many	fine	high	bad
7	few	small	many	rough	low	fair
8	few	small	many	rough	high	terrible
9	few	large	few	fine	low	fair
10	few	large	few	fine	high	bad
11	few	large	few	rough	low	fair
12	few	large	few	rough	high	terrible
13	few	large	many	fine	low	good
14	few	large	many	fine	high	bad
15	few	large	many	rough	low	fair
16	few	large	many	rough	high	bad
17	many	small	few	fine	low	fair
18	many	small	few	fine	high	bad
19	many	small	few	rough	low	fair
20	many	small	few	rough	high	terrible
21	many	small	many	fine	low	good
22	many	small	many	fine	high	bad
23	many	small	many	rough	low	fair
24	many	small	many	rough	high	bad
25	many	large	few	fine	low	good
26	many	large	few	fine	high	bad
27	many	large	few	rough	low	fair
28	many	large	few	rough	high	bad
29	many	large	many	fine	low	good
30	many	large	many	fine	high	ok
31	many	large	many	rough	low	good
32	many	large	many	rough	high	bad

the facility operation and provide a demonstration of the fuzzy control chart technique. The results of this study are being supplied to plant workers in order for them to provide input to further improve the technique.

Some actual beryllium exposure data were available for this study. These made it possible for us to develop some reasonably realistic simulations for each of the intermediate process steps for manufacturing the beryllium parts. The combination of plant data and simulation data were used in this study. The simulation-operator interaction process is iterative and is designed to enhance the beryllium exposure control techniques. A Shewhart-type control chart is used to measure the central tendency and the variability of the data. A process flow diagram of the plant simulation is shown in Figure 12.12.

The model has the following limitations or boundary conditions:

1. There are four machinists.

2. There are seven machines.

3. Machines 1 and 2 do rough cuts only.

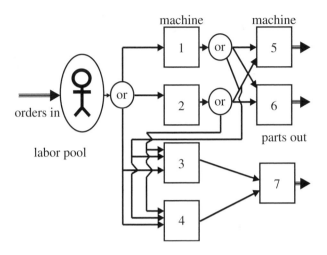

Figure 12.12. *The process flow diagram of the beryllium plant simulation.*

4. Machines 3 and 4 do both rough cuts and medium cuts.

5. Machines 5, 6, and 7 do only fine cuts.

6. Machine 7 accepts only work from machines 3 and 4.

7. Machines 5 and 6 accept only work from machines 1 and 2.

8. Each machinist does all of the work on one order.

9. All machinists have an equally likely chance of being chosen to do an order.

10. There are 10 possible paths through the plant. (At this point, all are equally likely.)

The simulation follows the algorithm below:

1. A random number generator determines how many orders will be processed on a given day (1 to 40).

2. Another random number generator chooses a machinist.

3. A third random number generator choose a part size.

4. A fourth random number generator chooses a path through the plant. For example, machine 1 to machine 3 to machine 7. (See Figure 12.12.)

5. The machine and path decide the type of cut (rough, medium, or fine). Machines 1 and 2 are for rough cuts only; machines 5, 6, and 7 are for fine cuts only; and machines 3 and 4 do rough cuts if they are the first machines in the path and medium cuts if they are the second machines in the path.

6. A random number generator chooses the number of setups for each machine on the path.

7. Another random generator chooses the beryllium exposure for the operator at each step.

The above procedure is carried out for each part, each day. The entire procedure is repeated the following day, until the required number of days has passed. For this study, the procedure was run for 30 days to generate some sample control charts. A description of the fuzzy control chart construction follows.

12.3.4   Example 4: Establishing fuzzy membership values

This example will follow each step of the process for a specific machinist for a given day. The description will then be extended to the work for the entire day for all machinists. From the simulation, on day 1, 13 part orders were placed. Machinist 2 processed four of these orders, machinist 1 processed three orders, machinist 3 processed four orders, and machinist 4 processed two orders. Machinist 2 will be used to demonstrate the fuzzy system.

The cumulative size of the four parts that machinist 2 processed on day 1 was calculated to be 64.59. The number of setups that he/she performed was 45. The numeric value for the type of cuts he/she performed on that day was 1.63. Finally, the machinist's beryllium exposure was $0.181 \mu g/m^3$ for that eight-hour period.

First, we note that since all input variables are binary, in almost every case, we will fire all 32 rules. Upon inserting the input values into the membership functions shown in Figure 12.10, the following values are obtained: For number of parts = 4, from Figure 12.10, the membership in many is 0.4 and the membership in few is 0.6. For size of parts = 64.59, from Figure 12.10, the membership in small is 0.52 and the membership in large is 0.48. For number of setups = 45, from Figure 12.10, the membership in many is 0.35 and the membership in few is 0.65. For type of cuts = 1.63, from Figure 12.10, the membership in rough is 0.16 and the membership in fine is 0.84. The beryllium exposure is 0.181. From Figure 12.10, the membership in high is 0.45 and the membership in low is 0.55. The variable values for set of inputs from Figure 12.10 for all rules are listed in Table 12.8, along with the variable values for the rule outputs.

In this study, the min–max technique was used to resolve the "and–or" nature of the rules and the centroid method was used for defuzzification. For example, rule 1 (Table 12.8) is fired with the following weights:

- number of parts: few = 0.6;

- size of parts: small = 0.52;

- number of setups: few = 0.65;

- type of cut: fine = 0.84;

- beryllium exposure: low = 0.55.

The rule consequent "fair" takes the minimum value 0.52. Observation of the last column in Table 12.8 reveals that the consequent fair appears 10 times with values ranging from 0.16 to 0.52. The min–max rule assigns the maximum value of 0.52 to the consequent fair. Similarly, the consequent "terrible" appears five times with a maximum value of 0.45. "OK" appears twice with a maximum value of 0.35. "Bad" appears 10 times with a maximum value of 0.45, and "good" appears five times with a maximum value of 0.4. The centroid defuzzification method when combined with the min–max rule "clips" the output membership functions at their maximum value. In this example, the membership functions are clipped as follows: good = 0.4, fair = 0.52, OK = 0.35, bad = 0.45, terrible = 0.45. The shaded area in Figure 12.13 shows the results of the clipping operation in this example. The defuzzified value is the centroid of the shaded area in Figure 12.13. In this case, the centroid is equal to 0.5036. Thus on day 1, machinist 2 had an OK "type of day" ($0.5036 \approx 0.5$).

The next step is to provide an average and a distribution for the entire day based on the results from each machinist. The procedure outlined above can be followed for each machinist for day 1. The results, with the average and the range for the day, are given in

Table 12.8. *Rules with membership values for machinist 2, day 1 from the simulation study. The rule input values are as follows: Number of parts* = 4.0, *size of parts* = 64.59, *number of setups* = 45, *type of cut* = 1.63, *and beryllium exposure* = 0.181. *Legend: F = few, M = many, S = small, L = large, Fi = fine, R = rough, Lo = low, H = high, Fa = fair, OK = okay, T = terrible, B = bad, G = good.*

Rule number	Number of parts	Size of parts	Number of setups	Type of cut	Beryllium exposure	Type of day
1	F = 0.60	S = 0.52	F = 0.65	Fi = 0.84	Lo = 0.55	Fa = 0.52
2	F = 0.60	S = 0.52	F = 0.65	Fi = 0.84	H = 0.45	T = 0.45
3	F = 0.60	S = 0.52	F = 0.65	R = 0.16	Lo = 0.55	OK = 0.16
4	F = 0.60	S = 0.52	F = 0.65	R = 0.16	H = 0.45	T = 0.16
5	F = 0.60	S = 0.52	M = 0.35	Fi = 0.84	Lo = 0.55	Fa = 0.35
6	F = 0.60	S = 0.52	M = 0.35	Fi = 0.84	H = 0.45	B = 0.35
7	F = 0.60	S = 0.52	M = 0.35	R = 0.16	Lo = 0.55	Fa = 0.16
8	F = 0.60	S = 0.52	M = 0.35	R = 0.16	H = 0.45	T = 0.16
9	F = 0.60	L = 0.48	F = 0.65	Fi = 0.84	Lo = 0.55	Fa = 0.48
10	F = 0.60	L = 0.48	F = 0.65	Fi = 0.84	H = 0.45	B = 0.45
11	F = 0.60	L = 0.48	F = 0.65	R = 0.16	Lo = 0.55	Fa = 0.16
12	F = 0.60	L = 0.48	F = 0.65	R = 0.16	H = 0.45	T = 0.16
13	F = 0.60	L = 0.48	M = 0.35	Fi = 0.84	Lo = 0.55	G = 0.35
14	F = 0.60	L = 0.48	M = 0.35	Fi = 0.84	H = 0.45	B = 0.35
15	F = 0.60	L = 0.48	M = 0.35	R = 0.16	Lo = 0.55	Fa = 0.16
16	F = 0.60	L = 0.48	M = 0.35	R = 0.16	H = 0.45	B = 0.16
17	M = 0.40	S = 0.52	F = 0.65	Fi = 0.84	Lo = 0.55	Fa = 0.40
18	M = 0.40	S = 0.52	F = 0.65	Fi = 0.84	Hi = 0.45	B = 0.40
19	M = 0.40	S = 0.52	F = 0.65	R = 0.16	Lo = 0.55	Fa = 0.16
20	M = 0.40	S = 0.52	F = 0.65	R = 0.16	H = 0.45	T = 0.16
21	M = 0.40	S = 0.52	M = 0.35	Fi = 0.84	Lo = 0.55	G = 0.35
22	M = 0.40	S = 0.52	M = 0.35	F = 0.84	H = 0.45	B = 0.35
23	M = 0.40	S = 0.52	M = 0.35	R = 0.16	Lo = 0.55	Fa = 0.16
24	M = 0.40	S = 0.52	M = 0.35	R = 0.16	H = 0.45	B = 0.16
25	M = 0.40	L = 0.48	F = 0.65	Fi = 0.84	Lo = 0.55	G = 0.40
26	M = 0.40	L = 0.48	F = 0.65	Fi = 0.84	Hi = 0.45	B = 0.40
27	M = 0.40	L = 0.48	F = 0.65	R = 0.16	L = 0.55	Fa = 0.16
28	M = 0.40	L = 0.48	F = 0.65	R = 0.16	H = 0.45	B = 0.16
29	M = 0.40	L = 0.48	M = 0.35	Fi = 0.84	Lo = 0.55	G = 0.35
30	M = 0.40	L = 0.48	M = 0.35	Fi = 0.84	H = 0.45	OK = 0.35
31	M = 0.40	L = 0.48	M = 0.35	R = 0.16	Lo = 0.55	G = 0.16
32	M = 0.40	L = 0.48	M = 0.35	R = 0.16	H = 0.45	B = 0.16

Table 12.9. Figures 12.14 and 12.15 are the fuzzy X bar and R charts, respectively, for the 30-day run in this example. For the same 30-day run, the daily average beryllium exposure and beryllium exposure ranges were plotted in the form of X bar–R charts. These plots are presented in Figures 12.16 and 12.17, respectively.

The motivation for this study was that beryllium exposure is highly task dependent. In the new Los Alamos beryllium facility, daily tasks will be highly variable. This means that a standard single-variable Shewhart X bar–R chart based only on daily beryllium exposure probably will not provide the control information required. In fact, previous beryllium exposure data show that these control charts do not supply the needed information. The 30-day simulator run that produced the data for Figures 12.14 and 12.15 was run in a totally random mode. This means that, as far as the simulator reproducing the real plant, this 30-day run was normal and should be entirely in control. The fuzzy X bar–R charts, Figures 12.14 and 12.15, based on the "type of day" (the type of day based on the daily circumstances)

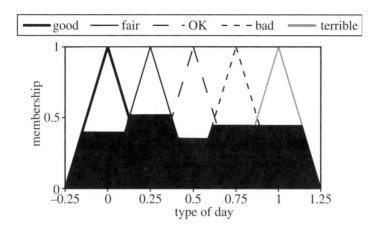

Figure 12.13. *"Clipped" output membership functions for Example* 4.

Table 12.9. *Computed value for day* 1 *from Example* 4.

Description	Value
machinist 1 (type of day)	0.4041
machinist 2 (type of day)	0.5036
machinist 3 (type of day)	0.4264
machinist 4 (type of day)	0.4088
average (*X* bar)	0.4357
range (*R*)	0.0995

show that the plant is entirely in control. Conversely, the X bar chart for beryllium exposure only (Figure 12.16) shows the plant as being out of control on days 5 and 24. The R chart (Figure 12.17), however, does not show control problems within the data sets. This is what one would expect because the perturbations come on a daily basis. Thus in this case the fuzzy control charts performed better than the purely statistical charts based only on beryllium exposure data. This is because the fuzzy charts have the ability to easily take into account varying circumstances, like heavy workdays and large parts, etc. All four charts, and the remaining charts in this section and section 12.4, use control limits based on (12.1)–(12.4). In this case, the sample size, n, was equal to 4, and the parameters A_2 and D_4 were 0.729 and 2.282, respectively.

There are statistical techniques that can take the daily plant variation into account. They will be discussed in section 12.4. Three other observations need to be made here:

- The beryllium exposure shown in Figure 12.16 is often above the $0.2\mu g/m^3$ limit. This reflects the fact that most of the data that was used to construct the simulator came from work performed before the new limits were put into effect.

- Other information such as the number of setups and size of parts might seem inconsistent with information that a machinist might be familiar with. Unfortunately, even though nothing mentioned here is secret information, a great deal of the information at Los Alamos is only available on a need-to-know basis. Consequently, in many publications we are unable to use "real numbers." The model used here is totally consistent and "not quite real numbers" will not affect the results.

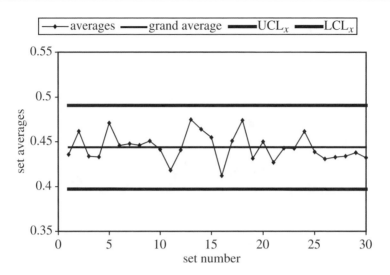

Figure 12.14. *Fuzzy X bar chart for a "normal" 30-day beryllium plant run.*

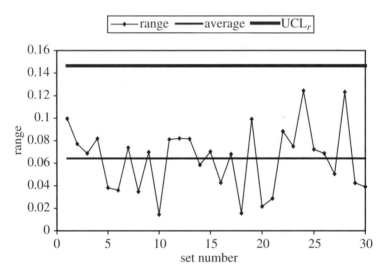

Figure 12.15. *Fuzzy range (R) chart for a "normal" 30-day beryllium plant run.*

- Because of the lack of available information, we are not addressing certain questions, such as "Are the tolerance limits inside or outside the natural process limits?" until we are able to work with actual plant data.

The following is a discussion of how well the fuzzy X bar–R charts detect out-of-control situations.

12.3.5   Example 5: Fuzzy control

For this case, it is assumed that the plant is running for three days (5, 6, and 7) with the ventilating system working at a diminished capacity without the operators' knowledge. Days 5, 6, and 7 are reasonably normal days otherwise, with the workload ranging from 29 parts

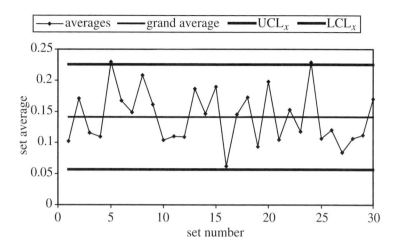

Figure 12.16. *Beryllium exposure X bar chart for a "normal" 30-day beryllium plant run.*

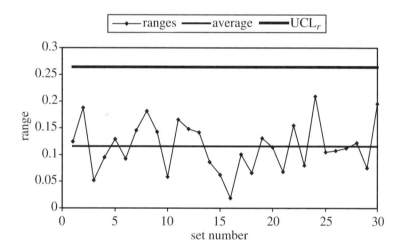

Figure 12.17. *Beryllium exposure range (R) chart for a "normal" 30-day beryllium plant run.*

on day 5 to 19 parts on day 7. The simulator is programmed to produce reasonably high beryllium readings on these days. The fuzzy X bar and R charts for a 30-day run covering these three days are shown in Figures 12.18 and 12.19, respectively.

The fuzzy X bar chart shown in Figure 12.18 signals that days 5, 6, and 7 are out of control. In an actual operating situation, Figure 12.18 probably would be just an extension of a "running" chart like Figure 12.14 with the control limits already in place, so the out-of-control situation would be caught immediately. The fuzzy R chart shown in Figure 12.19 shows that the set of ranges for the three days in question are about normal. This is what would be expected if all readings were high for every machinist on a given day. Figures 12.20 and 12.21 show only the beryllium exposure X bar and R charts for the same 30-day run. The standard single-variable X bar chart shown in Figure 12.20 shows, equally as well as the fuzzy X bar chart, that days 5, 6, and 7 are out of control. In this case, the recomputed

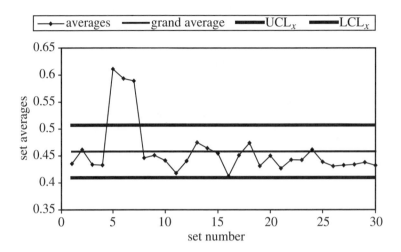

Figure 12.18. *Fuzzy X bar chart showing results for a ventilating system that worked poorly for three days during a 30-day beryllium plant run.*

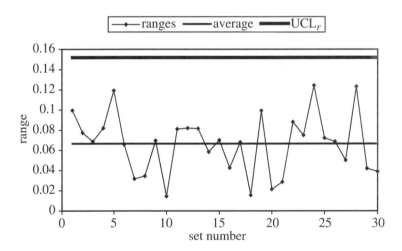

Figure 12.19. *Fuzzy range (R) chart showing results for a ventilating system that worked poorly for three days during a 30-day beryllium plant run.*

control limits are raised slightly because of the higher beryllium exposure readings. Day 24, which was out of control before, is now barely in control.

Day 16 is now below the lower control limit because of the movement in the control limits, but this is acceptable since low beryllium readings are good. The standard R chart (Figure 12.21) shows nothing unusual as far as dispersion within the sets is concerned. This is what is expected for the "bad ventilation system" problem. Next, the issue of how well the fuzzy X bar–R charts handle high beryllium exposure conditions that are justified will be discussed.

12.3.6   Example 6: A campaign

In this case, it is assumed that the plant, for three days (13, 14, and 15), is running a campaign, trying to process a large number of similar parts for a special project. In this

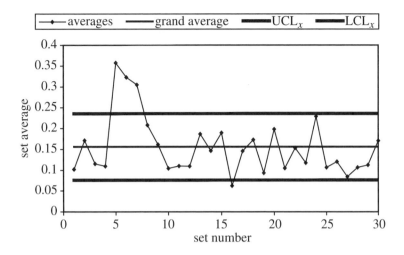

Figure 12.20. *Standard X bar chart (beryllium exposure only) showing results for a ventilating system that worked poorly for three days during a 30-day beryllium plant run.*

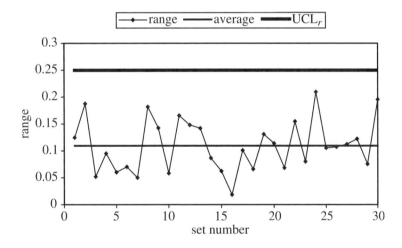

Figure 12.21. *Standard range (R) chart (beryllium exposure only) showing results for a ventilating system that worked poorly for three days during a 30-day beryllium plant run.*

case, the parts are large, require a precision finish with lots of fine cuts, and require more than a normal number of setups. All machinists receive relatively large beryllium exposure (an average of about $0.3\mu g/m^3$). In this case, the high beryllium exposures are expected and considered normal. The simulator is programmed to produce the above conditions for the designated three days. The fuzzy X bar and R charts for a 30-day run covering these three days are shown in Figures 12.22 and 12.23, respectively. Figures 12.24 and 12.25 show only the beryllium exposure X bar and R charts for the same 30-day run. The fuzzy X bar chart, Figure 12.22, handles the high beryllium exposure that is due to a high workload or high capacity days, as intended. The plant is not out of control because it is operating as expected for a high workload. The standard single-variable X bar chart (Figure 12.24), based on beryllium exposure only, signals that the plant is out of control for the days that

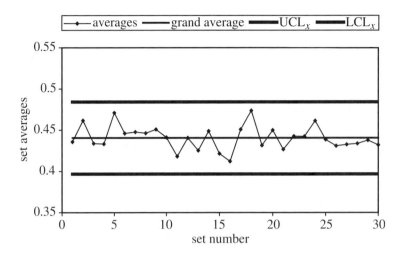

Figure 12.22. *Fuzzy X bar chart showing results for a high-capacity day with high beryllium exposure readings for three days during a 30-day beryllium plant run.*

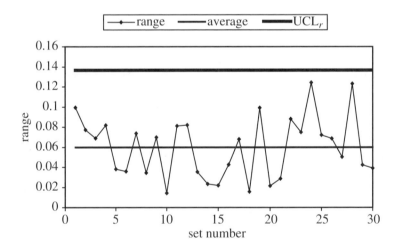

Figure 12.23. *Fuzzy range (R) chart showing results for a high-capacity day with high beryllium exposure readings for three days during a 30-day beryllium plant run.*

high beryllium readings are obtained. This is not the desired result. Both range charts, Figures 12.23 and 12.25, show no unusual dispersion in the in-set data, as expected.

The next section discusses an SPC technique, using regression that can be applied to this type of problem.

12.4 SPC techniques for measurement data requiring correlation and using regression

The beryllium exposure control problem can be solved without fuzzy logic. One can regress all the data from a 30-day run and find a relationship between the expected beryllium exposure (y_e) and the variables, number of parts (x_1), size of parts (x_2), number of setups (x_3), and

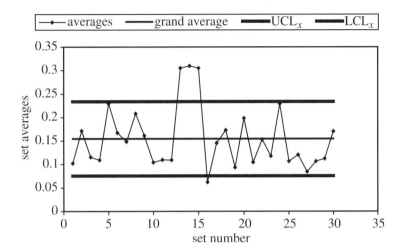

Figure 12.24. *Standard X bar chart (beryllium exposure only) showing results for a high-capacity day with high beryllium exposure readings for three days during a 30-day beryllium plant run.*

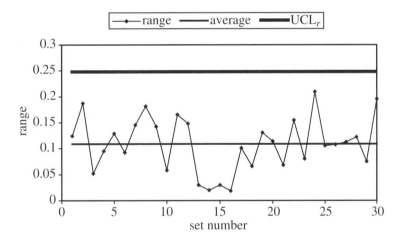

Figure 12.25. *Standard range (R) chart (beryllium exposure only) showing results for a high-capacity day with high beryllium exposure readings for three days during a 30-day beryllium plant run.*

the type of cut (x_4). The regression equation form that was used for this study is given as

$$y_e = a_0 + a_1 * x_1 + a_2 * x_2 + a_3 * x_3 + a_4 * x_4. \tag{12.9}$$

Obviously, more sophisticated forms can be used, but the user is cautioned not to over-fit the data because it would be easy to turn out-of-control points into expected points. Although our search was not exhaustive, in all the reference material listed at the end of the chapter, this technique is only briefly mentioned in Western Electric Company, Inc. (1995). It is not surprising, therefore, that the beryllium plant operators were not using this technique. Statisticians tend to automatically migrate toward an approach like this, however. The procedure for this technique is as follows:

- Choose a form for the regression equation, e.g., (12.9).

- Gather all the individual data for the entire run (in this case, 30 days). These data include the measured beryllium exposures (y_m) and the variables x_1 through x_4.

- Regress the data and find the values of a_0 through a_4.

- Use the regression equation (12.9) to find the expected beryllium exposures and compare them to the measured exposures. Compute the deviation (D). (In this case, $D = y_e - y_m$.)

- Compute the average deviation and the deviation range for each data set or each day.

- Compute the grand average deviation and the average deviation range.

- Plot this information on standard X bar–R charts using the same equations and tables as before to compute the control limits.

Examples 4–6 will be reworked as Examples 7–9 using the regression technique in order to compare the results with the fuzzy technique. Example 7 is the normal run, Example 8 is the bad ventilating system run, and Example 9 is the high capacity days run. Table 12.10 gives the regression coefficients obtained for Examples 7–9.

Table 12.10. *Regression coefficients for Examples 7–9.*

Example number	a_0	a_1	a_2	a_3	a_4
7	−0.00723695	0.0175316	0.0304519	−0.01306050	0.0379174
8	−0.00852286	0.0317977	0.0340590	−0.02612060	0.0482558
9	−0.0144913	0.0135056	0.0233584	−0.00861036	0.0771309

12.4.1   Example 7: Regression on Example 4

This is the normal 30-day run; nothing should be out of control. These are the same operating conditions as in Example 4. Figures 12.26 and 12.27 are the X bar and R charts for this example using regression.

The X bar chart for this example, Figure 12.26, shows no out-of-control sets, as expected. Day 24 does show a dip toward the control limit line. If we observe Figure 12.16, the X bar chart for beryllium exposure only, for a normal run, we see that day 24 was an "out-of-control" day. Figure 12.26 dips in the opposite direction from Figures 12.4–12.18 because of the way deviation was defined ($D = y_p - y_m$). Figure 12.14, the fuzzy X bar chart for the same run, shows a fairly normal point for day 24, indicating that the fuzzy rules explained the higher beryllium exposure reading adequately. The R chart for this example shows too much dispersion within the set on day 11. This is a range out-of-control condition. The fuzzy range chart for this run, Figure 12.15, shows a day with a normal range. The beryllium exposure only range chart for this run, Figure 12.17, shows a normal range as well for day 11. The deviations from the regression surface for day 11 show one point nearly on the surface, one point about an average range value from the surface, and two points about two average range values from the surface. This provides an extremely large deviation range for that day. However, it apparently has little or no physical meaning and can be listed as a false alarm.

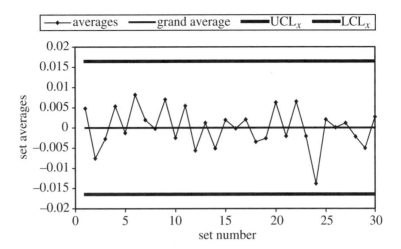

Figure 12.26. *Regression X bar chart showing results for a normal 30-day beryllium plant run.*

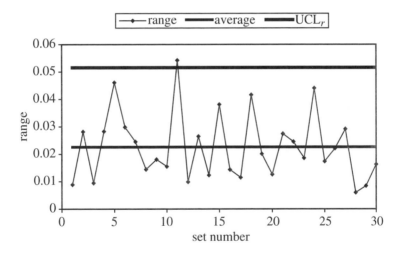

Figure 12.27. *Regression range (R) chart showing results for a normal 30-day beryllium plant run.*

12.4.2   Example 8: Regression on Example 5

In this 30-day run, days 5, 6, and 7 are days where the ventilating system is not working well. These three days should be out of control. These are the same operating conditions as in Example 5. Figures 12.28 and 12.29 are the X bar and R charts for this example using regression.

The X bar chart for this example does indeed show the plant to be out of control on days 5, 6, and 7, as is the case. Again the deviation moves in a negative direction because of the way it is defined. It is interesting to note how the regression pushed all the "in-control" points either onto the zero deviation line or onto the positive side of it. This is a negative point about the regression technique. If there is a strong out-of-control condition when the

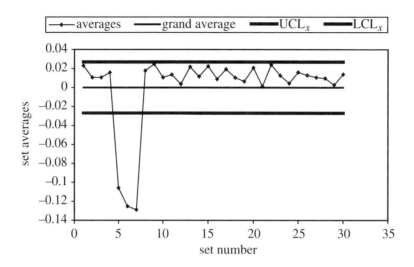

Figure 12.28. *Regression X bar chart showing results for a ventilating system that worked poorly for three days during a 30-day beryllium plant run.*

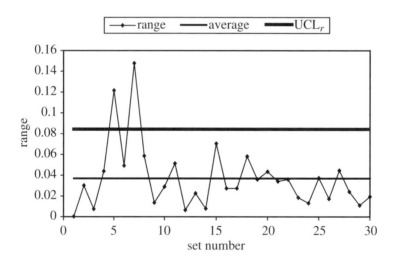

Figure 12.29. *Regression range (R) chart showing results for a ventilating system that worked poorly for three days during a 30-day beryllium plant run.*

regression line and the control limits are being established, the out-of-control points will have a strong influence on the limits. This would not happen if the regression equation was established during in-control conditions and the out-of-control deviation points were just computed from that regression surface. The range chart, Figure 12.29, shows that days 5 and 7 have too much dispersion in the data sets. Neither the fuzzy range chart for this run, Figure 12.19, nor the standard R chart (beryllium exposure only) shows unusual deviations. The range of deviations shown in Figure 12.29 represents excursions from the regression surface and has no physical significance. These are false alarms.

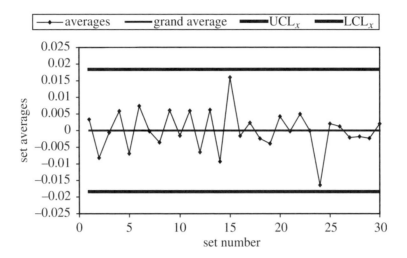

Figure 12.30. *Regression X bar chart showing results for a high-capacity day with high beryllium exposure readings for three days during a 30-day beryllium plant run.*

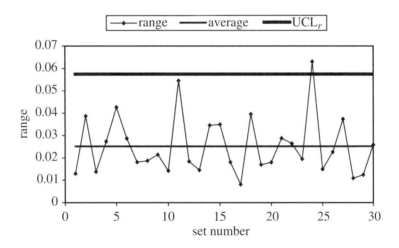

Figure 12.31. *Regression range (R) chart showing results for a high-capacity day with high beryllium exposure readings for three days during a 30-day beryllium plant run.*

12.4.3   Example 9: Regression on Example 6

In this 30-day run, days 13, 14, and 15 are high-capacity days. These three days should not be out of control. These are the same operating conditions as in Example 6. Figures 12.30 and 12.31 are the X bar and R charts for this example, using regression.

The X bar chart for this example shows the plant to be in control on days 13, 14, and 15, as is the case. The range chart, Figure 12.31, shows day 24 to have too much dispersion in the data set. The fuzzy range chart for this run, Figure 12.23, and the standard R chart (beryllium exposure only), Figure 12.25, both show reasonably high deviations for day 24, but not out-of-control conditions. The range of deviations shown in Figure 12.31 does have some physical significance, but the chart is still giving a false alarm.

12.4.4   Comments on regression

Although this example is not exceptionally good for the regression method, it does work very well if there is a large amount of data and if the original analysis is based on "in-control" conditions. The fact that the regression is based on the sum of the squared errors causes larger than desired distortion in the control chart if it is constructed with data that has out-of-control points in it. Admittedly, (12.9) is a very simple regression equation. A more sophisticated equation using a squared term or cross-products of $x_i * x_j$ probably would do a better job, but only if the regression is based on in-control data, such as in Example 7. By the same token, the fuzzy membership functions and rules used in this example also can be more sophisticated. The binary membership functions in Figure 12.10 and the simple basis for the rules in Table 12.7 can and probably will be enhanced in the future. Given the simple basis, the fuzzy system works very well. A final comment on regression versus fuzzy is that the "type of day" concept used by the fuzzy system is a little more meaningful to plant workers than the concept of deviation from the expected value used in the regression technique.

12.5   SPC techniques for attribute data

Most attribute data have only two values, for example, meets specifications or does not meet specifications, is useful or should be discarded, etc. The four most common types of control charts used for attribute data are the following:

1. The p-chart is used to record the proportion of discards, rejects, or "does not meets" per sample set. For the p-chart, the sample size can vary.

2. The np-chart is similar to the p-chart, except that the sample size must be constant and the number of rejects, etc. are recorded as opposed to the proportion.

3. The c-chart is used to record the flaws or less-than-perfect conditions for a sample of constant size. This type of chart is usually used to measure things like the number of devices on a circuit board that do not meet specifications or the number valves on a valve tray that do not open properly. The circuit boards and the valve trays must be the same size.

4. The u-chart is similar to the c-chart except that the sample size can vary. In this case the valve trays or circuit boards can be different sizes.

All of these charts are similar, but for the sake of brevity only the p-chart will be discussed here. For a good and quick discussion of all these techniques, see Mamzic (1995). Example 1 will be recast in order to demonstrate the use of the p-chart. The data will be changed to the number of rejects per sample instead of the actual average measurement of the variable (the length of the log). Since the sample size ($n = 5$) is constant, this example could be done as an np-chart, but the p-chart approach fits nicely into the fuzzy examples in the next section.

12.5.1   Example 10: Log cabin revisited

In this example, the log cabin building problem is revisited. The goal still is to produce 20-foot or 240-inch logs, ±2.5 inches. Logs with lengths greater than 242.5 inches and less than 237.5 inches will be rejected. The random data from Example 1, shown in Table 12.1, are altered to reflect the fraction of rejects for each set of sample size 5. These data are shown in Table 12.11.

Table 12.11. *Proportions of rejections per sample of sets of five logs (first try).*

Set number	Proportion of set rejected (p)
1	0.6
2	0.8
3	0.4
4	0.6
5	0.6
6	1.0
7	0.8
8	0.8
9	0.6
10	1.0
11	1.0
12	1.0
13	0.6
14	0.8
15	0.8
16	0.8
17	1.0
18	0.6
19	0.8
20	0.8

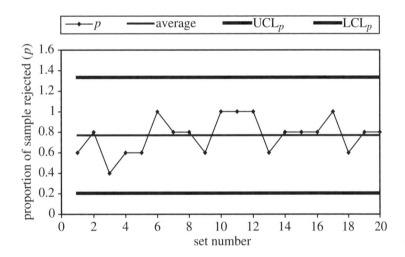

Figure 12.32. *The p-chart for the log cabin problem (first try).*

Figure 12.32 is the p-chart for this run. This chart shows the individual p-values (the fraction of rejections per set), the average p-value for the entire run ($\bar{p}$), and the upper and lower control limits, UCL_p and LCL_p. This chart also provides information regarding patterns in the central tendency of the data. The upper and lower control limits are defined by

$$\mathrm{UCL}_p = \bar{p} + 3\sqrt{\frac{\bar{p}(1-\bar{p})}{\bar{n}}}, \tag{12.10}$$

$$\mathrm{LCL}_p = \bar{p} - 3\sqrt{\frac{\bar{p}(1-\bar{p})}{\bar{n}}}, \tag{12.11}$$

where $\bar{p}$ is the average p-value for the run described above and $\bar{n}$ is the average sample set size. In this case, $n = \bar{n} = 5$. Often these formulas will replace $\bar{n}$ with n_i, where n_i is the individual set size and each individual p-value will have individual control limits based on the individual set size. The reader may recognize the radical term in (12.10) and (12.11) as one standard deviation (1σ) for a binomial distribution. Therefore, for the p-chart the upper and lower control limits are the $\pm 3\sigma$ limits for a binomial distribution. Some authors (Box and Luceno (1997)) use the $\pm 2\sigma$ limits as warning markers in addition to the control limits.

The p-chart in Figure 12.32 tells essentially the same story as the X bar and R charts in Figures 12.1 and 12.2. The process is not out of control. The problem is common-cause, not special-cause. In order to meet specifications, the process—not the operation techniques—must be changed. Example 11 is the p-chart version of Example 2. The process has been changed to eliminate the common-cause problems, but a special-cause problem still exists.

12.5.2   Example 11: The log cabin with a special-cause problem

Example 10 is repeated, this time using a saw instead of an axe to eliminate the common-cause problem. In Example 2, the saw fixed the problem so well that no rejections occurred even though there was an out-of-control situation. A p-chart with no rejections says nothing about the problem, so the problem will be slightly modified. Suppose that we feel we need to cut the tolerance to ± 1 inch. The new logs must have a length between 239 and 241 inches. Logs outside these limits, but within the old limits, are considered seconds and can be used to construct the garage. Logs outside the old limits are culls or must be used for firewood. Since the saw was introduced, no logs fall outside of the old limits so no culls are produced. For all practical purposes, this still is a binomial problem, with the seconds considered rejects. Note that Figure 12.6 shows that the new tolerance limits fall inside the natural process limits. The special-cause problem still exists because we are using each newly cut log as a template for each proceeding cut. With the exception of the new saw, the process is run as before, taking a sample of five logs at regular intervals. After 20 intervals, a new p-chart is constructed. Data for this run are presented in Table 12.12. Figure 12.33 is the p-chart for this run. This chart shows, as in Figure 12.4, that near the end of the run the number of rejections starts to move above the upper control limit, signaling a special-cause event. This is because the length of the logs is starting to grow because a different template is being used every time. The p-chart provided an out-of-control signal. (Note that no lower control limit is shown since it would be less than 0.)

The idea of good logs, seconds, and culls presented in Example 11 brings up an interesting problem. How do we develop a p-chart for a multinomial problem since we cannot rely on a binomial distribution to define the control limits? This provides an opportunity to use fuzzy logic to solve the problem. The SPC community favors a generalized p-chart based on the chi-square statistic for the multinomial processes (Duncan (1950)). The next section is devoted to fuzzy techniques for solving multinomial problems. The generalized p-chart will be discussed in section 12.7.

12.6   Fuzzy techniques for attribute data

Raz and Wang (1990) and Wang and Raz (1990) presented the fuzzy solution to multinomial attribute problems. Laviolette and Seaman (1992, 1994) and Laviolette et al. (1995) added more detail to the work of Wang and Raz and then criticized it. One problem with the work presented by Laviolette and Seaman (1992, 1994) and Laviolette et al. (1995) is that the

Table 12.12. *Proportions of rejections per sample of sets of five logs, with modified tolerances (second try).*

Set number	Proportion of set rejected (p)
1	0.0
2	0.0
3	0.4
4	0.2
5	0.2
6	0.0
7	0.2
8	0.2
9	0.2
10	0.2
11	0.2
12	0.0
13	0.0
14	0.2
15	0.0
16	0.0
17	0.2
18	0.0
19	0.6
20	0.8

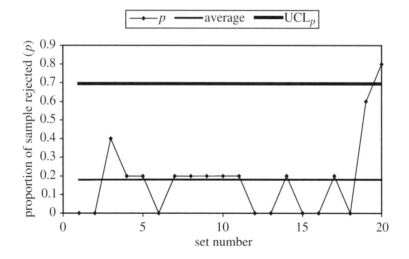

Figure 12.33. *The p-chart for the modified pellet problem (second try).*

membership functions used in their examples are not well defined. For example, the log cabin problem described in Example 11 has three membership functions, premium logs, seconds, and culls. In this problem, the membership functions are "crisp." One could say that all good logs are in the set containing lengths between 239 and 241 or in the set with an absolute error bound of 1. So the set of good logs is in the range that goes from 0 to 1. In a similar fashion the seconds are in the range that goes from 1 to 2.5. If it is assumed that logs are *never* farther than ±5 inches from the specified value, then the culls are in the range that goes from 2.5 to 5. In the real world the bounds cannot always be that clearly defined. In order to make the number come out even, assume that the measuring device (tape measure) is accurate only to ± 0.25 inches. Now good logs can be found in the range from 0 to 1.25. Seconds can be found in the range from 0.75 to 2.75, and culls in the range from 2.25 to 5.0. Finally, in this type of problem, the universe of discourse usually goes from 0 to 1. A score of 0 means perfect, or exactly 240 inches, and a score of 1 is terrible, or an error of 5 inches. If the ranges are normalized, then good logs range from 0 to 0.25, seconds from 0.15 to 0.55, and culls from 0.45 to 1.0. A triangular description of the type of logs produced is shown in Figure 12.34.

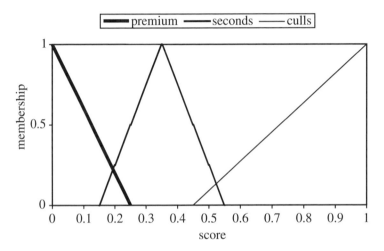

Figure 12.34. *Fuzzy membership functions for log types.*

Laviolette and Seaman (1992, 1994) and Laviolette et al. (1995) base most of their arguments on a brick problem described by Marcucci (1985). This problem involves the manufacture of bricks. Bricks from this process come in three forms: standard, chipped-face, and culls. The standard brick is classified as suitable for all purposes, the chipped-face brick is structurally sound but not suitable for every purpose, and the culls are unacceptable. Obviously, the membership functions for this situation cannot be as clearly defined as those for the log cabin problem, but there has to be some physical justification for assigning the membership functions. Although the connection is mentioned, it is not clear how the membership functions are connected to the physical world. The results can be no better than the model.

The fuzzy approach to developing a multinomial p-chart is as follows:

- Choose a set size and take a sample.

- Count the number in each category (e.g., premium, second, cull) in the sample set.

- Use the fraction in each category to define a fuzzy sample set in terms of the member-ship functions that define the system (e.g., Figure 12.34).

- Defuzzify the fuzzy sample in order to get one representative value for the sample set to use in constructing a p-chart.

This procedure is best described with an example. The example is an extension of the beryllium plant problem described in Examples 4–6 and Figure 12.12. The plant simulation works as before, but random number generators have been added to simulate scratches of various lengths and depths on the surfaces of the parts. The *p*-chart problem is a quality control problem. This is a different problem than the exposure control problem that was investigated earlier.

12.6.1   Example 12: Fuzzy attribute data

For this problem, all parts going through the beryllium plant are inspected twice. In the first inspection, part dimensions are measured with a machine to see if they fall within the desired tolerance times. This machine is a precision instrument, so parts are assigned to categories in a crisp manner. There are three categories: firsts, seconds, and culls. There is a reasonably large need for the seconds, and a tight tolerance for the firsts, so enough seconds are generated by this procedure to satisfy demand. The firsts are assigned a score of 0.0, the seconds a score of 0.5, and the culls a score of 1.0. A second inspection is performed to examine the finish of the part, or the surface area. The inspector visually checks for scratches, checks for the number of scratches, the average length of the scratches, and the average depth of the scratches. If the deepest scratch is more than three times the average scratch depth, then the deepest value is used rather than the average. A similar rule is used to document scratch length. In the actual plant inspection, the inspector does most of this visually and the recorded values are just good estimates. In the simulation, exact numbers are used. The numbers used in this example are not the real values but are representative values.

After the number, depth, and length of the scratches are recorded, a fuzzy rule-based system determines how much to "downgrade" the part from the first inspection category. Because beryllium metal is expensive, two additional categories have been added to the type of product that is obtained after both inspections. They are "possible rework to a first" and "possible rework to a second." This is because some finishes can be reworked without changing the dimensions beyond the tolerance limits. The fuzzy membership functions that describe the beryllium parts after the inspections are shown in Figure 12.35.

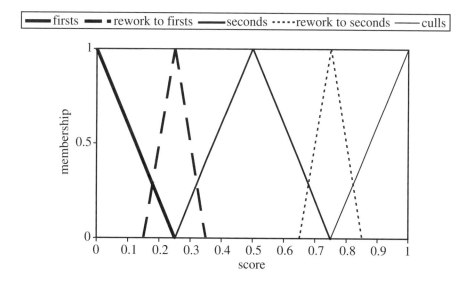

Figure 12.35. *Fuzzy membership functions describing beryllium parts after inspections.*

12.6.2   The fuzzy system used to downgrade parts

This fuzzy system consists of three input variables and one output variable. Each input variable has three membership functions and the output variable has five membership functions. The input and the output are connected by 27 rules. The three input variables are as follows:

1. Number of scratches: with a range of 0 to 9 and membership functions

    (a) small,

    (b) medium,

    (c) large.

2. Length of scratches: with a range of 0 to 2.4 centimeters and membership functions

    (a) short,

    (b) medium,

    (c) long.

3. Depth of scratches: with a range of 0 to 10 microns and membership functions

    (a) shallow,

    (b) medium,

    (c) deep.

The output variable is as follows:

1. Amount of downgrade: with a range of 0 to 0.4 and membership functions

    (a) very small,

    (b) small,

    (c) medium,

    (d) large,

    (e) very large.

Figure 12.36 shows the input membership functions and Figure 12.37 shows the output membership functions. The rules are given in Table 12.13 and the form of the rules is as follows:

> If (number of scratches) is. . . and if (depth of scratches) is. . . and if (length of scratches) is. . . , then the amount of downgrade is. . . .

The input variables are determined by measurements, or estimates, of the second inspector. The appropriate rules are fired. The min–max centroid approach is used to defuzzify the result, as in Example 4. The defuzzified value computed from this procedure is added to the score from the first inspection, giving a final score. The final score is applied to Figure 12.35 to place the part in its final category (firsts, rework to firsts, seconds, rework to seconds, culls). Entering the final score at the corresponding point on the abscissa of Figure 12.35 and then choosing the category with the highest membership function accomplishes this. Before continuing with the explanation of the procedure, it is appropriate to illustrate what has been done up to this point with some actual numbers.

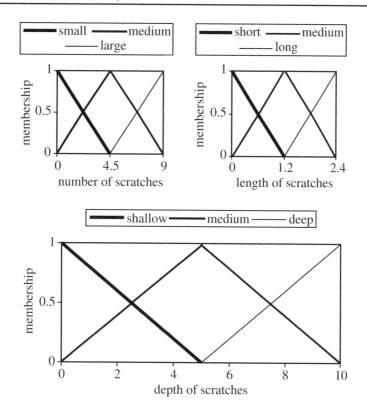

Figure 12.36. *Input membership functions for rules to downgrade beryllium parts.*

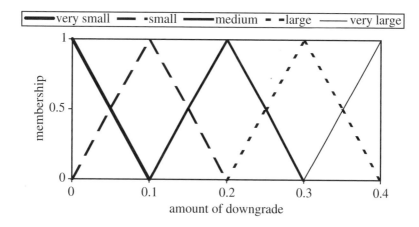

Figure 12.37. *Output membership functions for rules to downgrade beryllium parts.*

On the first day of the run, 13 parts were manufactured. The first inspector inspected part number 7 by comparing the dimensions to the specifications and assigning an initial score of 0.5. The second inspector checked the surface area of the part and found seven scratches. The average scratch length was 0.56cm. The average depth was small, but one scratch had a depth of about 7.5μ. Since this was greater than three times the average depth, the maximum, 7.5μ is used instead of the average. This inspection can be done visually with

Table 12.13. *Fuzzy rules for beryllium part downgrade.*

Rule number	Number of scratches	Length of scratches	Depth of scratches	Amount of downgrade
1	small	short	shallow	very small
2	small	short	medium	small
3	small	short	deep	large
4	small	medium	shallow	small
5	small	medium	medium	medium
6	small	medium	deep	large
7	small	long	shallow	large
8	small	long	medium	large
9	small	long	deep	very large
10	medium	short	shallow	small
11	medium	short	medium	medium
12	medium	short	deep	large
13	medium	medium	shallow	medium
14	medium	medium	medium	medium
15	medium	medium	deep	large
16	medium	long	shallow	large
17	medium	long	medium	large
18	medium	long	deep	very large
19	large	short	shallow	large
20	large	short	medium	large
21	large	short	deep	very large
22	large	medium	shallow	large
23	large	medium	medium	large
24	large	medium	deep	very large
25	large	long	shallow	very large
26	large	long	medium	very large
27	large	long	deep	very large

these numbers being only estimates. However, this material is quite expensive and with only a few parts to inspect each day the inspector can actually be quite meticulous.

These values are applied to the input membership functions in Figure 12.36. From this figure, for the number of scratches = 7, the membership in large is 0.56, the membership in medium is 0.44, and the membership in small is 0.0. The memberships for the length of scratches = 0.56, from the same figure, are 0.0 for long, 0.47 for medium, and 0.53 for short. The memberships for the depth of scratches = 7.5 are 0.5 for deep, 0.5 for medium, and 0.0 for shallow. Unlike Example 4, these input membership functions are not binary; therefore, in this case, only eight of the 27 rules will fire. Rules 1 through 9 will not fire because the variable "number of scratches" has no membership in the set small. Rules 16 through 18 and rules 25 through 27 will not fire because the variable "length of scratches" has no membership in the set long. Rules 10, 13, 19, and 22 will not fire because the variable "depth of scratches" has no membership in the set shallow. The eight rules left to fire are 11, 12, 14, 15, 20, 21, 23, and 24.

For rule 11 from Table 12.13, the inputs

- number of scratches: medium = 0.44;

- length of scratches: short = 0.53;

- depth of scratches: medium = 0.5

Table 12.14. *Fuzzy rules "fired" in Example* 12 *to determine the beryllium part downgrade.*

Rule number	Number of scratches	Length of scratches	Depth of scratches	Amount of downgrade
11	medium = 0.44	short = 0.53	medium = 0.50	medium = 0.44
12	medium = 0.44	short = 0.53	deep = 0.50	large = 0.44
14	medium = 0.44	medium = 0.47	medium = 0.50	medium = 0.44
15	medium = 0.44	medium = 0.47	deep = 0.50	large = 0.44
20	large = 0.56	short = 0.53	medium = 0.50	large = 0.50
21	large = 0.56	short = 0.53	deep = 0.50	very large = 0.50
23	large = 0.56	medium = 0.47	medium = 0.50	large = 0.47
24	large = 0.56	medium = 0.47	deep = 0.50	very large = 0.47

lead to the conclusion that the consequent of the rule "amount of downgrade" = medium has a value of 0.44. This is a consequence of the minimum portion of the min–max rule. The "fired" rules and their consequent values are shown in Table 12.14.

Using the maximum portion of the min–max rule and the values in the last column of Table 12.14 to resolve the "or" nature of the output membership functions, the following values are obtained:

- very small = 0.0;

- small = 0.0;

- medium = 0.44;

- large = 0.50;

- very large = 0.50.

The output membership functions shown in Figure 12.37 are "clipped" at these values, as shown in Figure 12.38. The centroid of the shaded area in Figure 12.38 is computed and is used as the defuzzified value for the amount of downgrade for part number 7.

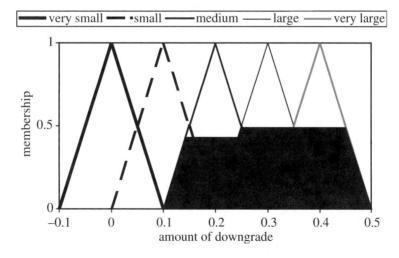

Figure 12.38. *"Clipped" output membership functions for Example* 12.

The centroid of the shaded area in Figure 12.38 is ≈ 0.304. This is the amount of downgrade applied to part 7 by the second inspector. The downgrade is added to the initial score, in this case 0.50, to obtain a final score of 0.804. This final score is the value on the abscissa of Figure 12.35 that is used to categorize part number 7. A line perpendicular to the score axis of Figure 12.35, at 0.804, is projected upward, crossing the membership functions "cull" and "rework to seconds" at ≈ 0.22 and 0.46, respectively. These values represent the membership of the final score in the respective sets; the values are read from the ordinate of the plot in Figure 12.35 at the points where the perpendicular line drawn from the score 0.804 crosses the respective membership function triangles. Since the membership in "rework to seconds" is greater than the membership in "cull," part number 7 is assigned the "rework to seconds" category. All 13 parts manufactured on day 1 are categorized in the same manner as part 7. At the end of day 1 the following results are obtained:

- number of parts categorized as firsts = 6;

- number of parts categorized as rework to firsts = 2;

- number of parts categorized as seconds = 2;

- number of parts categorized as rework to seconds = 1;

- number of parts categorized as culls = 2.

Table 12.15 shows the categorization for all parts for a 30-day run.

In this case, there is no choice of a sample size for each set. The sample size will vary from day to day since all parts produced daily are in the sample set. The obvious choice for this example is the p-chart because sample sizes vary.

The next step is to provide a fuzzy representation for each day based on the categorization of each part for that day. This is easily done using the *extension principle* from fuzzy logic and the concept of a triangular fuzzy number (TFN) as described by Kaufmann and Gupta (1985). A TFN is completely described by a triplet, $T = (t_1, t_2, t_3)$, or, in this case, the vector $[t_1, t_2, t_3]^T$. The values t_1, t_2, and t_3 are the x-values of the x–y pairs representing the corners of a triangle with the base resting on the x-axis ($y = 0$) and the apex resting on the line $y = 1$. Such a triangle can be described by the three points in the x–y plane $(t_1, 0)$, $(t_2, 1)$, and $(t_3, 0)$. For example, the triangular membership function "seconds" in Figure 12.35 can be described as a TFN with $t_1 = 0.25$, $t_2 = 0.5$, and $t_3 = 0.75$, or $[0.25, 0.5, 0.75]^T$. The other four output membership functions are described as follows:

- firsts = $[0.0, 0.0, 0.25]^T$;

- rework to firsts = $[0.15, 0.25, 0.35]^T$;

- rework to seconds = $[0.65, 0.75, 0.85]^T$;

- culls = $[0.75, 1.0, 1.0]^T$.

A matrix, called the A matrix, can be constructed with columns comprised of the five-output membership function TFNs. For this example, the A matrix is

$$A = \begin{bmatrix} 0.0 & 0.15 & 0.25 & 0.65 & 0.75 \\ 0.0 & 0.25 & 0.5 & 0.75 & 1.0 \\ 0.25 & 0.35 & 0.75 & 0.85 & 1.0 \end{bmatrix}.$$

Table 12.15. *The categorization for all parts for the 30-day run from Example* 12.

Day	Number of firsts	Number of rework to firsts	Number of seconds	Number of rework to seconds	Number of culls	Total number of parts
1	6	2	2	1	2	13
2	12	2	6	0	2	22
3	9	6	3	1	0	19
4	10	2	4	1	1	18
5	12	5	6	1	5	29
6	14	3	5	2	0	24
7	9	5	3	2	0	19
8	15	3	7	2	2	29
9	12	3	7	2	0	24
10	6	2	4	2	0	14
11	12	3	2	2	1	20
12	9	5	3	0	0	17
13	7	4	6	4	0	21
14	8	5	2	1	2	18
15	15	4	6	2	1	28
16	4	0	2	4	1	11
17	11	4	6	1	0	22
18	10	4	6	3	1	24
19	5	3	5	0	1	14
20	13	7	7	2	1	30
21	12	4	4	0	1	21
22	13	3	5	1	1	23
23	6	5	2	3	1	17
24	20	4	6	1	1	32
25	5	2	4	1	3	15
26	15	0	2	3	0	20
27	9	0	2	0	0	11
28	8	1	7	0	0	16
29	11	3	2	3	0	19
30	15	6	5	1	0	27

Next, a five-element vector called B is constructed. The first element of the B vector is the fraction of the daily readings that were firsts. The second element is the fraction of the readings that were rework to seconds, and so on. A B vector can be constructed for every day of the run from the values in Table 12.15. For example, for day 1, using row 1 from Table 12.15, the following value is obtained for the B vector:

$$B = \left[\frac{6}{13}, \frac{2}{13}, \frac{2}{13}, \frac{1}{13}, \frac{2}{13} \right]^T = [0.462, 0.154, 0.154, 0.077, 0.154]^T.$$

The product AB is a TFN that represents the fuzzy distribution for the day. For day 1 of this 30-day run, the TFN $AB \approx [0.227, 0.327, 0.504]^T$. This is a triangular distribution that is approximately halfway between "rework to firsts" and "seconds." Figure 12.39 shows how day 1 is distributed on a "score chart" like Figure 12.35. The shaded area is the TFN or fuzzy distribution for day 1. A different distribution is obtained every day. In order to construct a control chart, values for both a centerline and control limits must be determined. There are several metrics that can be used to represent the central tendency of a fuzzy set (Raz and Wang (1990), Wang and Raz (1990), Laviolette and Seaman (1992, 1994), Laviolette et al. (1995)). Two of them will be discussed here. The first metric to be discussed is the

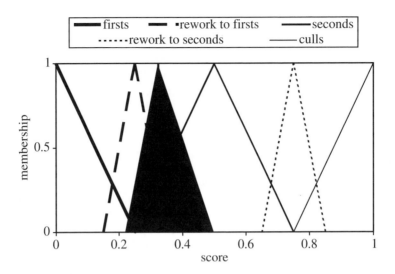

Figure 12.39. *The fuzzy distribution for day* 1 *shown on the score chart. The shaded area is the distribution for day* 1.

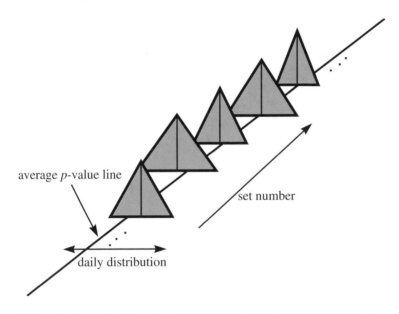

Figure 12.40. *Daily fuzzy distributions spread over the average p-value line.*

α-level fuzzy midrange (Laviolette and Seaman (1992, 1994), Laviolette et al. (1995)). The α-level fuzzy midrange for day 1 in the example is 0.346154. This is the midrange of the shaded triangle in Figure 12.39. The mean or the centerline for the 30-day run has a value of 0.275324. Figure 12.40 depicts the daily fuzzy distributions relative to the centerline or mean.

The α-level fuzzy midrange is defined for a fuzzy set as the midpoint of the crisp interval that divides the set into two subsets. One subset contains all the values that have a membership greater than or equal to α in the original set. The other subset contains all the values with memberships less than α. This interval is called the α-cut. Laviolette and Seaman (1992) showed that a central tendency of the TFN AB described above can be well represented by

$$R = \underline{\alpha}^T AB, \tag{12.12}$$

where R is the α-level fuzzy midrange and $\underline{\alpha}$ is a vector defined by

$$\underline{\alpha} = \left[\frac{1-\alpha}{2} \quad \alpha \quad \frac{1-\alpha}{2} \right]^T, \tag{12.13}$$

where α is the scalar value chosen for the α-cut. It is common to choose α equal to 0.5 (Laviolette and Seaman (1992)), in which case the vector $\underline{\alpha}$ becomes

$$\underline{\alpha} = [0.25, 0.5, 0.25].$$

Laviolette et al. (1995) point out that for sufficiently large sample size n, the vector B constitutes an observation from a multivariate normal distribution with a rank of $c - 1$, where c is the number of categories in the problem—in this case $c = 5$. The scalar R is then an observation from a univariate normal distribution with a mean $\mu = \underline{\alpha}^T A\pi$ and a variance $\sigma^2 = \underline{\alpha}^T A\Sigma A^T \underline{\alpha}$, where π and Σ are the respective mean vector and covariance matrix of the set of B vectors. The covariance matrix Σ is defined by

$$\Sigma = [\sigma_{ij}] = \begin{cases} \dfrac{\pi_i(1 - \pi_i)}{n}, & i = j, \\ \dfrac{-\pi_i \pi_j}{n}, & i \neq j. \end{cases} \tag{12.14}$$

This is convenient because it provides upper and lower control limits, UCL_{pf} and LCL_{pf}, for the fuzzy p-chart:

$$UCL_{pf} = \mu + z_c\sigma, \tag{12.15}$$
$$LCL_{pf} = \mu - z_c\sigma. \tag{12.16}$$

The factor z_c is a function of the confidence level for the normalized Gaussian random variable. These factors are given in Table 12.16, taken from Williams (1991).

Table 12.16. *Confidence levels as a function of the factor z_c for the normalized Gaussian random variable.*

z_c	Confidence level: $100 * (1 - \gamma)\%$
1.645	90.00
1.960	95.00
2.0 (2σ)	95.45
2.326	98.00
2.576	99.00
3.0 (3σ)	99.73
3.291	99.90
3.891	99.99

The symbol $\frac{\gamma}{2}$ represents the area under the normalized Gaussian curve in each tail beyond the values of $\pm z_c$ that represents the area of the desired confidence level. In this example, the desired confidence limit is 95%. Ninety-five percent of the area under the

normalized Gaussian curve lies between $\pm z_c = 1.96$ (nearly 2σ). The control limits for
each sample will be a function of n, the sample size. In the next section, we will address how
large n must be to have confidence that the sample vector B came from a multivariate normal
distribution. The example is continued using this α-level fuzzy midrange with $\alpha = 0.5$ to
construct the fuzzy p-chart.

The p-chart generated from the 30-day simulation run for Example 12 is shown in
Figure 12.41. The data from this run are presented in Table 12.17. Figure 12.41 and Table 12.17 show the fuzzy p-values for each set number. These values are computed as in the
preceding example. Also shown is the set mean, similar to the grand average used with X
bar–R charts. The chart and table also show the upper and lower control limits, UCL_{pf} and
LCL_{pf}, computed using (12.15) and (12.16). These control limits were not calculated in
the usual p-chart manner using (12.10) and (12.11) since the fuzzy p-chart is not binomial.
On this chart the control limits are approximately equal to the $\pm 2\sigma$ limits (95% confidence
limits). These limits normally are called the warning limits, as discussed in Example 10. The
manufacturers, however, can use whatever control limits they wish. In this case, they are
looking for very tight control, so the 95% confidence limits were used. Figure 12.41 shows
that two points are beyond the 95% confidence limit. These are points 16 and 25. Point 27
is inside the confidence limits. Twenty-eight of 30 points are within the control limits. This
is 93.33%, very close to 95%.

Table 12.17. *Results from Example* 12 *using α-level fuzzy midrange with $\alpha = 0.5$.*

Set number	Sample size, n	p-value	Mean, μ	UCL_{pf}	LCL_{pf}
1	13	0.346154	0.275324	0.418928	0.13172
2	22	0.278409	0.275324	0.385713	0.164935
3	19	0.226974	0.275324	0.394109	0.156539
4	18	0.267361	0.275324	0.397364	0.153284
5	29	0.359914	0.275324	0.371472	0.179177
6	24	0.234375	0.275324	0.381014	0.169635
7	19	0.253289	0.275324	0.394109	0.156539
8	29	0.295259	0.275324	0.371472	0.179177
9	24	0.270833	0.275324	0.381014	0.169635
10	14	0.312500	0.275324	0.413704	0.136944
11	20	0.246875	0.275324	0.391101	0.159547
12	17	0.194853	0.275324	0.400902	0.149746
13	21	0.354167	0.275324	0.388311	0.162337
14	18	0.298611	0.275324	0.397364	0.153284
15	28	0.263393	0.275324	0.373174	0.177475
16	11	0.471591	0.275324	0.431438	0.119210
17	22	0.247159	0.275324	0.385713	0.164935
18	24	0.325521	0.275324	0.381014	0.169635
19	14	0.321429	0.275324	0.413704	0.136944
20	30	0.283333	0.275324	0.369856	0.180793
21	21	0.223214	0.275324	0.388311	0.162337
22	23	0.250000	0.275324	0.383287	0.167361
23	17	0.341912	0.275324	0.400902	0.149746
24	32	0.216797	0.275324	0.366854	0.183794
25	15	0.425000	0.275324	0.409012	0.141636
26	20	0.209375	0.275324	0.391101	0.159547
27	11	0.142045	0.275324	0.431438	0.119210
28	16	0.265625	0.275324	0.404767	0.145881
29	19	0.246711	0.275324	0.394109	0.156539
30	27	0.210648	0.275324	0.374969	0.175679

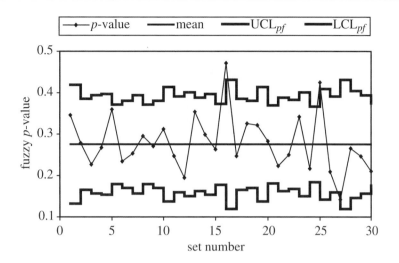

Figure 12.41. *Fuzzy p-chart for Example* 12 *using the α-level fuzzy midrange technique with* $\alpha = 0.5$.

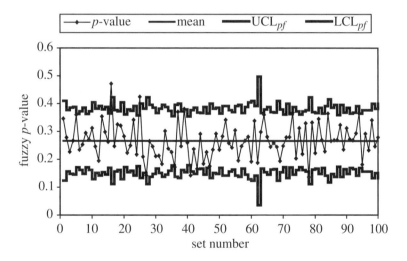

Figure 12.42. 100-*day fuzzy p-chart for Example* 12.

In order to conduct a second check of the control limits, a second run was made with 100 data sets, or over 100 days. These results are shown in Figure 12.42. The 100-day run produced five points that were outside the control limits. These points are set numbers 16, 25, 39, 41, and 94. This run demonstrates that exactly 95% of the points are within the 95% confidence limits. The data from this run were then divided into bins and the frequencies of these bins are plotted in Figure 12.43.

The data distribution shown in Figure 12.43 resembles a Gaussian distribution with an average, $\mu = 0.267786$, and standard deviation, $\sigma = 0.061848$. This indicates that the vector B is an observation from a multivariate normal distribution as assumed. Then the scalar R, plotted in Figures 12.41–12.43, probably is an observation from a univariate normal distribution, even with the sample sizes used in this problem.

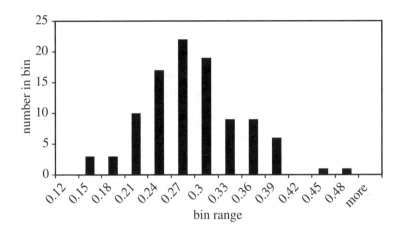

Figure 12.43. *The bin frequency for the data from the* 100-*day run, Example* 12.

The other metric that was used in this study to represent the central tendency of a fuzzy set is the fuzzy average, f_{avg}, defined by

$$f_{avg} = \frac{\int_0^1 x \mu_F(x) dx}{\int_0^1 \mu_F(x) dx}, \tag{12.17}$$

where $\mu_F(x)$ is the equation for the membership function or the fuzzy set. In Example 12, using this metric, from Figure 12.39, for day 1, $\mu_F(x)$ is given by the following set of equations:

$$\mu_F(x) = 0, \qquad\qquad\qquad x < 0.227,$$
$$\mu_F(x) = 10x - 0.227, \qquad 0.227 \le x \le 0.327,$$
$$\mu_F(x) = -5.65x + 2.85, \quad 0.327 < x \le 0.504,$$
$$\mu_F(x) = 0, \qquad\qquad\qquad x > 0.504.$$

The fuzzy average computed using (12.17) is 0.352564 compared to 0.346154 using (12.12), the α-level fuzzy midrange technique. Some authors prefer using the fuzzy average rather than the α-level fuzzy midrange technique because doing so makes it easier to visualize the physical meaning of the fuzzy average. The major problem is that it is more difficult to generate control limits for this technique. Wang and Raz (1990) have suggested using

$$UCL_{RW} = \mu + k\delta(T_{avg}), \tag{12.18}$$
$$LCL_{RW} = \mu - k\delta(T_{avg}), \tag{12.19}$$

where $T_{avg} = (t_1, t_2, t_3)$, the TFN for the average fuzzy triangle for the entire run. The constant k for the desired confidence level is generated from a simulation of the process.

The function $\delta(T_{avg})$ is defined by

$$\delta(T_{avg}) = \frac{t_3 - t_1}{2}. \tag{12.20}$$

A simulation can and will be done for this example, because there is a simulator available. However, this technique may not always be extremely useful since it depends on the availability and quality of the simulator.

First, it turns out that one can use the α-level fuzzy midrange to develop control limits for this case as well. If one picks $\alpha = \frac{1}{3}$, then (12.12)–(12.16), the mean $\mu = \underline{\alpha}^T A \pi$, and the variance $\sigma^2 = \underline{\alpha}^T A \Sigma A^T \underline{\alpha}$ can be used to generate a p-chart with the p-values generated by (12.12) being identical to those generated by (12.17) (where $\underline{\alpha} = [\frac{1}{3}, \frac{1}{3}, \frac{1}{3}]$). Example 12 is redone using this technique to generate a p-chart based on fuzzy averages. The data are presented in Table 12.18, and the p-chart is shown in Figure 12.44. The data in Table 12.18 can be compared to those in Table 12.17. The sample set size n is the same in each case. The p-chart using $\alpha = \frac{1}{3}$ is slightly different than the p-chart generated using $\alpha = \frac{1}{2}$. The results are the same for both p-charts. Data sets 16 and 25 have p-values outside the control limits. The point is that it probably does not make much difference which technique is used.

Table 12.18. *Results from Example* 12 *using α-level fuzzy midrange with $\alpha = \frac{1}{3}$.*

Set number	p-value	Mean, μ	UCL$_{pf}$	LCL$_{pf}$
1	0.352564	0.284981	0.422746	0.147217
2	0.287879	0.284981	0.390881	0.179081
3	0.236842	0.284981	0.398936	0.171026
4	0.277778	0.284981	0.402058	0.157904
5	0.364943	0.284981	0.377219	0.192743
6	0.246528	0.284981	0.386373	0.183589
7	0.263158	0.284981	0.398936	0.171026
8	0.304598	0.284981	0.377219	0.192743
9	0.281250	0.284981	0.386373	0.183589
10	0.321429	0.284981	0.417734	0.152228
11	0.258333	0.284981	0.396050	0.173912
12	0.205882	0.284981	0.405453	0.164510
13	0.361111	0.284981	0.393374	0.176589
14	0.305556	0.284981	0.402058	0.167904
15	0.273810	0.284981	0.378852	0.191110
16	0.477273	0.284981	0.434747	0.135215
17	0.257576	0.284981	0.390881	0.179081
18	0.333333	0.284981	0.386373	0.183589
19	0.327381	0.284981	0.417734	0.152228
20	0.291667	0.284981	0.375669	0.194293
21	0.234127	0.284981	0.393374	0.176589
22	0.260870	0.284981	0.388554	0.181409
23	0.348039	0.284981	0.405453	0.164510
24	0.229167	0.284981	0.372789	0.197173
25	0.427778	0.284981	0.413233	0.156729
26	0.225000	0.284981	0.396050	0.173912
27	0.159091	0.284981	0.434747	0.135215
28	0.276042	0.284981	0.409160	0.160802
29	0.258772	0.284981	0.398936	0.171026
30	0.222222	0.284981	0.380574	0.189388

Figure 12.45 shows the results of the simulation with a run of 1000 days. The results from this run were used to determine the value of k to be used with (12.18) and (12.19). The value of k that corresponded with the 95% confidence level is 0.9217. The $\pm 95\%$ confidence levels shown in Figure 12.45 are 0.403517 and 0.176457. Figure 12.46 is a p-chart using the same run data as Figure 12.44 but uses the 95% confidence limits of 0.403517 and 0.176457. This gives a slightly different—and probably better—result than using the computed k value and (12.18)–(12.20). The difference is that the overall means are slightly different for the 30-day run and the 1000-day run. The results are similar to the other p-charts for this run.

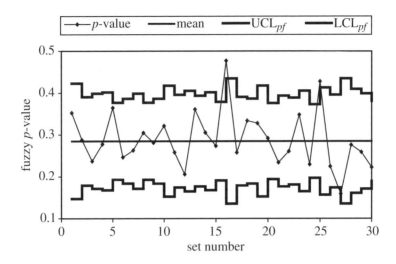

Figure 12.44. *Fuzzy p-chart for Example* 12 *using the α-level fuzzy midrange technique with* $\alpha = \frac{1}{3}$, *equivalent to the fuzzy average method.*

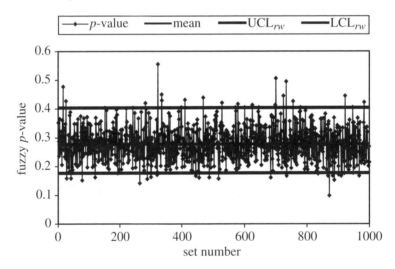

Figure 12.45. *Fuzzy p-chart for Example* 12 *developed from the* 1000-*day simulation showing the upper and lower* 95% *confidence levels.*

Data sets 16 and 25 have p-values outside the control limits. In this case, data set 27 is below the lower control limit.

This is obviously because the control limits generated using (12.18)–(12.20) do not depend on the sample size n. In the final analysis, the results are quite similar no matter which technique is used to determine the control limits.

12.7   Statistical techniques for multinomial attribute data

A common statistical solution to multinomial attribute problems that is sometimes criticized because of the possibility of losing information (Raz and Wang (1990), Wang and Raz

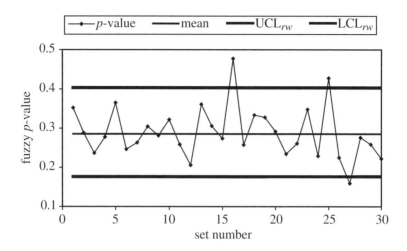

Figure 12.46. *Fuzzy p-chart for Example* 12 *using the* 95% *confidence limits developed from a simulation to determine the control limits.*

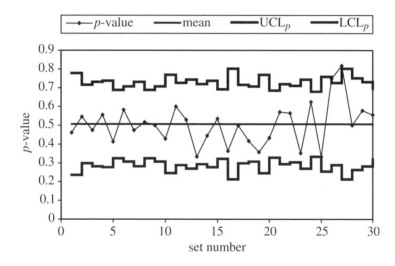

Figure 12.47. *Binomial p-chart for Example* 12, *"firsts" versus all other categories.*

(1990), Laviolette and Seaman (1992, 1994), Laviolette et al. (1995)) is to treat each category separately and produce a binomial *p*-chart for each category. For Example 12, five binomial *p*-charts are required. The five binomial *p*-charts developed from the data generated for Example 12 are shown in Figures 12.47–12.51. The upper and lower control limits for these charts were calculated using equations similar to (12.10) and (12.11) with the exception that the factor multiplying the radical (the standard deviation) was 1.96 instead of 3. The 1.96 factor represents the 95% confidence limits. Figure 12.47 is a binomial *p*-chart for "firsts" versus all other categories, or "firsts" and "nonfirsts." This figure shows that both set numbers 26 and 27 are beyond the upper control limit for this problem. Figure 12.48 is the binomial *p*-chart for "rework to firsts" versus all other categories. The lower control limit for set number 26 is barely above 0, although it is not perceptible in Figure 12.48. This means set number 26 is slightly below the lower control limit. Figure 12.49 is the binomial

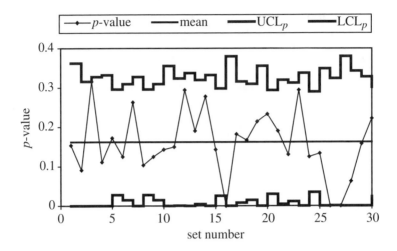

Figure 12.48. *Binomial p-chart for Example* 12, *"rework to firsts" versus all other categories.*

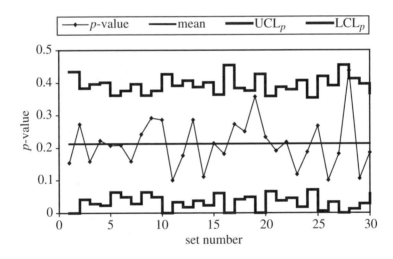

Figure 12.49. *Binomial p-chart for Example* 12, *"seconds" versus all other categories.*

p-chart for "seconds" versus all other categories. This chart shows set number 28 above the upper control limit. Figure 12.50 is the binomial *p*-chart for "rework to seconds" versus all other categories. This chart shows set numbers 13 and 16 above the upper control limit. Figure 12.51 isthe binomial *p*-chart for "culls" versus all other categories. This chart shows set numbers 5 and 25 above the upper control limit. These five binomial charts certainly tell a different story than the combined multinomial charts. They may be very useful in trying to determine something about the individual categories, but that is not the problem under investigation here. This may be the reason that Laviolette et al. (1995) claim that arbitrary combinations of data result in a loss of information and chart sensitivity.

Duncan (1950) developed a multinomial method based on the chi-square (χ^2) statistic. Marcucci (1985) later modified the technique somewhat and called it the generalized *p*-chart technique. This technique is based on a measure of the discrepancy that exists between

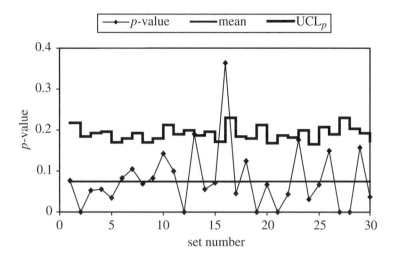

Figure 12.50. *Binomial p-chart for Example* 12, *"rework to seconds" versus all other categories.*

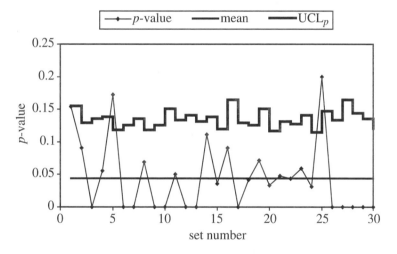

Figure 12.51. *Binomial p-chart for Example* 12, *"culls" versus all other categories.*

observed and expected frequencies as determined by the χ^2 statistic. The χ^2 statistic is given by

$$\chi_i^2 = \sum_{j=1}^{k} \frac{(o_{ij} - e_j)^2}{e_j},\tag{12.21}$$

where o_{ij} is the observed frequency for category j for set number i, and e_j is the expected frequency for that category. The upper index k is the number of categories—in our case k equals 5. The observed values for each category for each set number (day) for Example 12 are given in Table 12.15. The expected values are the averages (or the sum of each category column in Table 12.15 divided by the sum of the number of samples column in Table 12.15). That is,

- firsts: $e_1 = 0.507293$;

- rework to firsts: $e_2 = 0.162075$;

- seconds: $e_3 = 0.212318$;

- rework to seconds: $e_4 = 0.074554$;

- culls: $e_5 = 0.04376$.

Technically, this is not quite right because the user is supposed to know, or have a way to estimate, the expected values of the data before using (12.21). If the expected frequencies have to be computed from the sample statistics, the degrees of freedom will be reduced appropriately. Equation (12.21) can be used when building control charts with expected values known ahead of time. A correction to (12.21) for the case where the expected values are computed while building the control chart will be discussed shortly. For Example 12, the generalized p-chart or χ^2-chart is shown in Figure 12.52.

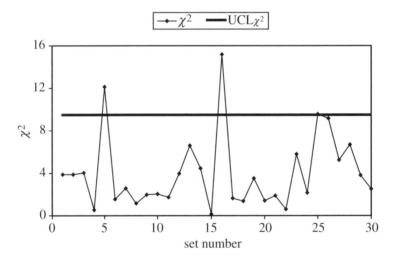

Figure 12.52. *Generalized p-chart for Example* 12 *using the* χ^2 *statistic.*

The upper control limit, UCL_{χ^2}, is taken directly from a χ^2 probability table. There is only one control limit because this is a one-tailed test. A portion of the χ^2 probability table is given in Table 12.19 taken from Williams (1991). The left-hand column of the table (df) contains the degrees of freedom $k-1$. In this example, df $= 4$. In order to use Table 12.19 to determine the 95% confidence level for the chi-square distribution, we want to determine the value of the abscissa, A, such that the probability that χ^2 is greater than A is 0.05. This value from the table, with df $= 4$, is 9.488. This is the value for the upper control limit in Figure 12.52.

Figure 12.52 shows three sets out of control, set numbers 5, 16, and 25. The fuzzy example had determined that only sets 16 and 25 were out of control. The next step is to look at the 100-day run for the generalized p-chart to see if that test provides additional information. The generalized p-chart for that test is shown in Figure 12.53. The control limit for this chart is determined as in the 30-day run. This chart shows six points out of control and "almost" a seventh point. The "almost" point is set number 25, which is out in the 30-day run but is not out in the 100-day run. This is due to the fact that the expected

Table 12.19. *Values of the abscissa A versus the probability $P(\chi^2 > A)$ for the χ^2 random variable.*

df	0.100	0.050	0.025	0.010	0.005	0.001
1	2.706	3.841	5.024	6.635	7.879	10.828
2	4.605	5.991	7.378	9.210	10.597	13.816
3	6.251	7.815	9.348	11.345	12.838	16.266
4	7.779	9.488	11.143	13.277	14.860	18.467
5	9.236	11.071	12.833	15.086	16.750	20.515

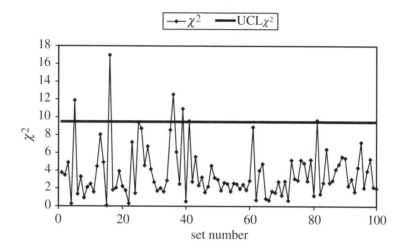

Figure 12.53. *Generalized p-chart for Example 12 using the chi^2 statistic for the 100-day run.*

values are different for the 100-day run than they are for the 30-day run. A remedy for this situation will be discussed shortly. The six to seven points out of control indicate that this test does not quite meet the 95% confidence level as the fuzzy test did.

The data from this 100-day run were divided into bins. The frequencies of these bins are plotted in Figure 12.54. The data distribution shown in Figure 12.54 does resemble a chi-square distribution. The 95% confidence level for Figure 12.54 is 9.5609 versus 9.488 for the true chi-square.

Equation (12.21) should not be used unless the expected values e_j are known a priori. Marcucci (1985) presents the Z^2 statistic defined by (12.22) below; this statistic can be used when the expected values are determined during some base set of observations, called set number 0, when the plant is in control:

$$Z_i^2 = n_i n_0 \sum_{j=1}^{k} \frac{(o_{ij} - e_j)^2}{X_{ij} + X_{0j}}, \qquad (12.22)$$

where Z_i^2 is the statistic for set number i. The index j represents the categories, in this case 1 to 5, where $k = 5$. The terms o_{ij} and e_j are the observed values for category j and set number i and expected values for category j, respectively. The terms X_{ij} and X_{0j} are the number of observations for category j and set number i and category j and the base set 0, respectively. The terms n_i and n_0 are the total number of observations for set number i and

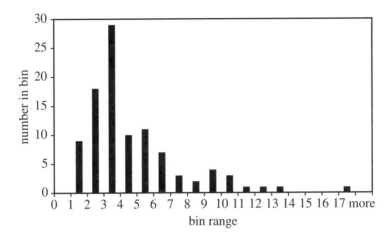

Figure 12.54. *The bin frequency for the data from the* 100-*day run, Example* 12, *using the chi^2 statistic.*

the base set number 0. It would be convenient to have the base set be the total 30-day run, but that would not be technically correct either. Instead the 100-day run was used as set 0 to determine the values of e_j. The values of e_j, the expected values, are as follows:

- firsts: $e_1 = 0.532728$;

- rework to firsts: $e_2 = 0.147635$;

- seconds: $e_3 = 0.20688$;

- rework to seconds: $e_4 = 0.0678452$;

- culls: $e_5 = 0.049116$.

Figure 12.55 shows the 30-day run from Example 12, using the Z^2 statistic and the 100-day run to determine the expected values.

This time the Z^2 value for set number 25 is 9.415, less than 9.488, so this chart shows only two points out of control. Obviously, if the user has to take a good deal of data in order to determine the expected value, the technique usually will be of less value than other control chart techniques. Figure 12.56 is a plot showing the same 30-day run using the Z^2 statistic and the expected values computed from the run itself. This is the normal control chart construction. Again, three points are out of control. Observation of Figures 12.52, 12.55, and 12.56 reveals that they all are quite similar. In fact with the exception of point 5 and sometimes 25, the generalized p-charts give results similar to the fuzzy charts shown in Figures 12.41, 12.44, and 12.46.

The reason for the difference is that the fuzzy chart looks more at the average value, and the chi-square technique will accentuate the dispersion due to using the sum of the square of the difference between the observed and the expected value. In the case of set number 5, the mean is approximately 0.36, where the 30-day mean is about 0.28 (depending on the fuzzy technique used). This is not very far out of line, but set number 5 contains five culls in a sample size of 29. This produces an observed value of 0.1724 to be compared with an expected value of 0.04376. When squared, this produces a large χ^2 or Z^2 value. This can be seen in Figure 12.51, the individual p-chart for culls. In Example 12, it is probably not good

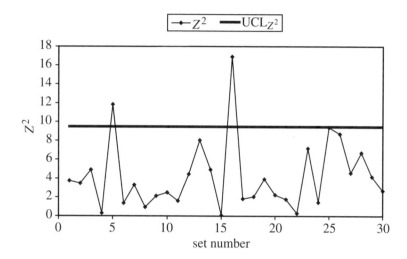

Figure 12.55. *Generalized p-chart for Example* 12, *using the* Z^2 *statistic and the* 100-*day run to determine the expected values.*

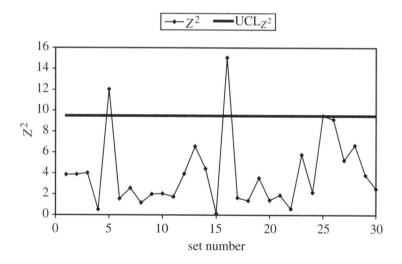

Figure 12.56. *Generalized p-chart for Example* 12, *using the* Z^2 *statistic and the run itself to determine the expected values.*

to have the generalized p-chart produce additional "out-of-control" values because all the out-of-control values in this example are false alarms. It is desirable to detect out-of-control conditions but it is also desirable to minimize the number of false alarms produced. In some cases, however, determining the amount of dispersion in the set might be very important.

One other issue is the minimum number of sample points that should be in a set before one can comfortably use the chi-square distribution or, for that matter, the Gaussian distribution used with the fuzzy technique. Cochran (1954) and later Yarnold (1970) have done a great deal of work to determine this minimum number. Cochran concluded that no less than 20% of the expected frequencies $n_i e_j$ in (12.21) and $\frac{n_i(X_{ij}+X_{0j})}{n_i+n_0}$ in (12.22) should be less than 5, and none of the expected frequencies should be less than 1. Yarnold's rules are a little less restrictive. If k is the number of categories and r is the number of expected frequencies less than 5, then for $k \geq 3$, the minimum expectation should be at least $\frac{5r}{k}$. Table 12.20 is a tabulation of $n_i e_j$ and $\frac{5r}{k}$ for Example 12 with the expectations computed with the run itself. A similar tabulation using the expectations from the 100-day run from formula (12.22) yields similar results.

Table 12.20. *Value of $e_i n_i$ and $\frac{5r}{k}$ for Example 12 with self-generated expectations.*

Set number	n_i	$e_1 n_i$	$e_2 n_i$	$e_3 n_i$	$e_4 n_i$	$e_5 n_i$	$\frac{5r}{k}$
1	13	6.595	2.107	2.760	0.969	0.569	4
2	22	11.160	3.566	4.671	1.640	0.963	4
3	19	9.639	3.079	4.034	1.417	0.831	4
4	18	9.131	2.917	3.822	1.342	0.788	4
5	29	14.711	4.700	6.157	2.162	1.269	3
6	24	12.175	3.890	5.096	1.789	1.050	3
7	19	9.639	3.079	4.034	1.417	0.831	4
8	29	14.711	4.700	6.157	2.162	1.269	3
9	24	12.175	3.890	5.096	1.789	1.050	3
10	14	7.102	2.269	2.972	1.044	0.613	4
11	20	10.146	3.242	4.246	1.491	0.875	4
12	17	8.624	2.755	3.609	1.267	0.744	4
13	21	10.653	3.404	4.459	1.566	0.919	4
14	18	9.131	2.917	3.822	1.342	0.788	4
15	28	14.204	4.538	5.945	2.088	1.225	3
16	11	5.580	1.783	2.335	0.820	0.481	4
17	22	11.160	3.566	4.671	1.640	0.963	4
18	24	12.175	3.890	5.096	1.789	1.050	3
19	14	7.102	2.269	2.972	1.044	0.613	4
20	30	15.219	4.862	6.370	2.237	1.313	3
21	21	10.653	3.404	4.459	1.566	0.919	4
22	23	11.668	3.728	4.883	1.715	1.006	4
23	17	8.624	2.755	3.609	1.267	0.744	4
24	32	16.233	5.186	6.794	2.386	1.400	2
25	15	7.609	2.431	3.185	1.118	0.656	4
26	20	10.146	3.242	4.246	1.491	0.875	4
27	11	5.580	1.783	2.335	0.820	0.481	4
28	16	8.117	2.593	3.397	1.193	0.700	4
29	19	9.639	3.079	4.034	1.417	0.831	4
30	27	13.697	4.376	5.733	2.012	1.182	3

Observation of the last column in Table 12.20 reveals that none of the sets used in this study satisfy the minimum size to comfortably use the chi-square test. The minimum set size should contain 46 samples. The calculation is as follows:

- $ne_1 = (46)(0.507293) = 23.335$;

- $ne_2 = (46)(0.162075) = 7.455$;

- $ne_3 = (46)(0.212318) = 9.767$;

- $ne_4 = (46)(0.074554) = 3.429$;

- $ne_5 = (46)(0.04376) = 2.013$;

- $r = 2$, the number of expectations less than 5;

- $\frac{5r}{k} = 2$, the minimum size for expectations less than 5;

- $n = 46$ is the lowest number to satisfy the criterion.

Furthermore, similar rules apply in order to obtain an accurate approximation to the normal distribution used with the fuzzy technique (Laviolette and Seaman (1994), Yarnold (1970)).

Obviously, the distributions shown in Figures 12.43 and 12.54 are not perfect. Technically, one should probably not try to solve this problem, but, unfortunately, the problem has to be solved. We cannot refuse to solve problems just because they do not provide ideal conditions. In Example 12, it is obvious that when using the smaller number of samples the results obtained are approximately what are expected—five false alarms out of 100 sets. This is not to say that one should use any of these techniques carelessly. The user should always take great care to make sure that the model, statistical or fuzzy, fits the problem as well as possible. This was the reason for spending so much effort to develop the fuzzy sets shown in Figure 12.35, used in Example 12.

Wheeler and Chambers (1992) make two points that are very useful to keep in mind when working with control charts. First, they demonstrated that Shewhart's X bar–R techniques works well for the following distributions:

- uniform distribution,

- right triangular distribution,

- normal distribution,

- Burr distribution,

- chi-square distribution with two degrees of freedom,

- exponential distribution.

The technique is very robust. The techniques used here for attribute data should prove to be robust also.

Second, they point out that there is nothing magical about the $\pm 3\sigma$ range for upper and lower control limits, even though the 3σ range is well founded and easily interpreted. When Shewhart was developing the control chart technique, he was looking for a range that the measured variable would not go beyond too often, causing too many false alarms and therefore excessive trouble shooting. The $\pm 3\sigma$ range fit his goals because it will produce only three false alarms in 1000 sets if the distribution is normal or Gaussian. In most cases, use of some empirical rules will work as well as strict use of the sigma limits. If the data set is homogeneous, the following three rules can be applied:

1. Roughly 60% to 75% of the data will be located within a distance of one sigma unit on either side of the centerline or average line.

2. Approximately 90% to 98% of the data will be located within a distance of two sigma units on either side of the centerline.

3. About 99% to 100% of the data will be located within a distance of three sigma units on either side of the centerline.

This means that there are several methods of solving problems with control charts. The user does not necessarily have to have a perfect distribution or set the control limits at exactly one of the sigma limits in order to meet his/her goals. This is not to suggest that the user should be sloppy about setting the tolerance limits, which will determine the quality of the products that will actually be sold. This is a different problem as explained earlier. If the tolerance limits are set correctly, then the control limits will signal a problem before the tolerance limits are violated. What the user wants to determine from the control chart is the following:

> "Is there a pattern or a control limit violation that shows that the process is no longer working normally? Do I need to shut the process down (costing money) in order to find the problem?"

or

> "Does the control chart indicate that process improvements have actually changed the process? Is there a data pattern that shows this?"

The correct choice of a control chart should not necessarily come from a contest between techniques based on which produces the most alarms. It should be based on which technique can answer the above questions most clearly.

12.8   Discussion and conclusions

The fuzzy "type of day" Shewhart-type X bar–R control chart has the potential to take into account the task dependency beryllium exposure for beryllium plant operations. Based on the studies completed up to this point, we believe these control charts will provide more realistic information than the standard single-variable X bar–R chart using only beryllium exposure information. Because of its ability to take into account task dependency, the "type of day" chart can be used to determine the significance of plant improvements as well as trigger "out-of-control" alarms. This fuzzy technique should work well with many other task dependent problems, as long as they are well defined semantically. The least squares approach also will work for this type of problem but in many cases will not be as descriptive as the fuzzy approach. The least squares approach can produce problems if the data used to develop a control chart contains many out-of-control points. This is because the technique squares the difference between the expected value and the measured value.

A fuzzy technique for dealing with multinomial attribute data has been presented and shown to work well if the problem is defined well. This technique will not work that well if the membership functions that define multinomial categories are assigned arbitrarily. Individual p-charts that deal with multinomial attribute data tend to be confusing. The chi-square technique for multinomial data works nearly as well as the fuzzy technique but has only one control limit. The single control limit is not a problem in the examples presented here because all efforts were directed toward keeping the process below the upper limit. If both upper and lower limits are important, additional work will be required to determine the meaning given by the chi-square chart.

Finally, the computer models of the beryllium plant operation described here were built from a semantic description of the process as were the fuzzy rule base and membership functions. Consequently, the correlation between the fuzzy model and the plant simulation was quite good. Both models come from the same description. It is important when developing a fuzzy model of a process that much care is taken to listen to the experts and obtain the best model possible. If the domain expert is knowledgeable, then the task usually is not very difficult, but it may require several iterations to get it right. The fuzzy control chart will only be as good as the fuzzy rules and membership functions that provide the input.

References

G. BOX AND A. LUCENO (1991), *Statistical Control by Monitoring and Feedback Adjustment*, Wiley, New York.

W. G. COCHRAN (1954), Some methods for strengthening the common χ^2 tests, *Biometrics*, 10, pp. 417–451.

COMMITTEE E-11 (1995), *Manual on Presentation of Data and Control Chart Analysis*, 6th ed., American Society for Testing and Materials, Philadelphia.

A. J. DUNCAN (1950), A chi-square chart for controlling a set of percentages, *Indust. Quality Control*, 7, pp. 11–15.

A. KAUFMANN AND M. M. GUPTA (1985), *Introduction to Fuzzy Arithmetic: Theory and Applications*, Van Nostrand Reinhold, New York.

M. LAVIOLETTE AND J. W. SEAMAN, JR. (1992), Evaluating fuzzy representations of uncertainty, *Math. Sci.*, 17, pp. 26–41.

M. LAVIOLETTE AND J. W. SEAMAN, JR. (1994), The efficacy of fuzzy representations of uncertainty, *IEEE Trans. Fuzzy Systems*, 2, pp. 4–15.

M. LAVIOLETTE, J. W. SEAMAN, JR., J. D. BARRETT, AND W. H. WOODALL (1995), A probabilistic and statistical view of fuzzy methods, *Technometrics*, 37, pp. 249–292.

C. L. MAMZIC, ED. (1995), *Statistical Process Control*, ISA Press, Research Triangle Park, NC.

M. MARCUCCI (1985), Monitoring multinomial processes, *J. Quality Tech.*, 7, pp. 86–91.

W. J. PARKINSON, S. P. ABLEN, K. L. CREEK, P. J. WANTUCK, T. ROSS, AND M. JAMSHIDI (2000), *Application of Fuzzy Set Theory for Exposure Control in Beryllium Part Manufacturing*, paper presented at the World Automation Conference, Maui, HI.

T. RAZ AND J. WANG (1990), Probabilistic and membership approaches in the construction of control charts for linguistic data, *Production Planning Control*, 1, pp. 147–157.

W. A. SHEWHART (1980), *Economic Control of Quality Manufactured Product*, American Society for Quality Control, Milwaukee, WI.

W. A. SHEWHART (1986), *Statistical Method from the Viewpoint of Quality Control*, Dover, New York.

J. WANG AND T. RAZ (1990), On the construction of control charts using linguistic variables, *Internat. J. Production Res.*, 28, pp. 477–487.

WESTERN ELECTRIC COMPANY, INC. (1958), *Statistical Quality Handbook*, 2nd ed., Indianapolis, IN.

D. J. WHEELER AND D. S. CHAMBERS (1992), *Understanding Statistical Process Control*, 2nd ed., SPC Press, Knoxville, TN.

R. H. WILLIAMS (1991), *Electrical Engineering Probability*, West Publishing Company, St. Paul, MN.

J. K. YARNOLD (1970), The minimum expectation in χ^2 goodness of fit tests and the accuracy of approximations for the null distribution, *J. Amer. Statist. Assoc.*, 65, pp. 864–886.

Chapter 13

Fault Tree Logic Models

Jonathan L. Lucero and Timothy J. Ross

Abstract. Combining, propagating, and accounting for uncertainties are considered in this new extension of reliability assessment. Fault trees have been used successfully for years in probabilistic risk assessment. This scheme is based on classical (Boolean) logic. Now this method is expanded to include other logics. Three of the many proposed logics in the literature considered here are Lukasiewicz, Boolean, and fuzzy. This is a step towards answering the questions: Which is the most appropriate logic for the given situation, and how can uncertainty and imprecision be addressed?

13.1   Introduction

The importance of a general logic-based methodology in system failure engineering is that it expands on classical failure analysis by allowing for the assessment of failure using various logic approaches. Classical (Boolean), Lukasiewicz, and fuzzy logic are three of the many theories that can be used in this general methodology to assess failure in mechanical or passive systems.

System failure engineering is the description and analysis of events and subevents leading to failure. This analysis is sometimes achieved by schematic models called fault trees. Fault trees diagram interactions of components as a flowchart with logical connectives between levels of the tree.

Our general methodology expands on fault trees by propagating probabilities and confidence measures from "bottom" level events to the "top" event, failure of the system using various logics. The probabilities are propagated using "t-norm" and "t-conorm" operators as aggregators for the connectives. These aggregators vary among the logical theories in the literature and are easily implemented using our general methodology. We illustrate the application of this methodology in structural engineering with the modeling of a simple truss system.

13.2   Objective

The primary objective of this chapter is to demonstrate the use of a general logic methodology to accurately assess safety in structural engineering. Accurately assessing safety means

325

accounting for degradation or breakdown of components leading to various types of system failure. Degradation can occur from weathering of steel or concrete. Breakdown of a component can be interpreted as any subservice functioning of a component, for example, below-service-level strengths of a load-bearing member. Using three logics—classical (Boolean), Lukasiewicz, and fuzzy—with our general methodology we attempt to show that certain logics can better describe a system's safety by interpretation and analysis of trends.

Fault tree analysis provides the groundwork for studying a system failure by progressively deducing levels of failure through its components and subcomponents. A truss, a wire rope, and a bridge are examples of structural systems with failures depending on the failure of its components. Using fuzzy logic, various degrees of failure, including the special cases of classical binary failure, can be assessed. Failure values between the binary extremes of failure and no failure can represent the uncertainties of the model or degradation of members where only partial failure can be noted. "When one considers that perhaps there can be various degrees of degradation between complete failure and no failure, and that many systems contain both random and ambiguous kinds of description, it becomes a natural step to consider the utility of fuzzy methodologies in describing this new paradigm" (Cooper and Ross (1997)).

Consider a truss with redundant members. Redundant members have a load-bearing capability that can be transferred to another member, and the system's strength remains unchanged. In this situation, the malfunction of a member does not render the system inoperable; it does leave the system less redundant and less safe. For example, corrosion of one strand of a wire rope means the rope is not as safe as a pristine rope (Nishida (1992)). Still, this rope functions. Moreover, if one member in a complex bridge shows cracks, it does not mean that the whole bridge will collapse. However, this too means that the system is less safe. Any of a number of general logic constructs may be used in whole or in part to assess system safety and failure.

One application of these logics lies in the use of triangular norms and triangular conorms. "The triangular norm (t-norm) and the triangular conorm (t-conorm) originated from studies of probabilistic metric spaces in which triangular inequalities were extended using the theory of t-norm and t-conorm" (Gupta and Qi (1991)). Karl Menger introduced the notion of t-norms in 1942 (Blumenthal and Menger (1970)). The exact distance between points x and y, $d(x, y)$, is unknown; only the probabilities of different values of $d(x, y)$ are known. Trying to represent the triangular inequality ($d(x, y) \leq d(x, z) + d(z, y)$), Menger discovered how to relate the probability of $d(x, y)$ to the probability of $d(x, z)$ AND $d(z, y)$.

Norms are used integrally in our general logic methodology as aggregators of the probabilities of components in the fault tree. Probabilities representing the potential for failure in the system components are aggregated to form a single value resulting in a final probability to be propagated and used in subsequent levels of analysis. These probabilities are aggregated according to the logical gates "AND" and "OR." These gates describe the interaction of components in the fault tree leading to failure.

In all logics, t-norms and t-conorms are used to calculate the intersection and union, respectively, of elements in the problem space. Moreover, the resulting aggregated probability values for failure are used in the calculations of the next level of components. Zadeh t-operators have been the premier choice in fuzzy logic controllers. "However, some theoretical and experimental studies seem to indicate that other types of t-operators may work better in some situations, especially in the context of decision-making processes" (Gupta and Qi (1991)). In this chapter, the norms from classical logic, Lukasiewicz logic, and fuzzy logic will be compared in a fault tree example.

Another feature of our general methodology is in the use of implication operators. Implication operators are used in this methodology in calculating the implication of subsequent events' confidence. Confidence is used here to account for the belief of the failure analyst in the probability values of events.

Therefore, two concepts characterize our general methodology: The first concept is to use a variety of logics through norm operators and implication operators, and the second concept is to extend traditional fault tree analysis with the use of confidence measures. These two concepts are described thoroughly in sections to follow.

13.3   Chapter overview

Here we describe the two methods of analysis used in this research, fault trees and a general logic methodology. Fault tree analysis, as a method of examining system failure, can be applied to a variety of fields including structural engineering (Adams and Sianipar (1997)). Structures can be represented as systems composed of load carrying members whose functions are interdependent. Fault trees describe this interdependence by progressively sectioning and simplifying this system's functions. Moreover, this method has been expanded to better quantitatively assess system failure by propagating probabilities of failure, each of which sometimes includes imprecision. Our expanded method is termed a general logic methodology, which can make use of fuzzy membership functions. Using these two ideas, we make comparisons of three different logical theories and make conclusions about their suitability.

This chapter consists of two parts, A and B. Part A (section 13.4) introduces the general methodology with background information such as fault tree construction and mathematical definitions. Part A approaches the probabilistic aspect of the general methodology. Part B (section 13.5) then discusses the general logic capabilities of our proposed method. Ideas such as aggregation of fuzzy sets and dependence relations are described to encompass the whole potential of this general methodology.

13.4   Part A: General methodology

Fault tree analysis is a tool that uses classical (Boolean) logic to deduce system failure. Schematic diagrams describe the system failure using events and gates. This procedure has been expanded into a general logic methodology.

The general logic methodology complements the classical logic used in traditional fault trees by providing a framework for the use of a variety of logics and enabling the analyzer to consider imprecision and uncertainty. Additionally, it provides the means to account for dependencies using rules of fuzzy logic, as will be discussed later in Part B.

13.4.1   Fault trees

"Fault tree analysis was developed by H. A. Watson of the Bell Telephone Laboratories in 1961–62 during an Air Force study contract for the Minuteman Launch Control System" (Henley and Kumamoto (1992)). Fault trees are a graphical approach to analyze system failure using Boolean algebra. Boolean algebra is a class of rules and manipulations in classical set theory. Using this and deductive reasoning, fault trees provide the causal relations of component failures in a system.

13.4.1.1   Fault tree construction

Fault trees are a diagram of consequential events. Beginning at the final failure and terminating at roots, fault trees are constructed by describing a variety of partitions necessary for failure of subsequent events. The tree starts at the top, the system failure of interest. This failure can be a structural collapse, for example, of a truss bridge. From this failure, the system is subdivided into simplified lower levels of failure. Using the bridge example, the lower levels may represent foundation, floor, or girder failures. The levels are described as either basic (failure data known or presumed) or intermediate (failure is calculated).

These levels are called "events" in the literature and are combined using logical connectives describing their interaction leading to the top event's failure. "Top" is the ultimate event to which all other failures lead. "Intermediate" is a level where preceding actions or events are required for failure. Similarly, "basic" is a level where no preceding actions or events are necessary to determine failure. Graphically this model is represented with the events and their "gates." Gates are the logical connectives describing the necessary interactions for failure in the forms of AND (intersection of events) and OR (union of events) or other logic gates. This interaction is based on the rules of Boolean algebra (classical logic). These rules are briefly covered in section 13.4.3.1 on set theory. As in a flow chart, the events and gates are combined into a sequence of events. This sequence terminates at the basic event. Figure 13.1 illustrates this process.

In this chapter, the fault tree is limited to the use of the logic gates AND and OR. Despite this limitation, these fault trees still can portray the system failure accurately. As an example, we show its use in structural engineering in the simple system (Figure 13.2) following the general fault tree construction (Figure 13.1).

In Figure 13.2, a fault tree describes the failure of a simple truss. This is a five-member statically determinate, simply supported truss. The failure of this truss represents the top event and is dependent on the failure of its components. These components are first the intermediate events, which are the failures of subtrusses. The two subtrusses are composed of members A, C, and D for the first sequence of events and members B, C, and E for the second events. Finally, the failure of each of these trusses depends on the failures of each of its members. These members represent the basic events of this system and termination of the tree. This example and discussion show qualitatively the use of fault trees in analyzing system dependencies. In the next section, we show how probabilities associated with each component are accounted for and used in determining a system failure potential through the use of a general logic methodology.

13.4.2 General logic methodology

In the paper by Cooper and Ross (1997), fault tree analysis is expanded to form a new paradigm for safety analysis. This paradigm, termed a general logic methodology, examines both passive and active systems (Part B) similar to the classical fault tree method. Additionally, uncertainties such as ambiguity, imprecision, and ignorance are considered. This topic is discussed in more detail in Part B. Briefly, accounting for this uncertainty is a method based on fuzzy sets. Fuzzy sets, which include classical binary sets, can be aggregated using a variety of t-norms and t-conorms through various logics, e.g., classical, Lukasiewicz, and fuzzy. In the lexicon of fuzzy logic, the sets are represented by membership functions. Membership values define an element's membership in fuzzy sets. Regardless of the logic used, norms calculate the intersection and union of the sets yielding a new value. This value is then propagated up the fault tree in a process called "implication." Attached to each probability value, and similarly propagated up the fault tree, is a confidence measure.

A confidence measure is a value that represents the analyst's belief in a selected probability for failure. This measure is propagated up the fault tree by first aggregating the measures with norm operators. Then with the use of an implication operator the truth of the

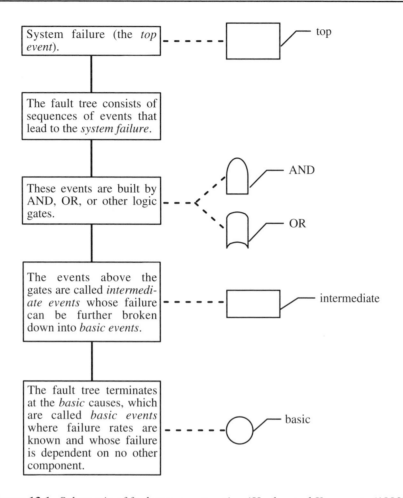

Figure 13.1. *Schematic of fault tree construction (Henley and Kumamoto (1992)).*

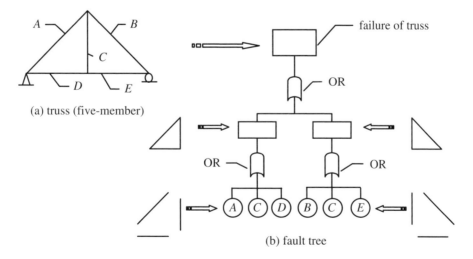

Figure 13.2. *Example fault tree of a five-member truss.*

causal relation is calculated. Implication operators are used in logic to assess truth values in inferences. A truth value is the measure of an inference. In the binary case, an inference is either true or false. In the general logic methodology, an inference can have an intermediate value between true and false. Again, this value is the belief in the propagated probabilities. A comparison of implication operators with norm operators for the three logics is shown in Lucero (1999).

13.4.2.1   General logic methodology procedure

General logic methodology begins with the construction of a fault tree similar to the preceding fault tree analysis. The difference between this general logic paradigm and the traditional fault tree analysis lies in the representation and calculation of probabilities for failure.

For now, the probabilities are represented as scalar singletons or discrete point values throughout the tree. These probabilities are aggregated using various logics after their interdependence has been determined.

Basic events can be either dependent on each other's function or independent. In the case of independent basic events, the probability of failure for each basic event is simply aggregated into a matrix of values. However, if the basic events are dependent, then calculation of this interdependence follows the traditional methods as described in Henley and Kumamoto (1992). After the relation between basic events is determined, aggregation of the probabilities is the next main step in the general logic methodology.

Triangular norm operators (t-norm, t-conorm) from various logics are used in this step to aggregate the values. The resulting single value is augmented with independent values of the next level forming a new set of probabilities. This set is then aggregated again using the norm operators (section 13.4.3.2) forming another single value to be grouped with probabilities of successive levels until the top is reached. This final value represents the probability for system failure. To summarize, the calculation of the propagating probabilities requires the determination of an initial probability set. These values are then successively aggregated and augmented with the higher levels using triangular norm operators until the final probability of failure for the top event is reached.

The following equation describes the operations involved between levels of the fault tree.

Successive levels, k:

B_k: 2×1 resulting matrix of confidence measures and probability values after norm and implication operations, where

$$B_k = \begin{bmatrix} T \\ P \end{bmatrix} \xleftarrow{\text{implication}} \begin{bmatrix} T^* \\ P^* \end{bmatrix} \overset{\text{norm}}{\underset{\text{norm}}{\longleftarrow}} [[A'] | [B_{k-1}]]; \qquad (13.1)$$

A' (input): $2 \times n$ matrix of basic event's truth and probability values (the first row is truth values, and the second row is probability values);

B_{k-1} (calculated): 2×1 matrix of aggregated basic event's truth and probability values at level $(k-1)$;

T (input):   input truth (confidence) value;

P (input/calculated):   input probability value of level (k);

$T*$ (calculated):   truth value (confidence) after norm operation;

$P*$ (calculated):   probability value after norm operation.

This process is repeated until the top level is reached. At this point, the value for the system probability for failure is determined. In the general logic methodology, the probabilities and the confidence measures (truth values) associated with each probability are propagated.

The propagation of the confidence measures up the fault tree is similar to the propagation of probabilities; however, one more step is included. This extra step is evident in (13.1). Confidence represents the degree to which the probability of the event is believed to be valid. This set of measures is first calculated and aggregated in the same way as the probabilities with the norms of section 13.4.3.2. As with the probability values, the input confidences from the first-level values of the tree have been calculated using the traditional methods found in Henley and Kumamoto (1992). The added step uses an implication operator (section 13.4.3.3). Implication is the term used in logic to describe a logical connective between two propositions (usually input and output). Each event (basic, intermediate, and top) is given a truth value. This value is used in the implication with the aggregated value of the previous step. This resulting truth value is then augmented and aggregated with the confidence of the subsequent basic event to yield a new value for the implication of the next level. This process is repeated until the top is reached, where a final truth value or confidence is calculated. This final value is the confidence in the probability for failure of the system.

Figure 13.3 shows the layout for the system in Cooper and Ross (1997). This is a generic model with basic events A, B, I, T, and F that represent an accident, A; bypass (an intentional activation rendering the system inoperable), B; and three safety subsystems, I, T, and F. The mutual dependence of I, T, and F is taken into account in this methodology using a relational matrix, R (section 13.5.2). The values of probability and confidence are then aggregated using norms and propagated as described earlier to end with final values for probability and associated confidence.

13.4.3   Mathematical operators

Of the many implication and norm operators, three are examined, one from each class of logic. The three norm operators are Lukasiewicz, probabilistic, and Zadeh (Ross (1995)). This choice reflects a comparison of three logics, Lukasiewicz, classical, and fuzzy. Because fuzzy logic is included, the notation will be based on the same notation used for fuzzy sets. However, in this study the sets used are scalar singletons on the probability interval. Similarly, the choice of implication operators reflects the same three logics.

13.4.3.1   Set theory

In this section, some basic concepts of Boolean algebra, or set theory, will be described. These concepts are intersection, union, and implication.

Intersection and union describe the location of an element in a universe X in the total space of information containing all the sets. In this research, each basic event is a set in the universe of probability with a range of real numbers from 0 to 1. An "element" is a component of failure whose location is described linguistically with "AND" and "OR" corresponding to intersection (t-norm) and union (t-conorm), respectively. An "element," also called a singleton, is also a set. For example, an intersection yields the set requiring all basic events to be included. Similarly, a union is the set requiring only one of the basic events to be included. Symbolically, this is represented as

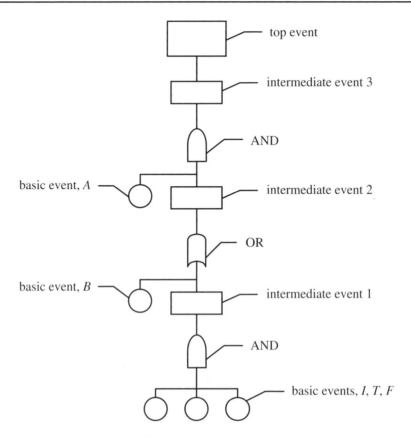

Figure 13.3. *Example fuzzy methodology fault tree (Cooper and Ross (1997)).*

intersection, "AND": $(B_1 \cap B_2) \to E \in X$;

union, "OR": $(B_1 \cup B_2) \to E \in X$, where

B_i: basic event, i;

E: resulting "element" from operation;

X: universe of probabilities of failure.

The next basic concept of set theory is implication. Implication is the operation in logic that determines the place of an element in the set corresponding to the inference of a hypothesis (antecedent) and a conclusion (consequent). In classical logic this is measured with binary truth values. These binary truth values are 0 for false and 1 for true. In other words, implication is stated as "if P, then Q." This can be decomposed into the union of two sets. In the literature, this is represented as

$$(P \to Q) \equiv (\bar{A} \cup B) \equiv (\text{either "not in } A\text{" or "in } B\text{")}.$$

In fuzzy logic, an element's place is not exclusively in any one set, but may reside partially in several sets. As a result, the implication operation is extended to give partial truth values ranging from 0 to 1.

Given in terms of fuzzy sets, the basic event confidence x implies the intermediate or top event's confidence y from the fuzzy sets A and B, respectively. In fuzzy logic, events' confidences are defined on membership grades $\mu_A(x)$ and $\mu_B(y)$ for events A and B, respectively. The resulting implication forms the membership function, $\mu_R(x, y)$, where R is the fuzzy relation yielding the truth value of the propagating confidence. This propagating confidence is the measure of truth for an event's failure (basic, intermediate, or top).

13.4.3.2 Norm operators

Triangular norm/conorm operators are used as aggregators of values corresponding to intersection and union. This section shows the operators that are compared along with general t-norm and t-conorm definitions that apply to each.

The following two definitions apply to t-norms and t-conorms, respectively, where x, y, and z represent basic event probabilities (Gupta and Qi (1991)).

Definition 13.1. *Let* $T : [0, 1] \times [0, 1] \to [0, 1]$. T *is a t-norm such that*

1. $T(0, 0) = 0$;

2. $T(1, 1) = 1$;

3. *for* $x, y \in [0, 1]$, $T(x, y) \le x$;

4. *for* $x, y, z \in [0, 1]$, $|T(x, y) - T(z, y)| \le |x - z|$;

5. *for* $x, y, z \in [0, 1]$, *if* $x \le z$, *then* $T(x, y) \le T(z, y)$ *(monotonicity)*;

6. *for* $x, y, z \in [0, 1]$, $T(x, T(y, z)) = T(T(x, y), z)$ *(associativity)*;

7. *for* $x, y \in [0, 1]$, $T(x, y) = T(y, x)$ *(commutativity)*.

Definition 13.2. *Let* $S : [0, 1] \times [0, 1] \to [0, 1]$. S *is a t-conorm such that*

1. $S(0, 0) = 0$;

2. $S(1, 1) = 1$;

3. *for* $x, y \in [0, 1]$, $S(x, y) \ge x$;

4. *for* $x, y, z \in [0, 1]$, $|S(x, y) - S(z, y)| \le |x - z|$;

5. *for* $x, y, z \in [0, 1]$, *if* $x \le z$, *then* $S(x, y) \le S(z, y)$ *(monotonicity)*;

6. *for* $x, y, z \in [0, 1]$, $S(x, S(y, z)) = S(S(x, y), z)$ *(associativity)*;

7. *for* $x, y \in [0, 1]$, $S(x, y) = S(y, x)$ *(commutativity)*.

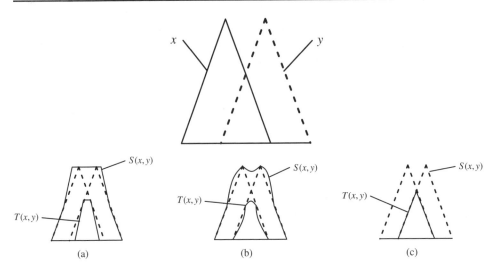

Figure 13.4. *Intersection (t-norm) and union (t-conorm) of fuzzy sets x and y for (a) Lukasiewicz, (b) probabilistic, and (c) Zadeh norm operators (Gupta and Qi (1991)).*

Lukasiewicz t-operators (Figure 13.4(a)), based on Lukasiewicz logic, are defined as follows:

$$\text{AND } (t\text{-norm}): \quad T(x, y) = \max(0, x + y - 1), \tag{13.2}$$

$$\text{OR } (t\text{-conorm}): \quad S(x, y) = \min(1, x + y). \tag{13.3}$$

Probabilistic t-operators (Figure 13.4(b)), based on classical logic, are defined as

$$\text{AND } (t\text{-norm}): \quad T(x, y) = xy, \tag{13.4}$$

$$\text{OR } (t\text{-conorm}): \quad S(x, y) = x + y - xy. \tag{13.5}$$

Zadeh's t-operators (Figure 13.4(c)), used in fuzzy logic, are defined as

$$\text{AND } (t\text{-norm}): \quad T(x, y) = \min(x, y), \tag{13.6}$$

$$\text{OR } (t\text{-conorm}): \quad S(x, y) = \max(x, y). \tag{13.7}$$

13.4.3.3   Implication operators

As previously explained, implication operators calculate the truth of an antecedent/consequent inference. This section gives the implication operators compared in this study along with common properties (Yager (1983)).

The following properties apply to the three implication operators.

Property 1:   The implications collapse to the two-valued operation defined as $(P \rightarrow Q) = (\text{not } A \text{ or } B)$.

Property 2:   Monotonicity of the antecedent is satisfied for classical and Mamdani's implication. Lukasiewicz implication for the case when $y = 1$ decreases as x increases from 0 to 0.5, then increases as x increases from 0.5 to 1.

Property 3:   Monotonicity of the consequent is satisfied for all three implication operators.

Property 4: All implications are continuous.

Property 5: All are "regular." "Regular" is defined as $\mu_R(x, y) = \mu_y$ when $x = 1$. Let $x, y \equiv$ antecedent, consequent (respectively).

Lukasiewicz's implication is defined as follows:

$$\mu_R(x, y) = \min\{1, [1 - \mu_A(x) + \mu_B(y)]\}. \qquad (13.8)$$

The classical implication operation has been extended to accommodate fuzzy membership functions:

$$\mu_R(x, y) = \max\{\min[\mu_A(x), \mu_B(y)], 1 - \mu_A(x)\}. \qquad (13.9)$$

Mamdani's implication operation typically used in system control is

$$\mu_R(x, y) = \min[\mu_A(x), \mu_B(y)]. \qquad (13.10)$$

13.4.4   Examples

The examples in this section include a simple generic failure analysis and a structural system representing a four-member, statically indeterminate truss. The generic system shows step-by-step procedures for implementing this methodology, assuming no interdependence of basic events. The structural system is a model taken from Lucero (1999) and is presented to show how a model of an actual system can be assessed as a fault tree. In Lucero (1999), other configurations and another system are analyzed and compared.

13.4.4.1   Example 1

This generic system (Figure 13.5) is composed of two intermediate events and the top event. A total of four basic events are accounted for, three as inputs to intermediate event 1 and one as an input to intermediate event 2. Using (13.1)–(13.10), final failure probabilities are calculated along with their respective confidence measures for the three logics, Lukasiewicz, classical, and fuzzy. These final values are presented as 2×1 matrices where the belief (truth) of the probability is in the first row and the probability is in the second row.

Input to intermediate event 1:

$$\begin{matrix} \text{truth} \\ \text{probability} \end{matrix} \begin{bmatrix} 0.76 & 0.98 & 0.83 \\ 0.55 & 0.72 & 0.94 \end{bmatrix}.$$

Norm operation:

	Lukasiewicz	probabilistic	Zadeh
truth	0.57	0.62	0.76
probability	0.21	0.37	0.55

Implication of basic events A, B, and C to intermediate event 1:
Implication operation:

Lukasiewicz	classical	Mamdani
1.0	0.62	0.63

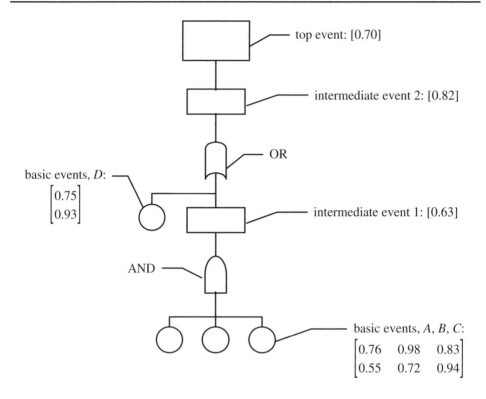

Figure 13.5. *Example* 1: *Generic fault tree.*

Aggregated values:

	Lukasiewicz	probabilistic	Zadeh
truth	1.0	0.62	0.63
probability	0.21	0.37	0.55

Input to intermediate event 2:

$$\text{basic event } D = \begin{matrix} \text{truth} \\ \text{probability} \end{matrix} \begin{bmatrix} 0.75 \\ 0.93 \end{bmatrix} \xrightarrow{\text{augment with}} \begin{bmatrix} 1.0 \\ 0.21 \end{bmatrix}, \begin{bmatrix} 0.62 \\ 0.37 \end{bmatrix}, \begin{bmatrix} 0.63 \\ 0.55 \end{bmatrix}.$$

Norm operation:

	Lukasiewicz	probabilistic	Zadeh
truth	1.0	0.91	0.75
probability	1.0	0.96	0.93

Implication of basic event D and intermediate event 1 to intermediate event 2:
Implication operation:

Lukasiewicz	classical	Mamdani
0.82	0.82	0.75

Aggregated values:

	Lukasiewicz	probabilistic	Zadeh
truth	0.82	0.82	0.75
probability	1.0	0.96	0.93

Input to top event:

$$\begin{matrix} \text{truth} \\ \text{probability} \end{matrix} \begin{bmatrix} 0.82 \\ 1.0 \end{bmatrix}, \begin{bmatrix} 0.82 \\ 0.96 \end{bmatrix}, \begin{bmatrix} 0.75 \\ 0.93 \end{bmatrix}.$$

Implication of intermediate event 2 to top event:
Implication operation:

Lukasiewicz	classical	Mamdani
0.88	0.70	0.70

Final failure of system:
Lukasiewicz:

truth	0.88
probability	1.0

Classic:

truth	0.70
probability	0.96

Fuzzy:

truth	0.70
probability	0.93

13.4.4.2 Truss system

The truss is composed of one level of intermediate events and one top event. In this structure, each component is axially loaded; the system is statically indeterminate, and supported with pins. Each intermediate event has components of basic events. These basic events represent members of the truss. Furthermore, the intermediate event represents the combination of members required to fail, leading to the failure of the truss. The top event is the truss failure composed of one level of intermediate events. As members are added to this system, more combinations of members are required and therefore additional intermediate events are necessary. These then combine at the top, resulting in the failure of the truss system.

The member numbers range from two to four in Lucero (1999). Here however, the truss has four members. The behavior of the norms is detected by varying the probability of one member, the basic event, of the intermediate events and aggregating it with various probabilities of the other members. The variation ranges from 0 to 1.0. The focus of this model is on capturing only the behavior of norm operators.

The truss system of Figure 13.6 is studied, where the corresponding fault trees are configured as shown in Figure 13.7.

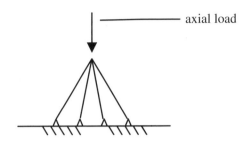

Figure 13.6. *Four-member axially loaded truss model.*

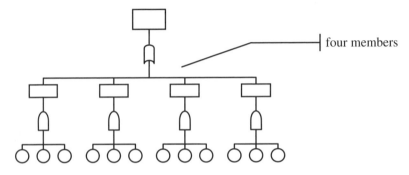

Figure 13.7. *Fault tree of four-member axially loaded truss.*

Comparison of the three logics is made by varying the probability for failure of one member from 0 to 1.0. These values are then propagated by the procedure previously discussed where only the probability is examined. They are propagated with three values for the other members 0.1, 0.6, and 0.8. The final, "top" probabilities for failure are shown in Figure 13.8 for the three logics Lukasiewicz, classical, and fuzzy.

13.4.5 Summary

The following discussion briefly describes the behavior of the norm operators with respect to the truss system. The results in Lucero (1999) constitute a more detailed analysis. Typical behavior is seen from a series of graphs. In the truss, however, the ordering from highest to lowest probabilities for failure becomes Zadeh, probabilistic, and Lukasiewicz.

In this summary, we will explain the use of the connectives in the fault tree of the truss system, describe this type of system in structural engineering, and finally conclude by discussing the use of the norms for structural systems. The fault trees can be described according to the type of connective aggregating the basic events. The fault tree for the truss connects its basic events with the "AND" (t-norm) gate. In this tree, failure of the subsequent event occurs if all these grouped basic events fail together. This same logic can be seen in actual structural systems.

The structural system that follows this fundamental logic is a continuous system. A continuous system has components serving more than one function. This function can be resisting both axial loads and bending moments, as in the case of concrete columns in buildings. Another example of multiple functions is redistribution of loads, as in the case of the truss. For example, if one member fails in a redundant truss system (Figure 13.6), then the other members take the additional load lost from the failing member.

comparison of axially loaded systems

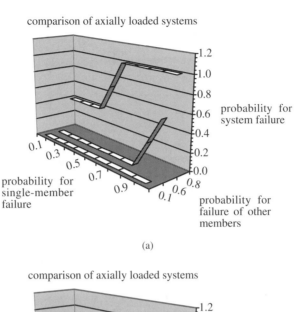

(a)

comparison of axially loaded systems

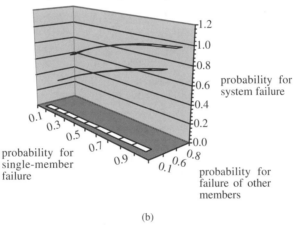

(b)

comparison of axially loaded systems

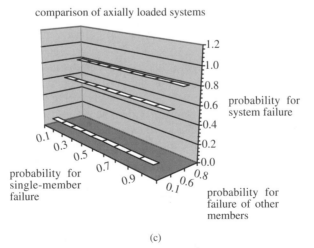

(c)

Figure 13.8. *Behavior of norms on truss system:* (a) *Lukasiewicz norms;* (b) *probabilistic norms;* (c) *Zadeh norms.*

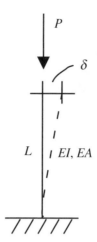

Figure 13.9. *Structure with independent failures.*

Based on the results of the research in Lucero (1999), the norms can be appropriately applied to the following types of structural systems: Zadeh to discrete systems, Lukasiewicz to continuous systems, and probabilistic to active independent systems.

Zadeh operators show probabilities consistent with intuition such as a parallel chain (Lucero (1999)) and can be better used for discrete structures. A parallel chain system is only as strong as the strongest weak link. This shows dependence among the chains in characterizing the failure of the system. For each case, this weak link was found and its probability carried for the system. For the case of the parallel chain, Zadeh gave the lowest values (most conservative results).

Lukasiewicz norms can be better utilized for continuous systems such as a concrete structure. These norms produce a lower bound to the probabilities when applied to the truss system. The truss system can be considered similar to a continuous system from a logical perspective in the sense that, for failure to proceed to subsequent levels, all the components must fail. All components failing is a characteristic of the "AND" (intersection) connective. Moreover, in a continuous system, all the components must fail. However, some resistance to failure is inherent since load-carrying mechanisms are shared. Lukasiewicz captures this quality of lower bound for "AND" gates by giving low values of propagating probabilities for failure.

Probabilistic aggregators of classical logic have been the means of calculating probabilities in fault tree analysis but may not be the most appropriate to use. In the results of the research in Lucero (1999), probabilistic aggregators gave values between those of Lukasiewicz and Zadeh. The two systems, parallel chain and truss, are examples of dependent systems. For example, the failure of the parallel chain is dependent on failures of each chain. Similarly, the failure of the truss is dependent on failures of adjoining members. Without knowing the degree of dependence of the system components, probabilistic t-operators cannot be appropriately applied. However, one system in structural engineering, where independence holds and probabilistic t-operators are better suited than Lukasiewicz and Zadeh, is shown in Figure 13.9.

For the axially loaded column in Figure 13.9, the three independent failure modes are due to static stress, static buckling, and dynamic buckling. For the failure due to static stress, the static stress must be greater than ultimate stress. For static buckling, the axial force must be greater than the allowable buckling load. Moreover, for dynamic buckling, the frequency of application of the load must be equal to the natural frequency of the system. All three cases are independent. As a result, the probabilistic t-operators can be used without assuming independence, since there is little interaction between these failure modes.

This study and Lucero (1999) reveal the characteristics of aggregators from the logics of Lukasiewicz, classical, and fuzzy. The Lukasiewicz, probabilistic, and Zadeh t-operators can give conservative, intermediate, or liberal probabilities depending on the situation. Whether a conservative or liberal view is chosen for a dependent or independent system, the goal of this research is to give an analyst more models to assess a system's probability for failure.

13.5 Part B: General methodology extended with fuzzy logic

Fuzzy logic allows for the modeling of uncertainty, ambiguity, or imprecision. This is achieved through the use of membership functions. "The essence of fuzzy logic is in its capability to model imprecision and vagueness by assigning membership functions to the variables. Membership functions (MFs) differ from probability density functions (PDFs) in that the PDFs describe the uncertainty in the future states of the variables while membership functions describe the vagueness in the definition of the state itself or the imprecision in the measurement or the representation of the variable used to characterize the future state" (Cooper and Ross (1997)).

Part A approached the probabilistic aspect of the general methodology where the probabilities are discrete (known) values.

In Part B, we show how fuzzy logic is used more specifically in this methodology. Membership functions can expand Part A to include uncertainty in the probability values. This concept is described in section 13.5.1. Moreover, the fuzzy logic approach allows for a different accounting of dependence among basic events with the use of rules and similarity relations. Section 13.5.2 covers this aspect of fuzzy logic as used here in the general methodology.

13.5.1 Membership functions

As mentioned previously, fuzzy set theory allows for uncertainty, ambiguity, and imprecision through the use of membership functions. The examples in Part A of this chapter cover the idea of precise probabilities for failure using scalar singletons or discrete point values to represent the probability values. These probabilities also can be represented with membership functions accounting for uncertainty, etc. However, they also can represent the failure description, where the imprecision is not on the future (probability) event but on the qualification of failure, e.g., partial failure. At this point, the probabilities or failure values are fuzzy numbers. Making the necessary calculations on these fuzzy numbers for this methodology requires knowledge of concepts such as *interval analysis* and *extended max and min*. These are covered thoroughly in Ross (1995).

13.5.2 Dependence

Many situations have components of failure at the basic level that are dependent on each other's function. Determination of their dependence has been expanded in this methodology from the traditional schemes. This new process involves expert knowledge or statistical data. In the case of passive systems, engineering judgment is used to form rules that govern the interrelation of these events. Additionally, statistical data on a component's operation is used to form similarity relations among basic events.

   This interdependence determination is the first step in the calculation and propagation of basic event probabilities and confidences in the general logic methodology. In Part A, the basic events were modeled as independent with no interdependence. In this section, basic events' interdependence are obtained as described above. The process begins with the calculation of a relational matrix. A relational matrix accounts for dependence among the basic events and is used to map the input values with their dependencies. This mapping is achieved by means of the composition operation. This operation uses a pairwise comparison of elements. One method used for a composition operation selects the maximum value from each set of pairwise minimum values. After the composition operation, the resulting set of values is the probability of failure and associated confidence. These values are then aggregated using various logics and propagated up the tree as described in Part A.

   The following describes the relational matrix formed from rules or statistical data, for a passive or active system, respectively. As previously mentioned, this represents the dependence among basic events and is used to calculate the event dependencies used at the first step of the tree.

Relational matrix:

$$R_{n \times n} = \begin{bmatrix} r_{11} & \cdots & \cdots & r_{1n} \\ \vdots & \ddots & & \vdots \\ \vdots & & \ddots & \vdots \\ r_{n1} & \cdots & \cdots & r_{nn} \end{bmatrix};$$

r_{ij} is the degree of dependence of basic event i on j $(i, j = 1, 2, \ldots, n)$. Once this relation is determined, the input basic event probabilities and confidences are mapped to their dependence values. This mapping takes place by means of the composition operation. Again, pairwise comparisons are made on elements from the relational matrix with elements from the input matrix as follows:

Input, the truth matrix, and probabilities for basic events at intermediate level 1:

$$A_{2 \times n} = \begin{bmatrix} T_1 & \cdots & \cdots & T_n \\ P_1 & \cdots & \cdots & P_n \end{bmatrix},$$

$$T_i = \text{confidence measure,} \qquad 0 \le T_i \le 1,$$
$$P_i = \text{probability value,} \qquad 0 \le P_i \le 1,$$
$$i = 1, 2, \ldots, n,$$
$$i = \text{basic event,}$$
$$n = \text{total number of basic events.}$$

Composition operation:

$$A'_{2 \times n} = A_{2 \times n} \circ R_{n \times n}, \tag{13.11}$$

where $A'_{2 \times n} = $ initial values for basic events, such that

$$A'_{2 \times n} = \max\{\min(A_{2 \times j}, R_{j \times i}), \ldots, \min(A_{2 \times j}, R_{n \times n})\}$$

for $i, j = 1, 2, n$.

The probabilities in the $2 \times n$ vector, A', of (13.11) are the values to be calculated from a variety of logics to determine the system failure probability. Rules or statistical information are used to form a relation $R_{n \times n}$ that describes the dependence among basic events for a passive or an active system. Sections 13.5.2.2 and 13.5.2.3 discuss how the matrix relation $R_{n \times n}$ is formed for such systems.

13.5.2.1 Example 2

This example illustrates the composition operation used to map the interdependence of basic events. This operation uses the input $2 \times n$ matrix of probabilities and confidences along with the $n \times n$ relational matrix. Again, the relational matrix is the matrix that defines the basic events' interdependence. How this (R) matrix is developed is discussed in sections 13.5.2.2 and 13.5.2.3.

Consider Example 1 of section 13.4.4.1. The determination of dependence affects the input to intermediate event 1. From Figure 13.5, we use the same input values of basic events A, B, and C. We map these values to a new 2×3 input matrix using the relational matrix R. "R" shows that basic event j relates to basic event i at a value of 1.0 when $j = 1$. This is an obvious observation since an event is entirely related to itself. The rest of the matrix values fill in a similar manner.

The resulting matrix A' after the composition operation is the new input values of probability and confidence to be used as input to "intermediate event 1."

Matrix of truth and probability values for basic events A, B, and C:

$$\begin{matrix} \text{truth} \\ \text{probability} \end{matrix} \begin{bmatrix} 0.76 & 0.98 & 0.83 \\ 0.55 & 0.72 & 0.94 \end{bmatrix}.$$

Relational matrix:

$$R = \begin{matrix} \quad \\ j \\ \vdots \\ \vdots \end{matrix} \begin{matrix} i \;\; \cdots \;\; \cdots \\ \begin{bmatrix} 1.0 & 0.5 & 0.3 \\ 0.5 & 1.0 & 0.4 \\ 0.3 & 0.4 & 1.0 \end{bmatrix} \end{matrix} \quad (i, j = 1, 2, n).$$

Composition operation from (13.11):

$$A' = \begin{bmatrix} 0.76 & 0.98 & 0.83 \\ 0.55 & 0.72 & 0.94 \end{bmatrix} \circ \begin{bmatrix} 1.0 & 0.5 & 0.3 \\ 0.5 & 1.0 & 0.9 \\ 0.3 & 0.9 & 1.0 \end{bmatrix} = \begin{matrix} \text{truth} \\ \text{probability} \end{matrix} \begin{bmatrix} 0.76 & 0.98 & 0.9 \\ 0.55 & 0.9 & 0.94 \end{bmatrix}.$$

Input to intermediate event 1:

$$\begin{matrix} \text{truth} \\ \text{probability} \end{matrix} \begin{bmatrix} 0.76 & 0.98 & 0.9 \\ 0.55 & 0.9 & 0.94 \end{bmatrix}.$$

13.5.2.2 Rules

Passive systems. Passive systems, as defined by Cooper and Ross (1997), are systems that "only require adherence to first principles of fundamental physical laws of nature and

Figure 13.10. *A simply supported bridge undergoing thermal expansion.*

production controls that assure no compromises to these principles in the actual implementation." These systems require no sensory detection and no reaction. For passive systems,
we propose that the relations among the basic events can be established through the use of
linguistic rules. Most structural engineering systems are passive systems, for example, a
bridge supported with pins and rollers. The pins brace the bridge as the rollers allow for the
movements of expansion and contraction during temperature changes (Figure 13.10).

For passive systems, a rule is a condition describing the behavior of components. Using
the bridge example, one rule might be "if temperature is high and pin doesn't move then
roller moves." Concepts like "high temperature," "pin doesn't move," and "roller moves"
are represented as fuzzy sets, whose membership functions are defined on appropriate scales.
An expert with sufficient knowledge of the basic events' failure dependencies forms a series
of these rules. Next, these rules are combined into matrices of ordered relations. These
matrices are then grouped with similar basic events and broken down using a variety of
defuzzification schemes. *Defuzzification* is the term used in the literature of fuzzy logic that
describes the consolidation of a fuzzy membership function into a single scalar value through
an averaging method. As a result of this operation, a matrix is formed (our R matrix) that
represents the dependence of basic events on one another for this failure component. This
matrix is then used to map these dependencies with input probabilities for failure to give
initial values for each basic event probability.

13.5.2.3 Similarity

Active systems.   Active systems are described in Cooper and Ross (1997) as systems where
"the safeguards depend on things that will function." This means that detection and reaction
are necessary. One example of this in structural engineering is vibration control (Preumont
(1997)). Either by means of cables or piezoelectric materials, sensors that detect critical
vibrations send input to controllers to react to the vibrations by contracting or expanding
members to alleviate the response (Figure 13.11). The relations among basic events in active
systems can be developed from statistical data on the interdependence of components.

The dependence of basic events in an active system usually comes from historical
statistical data. For example, with the vibration control of a bridge, one data value would be
that a motor controlling a tendon fails after 100,000 cycles of operation. A similarity relation
(again, our R matrix) is formed from this data relating dependence among the basic events
(Ross (1995)). This similarity relation is then mapped with input probabilities for failure
and confidences as in section 13.5.2 to give initial values of system failure. These values are
then propagated as in Part A.

13.5.3 Summary

Part B is a more comprehensive description of the general methodology. The general methodology uses concepts from fuzzy logic such as partial truths and partial failure. However, in

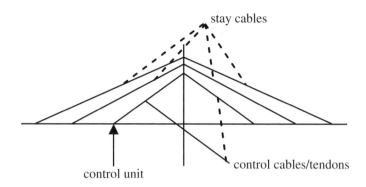

Figure 13.11. *Cable-stayed bridge with active tendons (Preumont (1997)).*

Part A, these concepts are applied to singleton values. In other words, the truths were presumed and the failure probabilities were presumed. Moreover, the interdependence of basic events was calculated using traditional methods of classical logic.

In Part B, these concepts were expanded to include fundamental characteristics of fuzzy logic. Membership functions describing the system variables now have uncertainty. Unlike Part A where input values are presumed, here the values are presumed imprecisely. Therefore, we alluded to tools to account for these "fuzzy" numbers.

In addition, Part B uses two more key aspects of fuzzy logic. These two aspects are rules and similarity relations. Applied to passive and active systems, these two aspects account for interdependence of basic events using rules or historical data, respectively. These supplement the traditional Boolean approaches to give more capacity to an analyst's decision making.

13.6   Conclusions

This chapter describes in two parts how fault tree analysis has been expanded to a general methodology. Part A describes the procedure for propagating singleton probability values and associated confidences. Three logics—Lukasiewicz, classical, and fuzzy—are used to demonstrate the effectiveness of this procedure. An example from structural engineering— an axially loaded truss—is included to demonstrate trends. This also shows that some other logics might be better suited to certain situations. This insight is due to the application of this method and shows its utility. In Part B, a fuzzy logic extension to the general method is described. Membership functions of fuzzy logic are described and have been proven to be useful in accounting for uncertainty. Next, dependencies using fuzzy logic are explained. Two types of dependencies are encountered in situations. These two types are due to passive and active systems. A brief description of these is included. With this and the procedure of Part A, one is able to assess system reliability in a new way.

References

T. M. ADAMS AND P. R. M. SIANIPAR (1997), Fault-tree model of bridge element deterioration due to interaction, *ASCE J. Infrastructure Systems*, 3, pp. 103–110.

L. M. BLUMENTHAL AND K. MENGER (1970), *Studies in Geometry*, W. H. Freeman, San Francisco.

J. A. COOPER AND T. J. ROSS (1997), Improved safety analysis through enhanced mathematical structures, in *Proceedings of the 1997 IEEE International Conference on Systems, Man, and Cybernetics*, IEEE, Piscataway, NJ, pp. 1656–1661.

M. M. GUPTA AND J. QI (1991), Theory of T-norms and fuzzy inference methods, *Fuzzy Sets Systems*, 40, pp. 431–450.

E. J. HENLEY AND H. KUMAMOTO (1992), *Probabilistic Risk Assessment*, IEEE, Piscataway, NJ.

J. LUCERO (1999), *A General Logic Methodology for Fault Tree Analysis*, M.Sc. thesis, University of New Mexico, Albuquerque, NM.

S. NISHIDA (1992), *Failure Analysis in Engineering Applications*, Butterworth–Heinemann, Oxford, UK.

A. PREUMONT (1997), *Vibration Control of Active Structures*, Kluwer Academic Publishers, Dordrecht, The Netherlands.

T. J. ROSS (1995), *Fuzzy Logic with Engineering Applications*, McGraw–Hill, New York.

R. R. YAGER (1983), On the implication operator in fuzzy logic, *Inform. Sci.*, 31, pp. 141–164.

Chapter 14

Uncertainty Distributions Using Fuzzy Logic

Ronald E. Smith, W. Jerry Parkinson, and Thomas R. Bement

Abstract. In this chapter, we look at the problem of developing a probability density function (PDF) with little or no numerical data and from that PDF generating a cumulative distribution function (CDF). We use the CDF to help us find a reliability estimate, or limits, on that estimate, for a determined value, for a given variable under study. We have used fuzzy logic and expert judgment to produce the required PDF. The technique used here is similar to the fuzzy control techniques described in Chapter 8. One major difference is that instead of utilizing only a single value that would normally be produced by defuzzification of the "truncated" output membership functions, we use all the information that we have generated. This includes the entire truncated output membership function that represents the variance of the knowledge generated by the fuzzy rule system. In this chapter, we also use the probability controller technique, described in Chapter 8, to analyze the same problem.

14.1 Introduction

The work in this chapter centers on the common problem of the need to have a quantitative value for a confidence level or the need to know the reliability of a piece of equipment. This information usually is required to make good decisions. If a reasonable amount of data exist, a probability density function (PDF) can be obtained. The cumulative distribution function (CDF) can be obtained by integrating the PDF. These two tools can be used to obtain desired answers, a confidence limit, or a reliability estimate. The problem arises when little or no data are available, there is no mathematical model to generate information from, or a mathematical model is inadequate.

The first solution to this problem is, if there is an expert, or even a reasonably knowledgeable person, available, to ask this person to give you a PDF for the situation, even if no data is available. The expert may or may not be able to do this. In addition, experts may not be knowledgeable about PDFs and may be embarrassed to admit it. Meyer and Booker

(2001) do not recommend asking experts for PDFs. If the expert can provide a PDF and everyone is comfortable with this solution, then the problem is solved. The statistician or engineer can then integrate the PDF, obtain a CDF, and then obtain limit estimates or reliability information. The problem is then solved and there is no need to read further. Unfortunately, it has been our experience that the experts, engineers or, in our case, scientists, do not feel comfortable doing this even though they feel comfortable with the subject matter. The three typical PDFs given by the experts are shown in Figure 14.1.

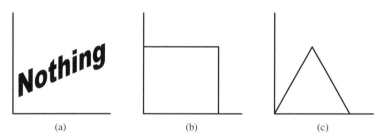

Figure 14.1. *Common PDFs given by experts:* (a) *nothing;* (b) *uniform distribution;* (c) *triangular distribution.*

A second solution is to use fuzzy rules and membership functions to generate what we might call a "comfort" function. This comfort function is generated by firing rules for a given situation and "clipping" the output membership functions according to some resolution law like the min–max method. This comfort function then can be normalized and treated like a PDF in lieu of better information. A PDF generated in this fashion usually is a better representation of what an expert really knows than a PDF produced by force. (The authors once attended a meeting where antifuzzy folks made the following comment about these ideas: "Don't use fuzzy logic. Just bend the expert's fingers back until he/she gives you a PDF.") This is probably because you are obtaining information from the expert in terms of ideas that he/she is comfortable with. From this point on, the researcher or the person who has been charged with finding the solution to this problem can use standard probabilistic techniques. The researcher can integrate the PDF to get a CDF and determine confidence limits or reliability values or whatever else is desired. This is one area where fuzzy logic and probability can work together, at least in the case of probabilistic risk assessment, probabilistic safety assessment, and reliability analysis—or any situation where a researcher needs to use a PDF or a CDF. A typical comfort function or PDF generated by this fuzzy technique is presented in Figure 14.2.

The authors wish to reiterate that there is no benefit in using fuzzy PDFs and CDFs when density functions and distribution functions can be produced from data or a mathematical

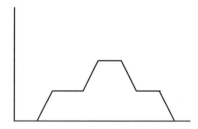

Figure 14.2. *Typical comfort function or PDF generated by this fuzzy technique.*

relationship. These functions will be better. The best time to use the fuzzy technique is when you have little numerical data and no mathematical model of the system. The fuzzy technique is most useful when using an expert who is knowledgeable about a system and who can build a semantic model using rules. (Although, in many problems the information that is available is all that is necessary. A combination of expert rules and data or a good model may give the best solution.) Fuzzy expert control systems are examples of a common and successful use of rules to describe a system or a relationship. The major difference between fuzzy control systems and the technique presented in this chapter is that fuzzy control systems require only a single "defuzzified" value. The technique presented here requires the entire fuzzy distribution. The idea is that there is more information available from the expert than a single value, and that knowledge should not be wasted.

This chapter is prompted by work done at Los Alamos National Laboratory in the mid 1990s. Statisticians needed PDFs and CDFs to develop relationships for the reliability of special components, but they had very little or no data. First, they tried to get PDFs from expert scientists, who really understood these components very well. They generally got poor results from these experts. (See Figure 14.1(a).) The statisticians decided to collaborate with some control engineers. Together they developed fuzzy rule-based comfort functions or PDFs, which were easily obtained from the expert scientists. The results they obtained were much better than before. They compared the results with what little data they had and obtained an excellent fit. This work was confidential and the specifics could not be published; however, a pseudoproblem was created using wear in a machine tool and the techniques used were demonstrated in that problem (Smith et al. (1997, 1998)).

The machine tool problem was noticed and the team was assigned another similar problem actually dealing with machine tools. This problem was not ideal, but it was actually a simpler problem than the one solved in the confidential work. A subject matter expert was available but there were no actual wear data. Several different fuzzy expert systems were designed, built, and tested. The major problem was that because there was no real wear information, it was very difficult to determine the best solution or even if the solution was valid. Later, an attempt was made to tie wear information to tool temperature data.

This was easier to quantify, but the expert only had a vague feeling for temperatures as compared to tool wear. He basically knew that there was a relationship between temperature and wear, but even if he touched the tool he would not have known what temperature it was. The expert was given the temperature ranges and asked to fit his fuzzy rule base and membership functions to temperature instead of wear. He did so, and, surprisingly, seven of ten data sets were fairly well represented by his system. This system can be quantified and give a good example of how this techniques works and, since temperature data are available, we can see what results might be expected from a method like this. Some of this work is presented in (Parkinson et al. (1999)).

14.2 Example 1: The use of fuzzy expertise to develop uncertainty distributions for cutting tool wear

Here we began developing an expert system with a researcher who had a good feeling for tool wear as a function of machine speed and forces applied to the tool and the work, all of which could be measured easily. The problem was changed in midstream when we were able to obtain data from a dissertation that related the machine speed and forces to cutting tool temperature. This work was being performed for an entirely different purpose. We felt fortunate to get the temperature data since it is extremely difficult and expensive to obtain.

Once the expert knew something about the temperature ranges in the problem, he felt that he could apply his knowledge of speed and forces versus wear to speed and forces versus temperature, then later relate temperature to wear. In most cases, workers are unlikely to have access to this information. We were fortunate to be able to compare our expert-produced CDFs to those produced by the data. In this case the expert was not an expert in temperatures, but he was given some idea of the temperature ranges involved and the ranges of speeds and forces. He then turned his "wear" rules into temperature estimates. The fuzzy temperature estimates were turned into fuzzy CDFs and compared with the CDFs produced from the actual experimental data. In this case, because we used temperatures instead of wear, the expertise was not as good as we would preferred, but the correlation was reasonable for seven of ten cases. Three cases were not as satisfactory, but considering that the rules and the membership functions were quite simple, these results were quite remarkable.

14.2.1 Demonstration of the technique

With dry machining becoming a reality, it is easy to destroy a work piece with a dull cutting tool. Current practice is to use a tool only once for finishing a piece. Suppose that it takes one minute to finish a piece. We have a tool that we believe will be good for three minutes of work under nominal operating conditions. We have used the tool twice (for two minutes) and need to determine the confidence that we can use it one more time.

The input variables are speed of rotation, thrust force on the piece, and cutting force exerted by the tool as illustrated in Figure 14.3. We developed the following set of input membership functions shown in Figures 14.4–14.6. (These are actually the ones used for the temperature problem but they are very similar to the ones developed for the tool wear problem.) Figure 14.7 shows the output membership functions for percent wear deviation.

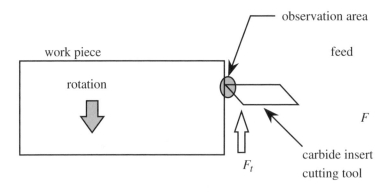

Figure 14.3. *Cutting tool, work piece, and associated forces.*

Table 14.1 shows the rules developed by the subject matter expert for tool wear. The last column in this table also shows the rules for temperature prediction since the input membership functions are the same for both cases.

These are the steps used to resolve the problem of interest:

1. Fire the rules that are involved in the problem.

2. Clip the resultant output membership function.

3. Normalize the resultant membership function. (This is the PDF.)

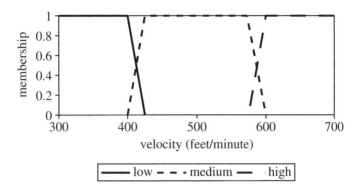

Figure 14.4. *Membership functions for speed of rotation.*

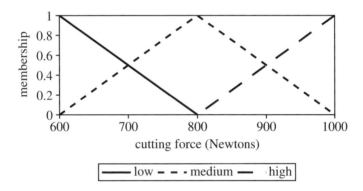

Figure 14.5. *Membership functions for cutting force.*

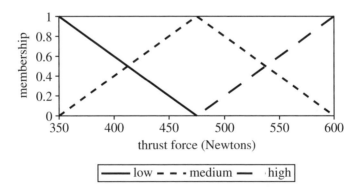

Figure 14.6. *Membership functions for thrust force.*

4. Repeat steps 1–3 for each set of conditions for the tool.

5. Apply a mixing rule if appropriate.

6. Use convolution to combine the resultant PDFs.

7. Integrate the combined PDFs to obtain a CDF.

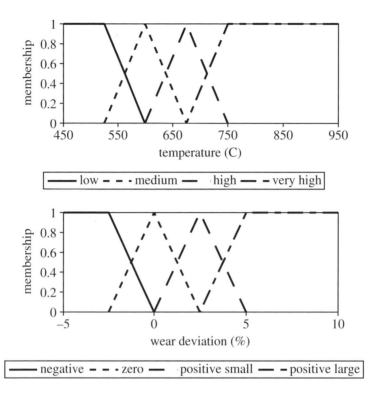

Figure 14.7. *Output membership functions for temperature and wear deviation.*

In our example problem, the tool was used twice. The first set of conditions were

- velocity = 351 feet/minute,

- cutting force = 893 Newtons,

- thrust force = 497 Newtons.

Table 14.2 shows the rules that were fired and their strengths (step 1).

Rules 5, 6, 8, and 9 were fired. The min–max rule was used to resolve the conflicts. The results of clipping the output membership functions (step 2) are shown in Figure 14.8. The PDFs resulting from normalizing the clipped membership function are shown in Figure 14.9. Figure 14.10 shows the resulting temperature CDF, the CDF from the experimental data, and the CDF of a normal distribution fit to the experimental data.

14.3 Example 2: Development of uncertainty distributions using fuzzy and probabilistic techniques

In this section, we describe a method for obtaining uncertainty distributions that was based on fuzzy control systems methodology and a probabilistic alternative. In particular, we considered the situation where the reliability of a specific system was based on whether or not performance fell within specifications. This section is designed to provide more background for the work described in section 14.2.

Our particular interest was in cases where the relationship between the condition of the system and its performance was not well understood, especially for some sets of possible

Table 14.1. *Rules relating input membership functions velocity, cutting force, and thrust force to both tool wear and temperature.*

Rule number	Velocity	Cutting force	Thrust force	Tool wear	Temperature
1	low	low	low	negative	low
2	low	low	medium	negative	low
3	low	low	high	negative	low
4	low	medium	low	negative	low
5	low	medium	medium	negative	low
6	low	medium	high	zero	medium
7	low	high	low	negative	low
8	low	high	medium	zero	medium
9	low	high	high	positive small	high
10	medium	low	low	negative	low
11	medium	low	medium	zero	medium
12	medium	low	high	zero	medium
13	medium	medium	low	zero	medium
14	medium	medium	medium	zero	medium
15	medium	medium	high	positive small	high
16	medium	high	low	zero	medium
17	medium	high	medium	positive small	high
18	medium	high	high	positive large	very high
19	high	low	low	zero	medium
20	high	low	medium	zero	medium
21	high	low	high	zero	medium
22	high	medium	low	positive small	high
23	high	medium	medium	positive small	high
24	high	medium	high	positive large	very high
25	high	high	low	positive small	high
26	high	high	medium	positive large	very high
27	high	high	high	positive large	very high

Table 14.2. *Rules fired in Example* 1.

Rule number	Velocity	Cutting force	Thrust force	Temperature
5	low (1)	medium (0.53)	medium (0.79)	low (0.53)*
6	low (1)	medium (0.53)	high (0.21)	medium (0.21)
8	low (1)	high (0.47)	medium (0.79)	medium (0.47)*
9	low (1)	high (0.47)	high (0.21)	high (0.21)*

*Indicates the rules that were actually used.

operating conditions, and where developing a better understanding was very difficult and/or expensive. For example, a system may have been developed and tested for a specific range of operating parameters, and it was now desired to characterize its reliability if it was used in untested or minimally tested parameter ranges. In applications, the manner in which uncertainty distributions were obtained ranges from being largely data driven to being highly subjective, depending upon where test results were available.

Because the actual application that motivated this work was confidential, an artificial problem description is used to illustrate the method.

14.3.1   Description and illustration of the fuzzy setup of the method

Consider a system with several components, each of which can influence the performance of the system. Each component was subject to wear, potentially degrading performance. One

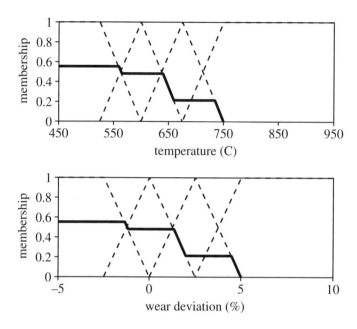

Figure 14.8. *Clipped output membership functions for temperature and wear deviation in Example* 1.

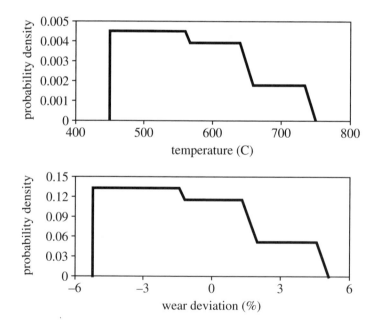

Figure 14.9. *PDFs for temperature and wear deviation in Example* 1.

example was an automobile fuel injector consisting of many components, each of which can influence the flow rate of the air/fuel mixture. Another was a catalytic converter which must meet certain performance specifications and which can degrade with use. For a given wear level, performance degradation was variable. Our objective was to characterize the

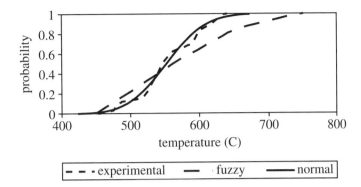

Figure 14.10. *Temperature CDFs: fuzzy, experimental, and normal (Example* 1).

relationship between performance and the condition of the components.

Assume the hypothetical system consists of four components, A, B, C, and D, where the relationship between the condition of the components and the performance of the system is not well understood. The understanding of the relationship was better for some conditions than for others.

The system experts classified component wear according to four levels or categories: negligible wear, slight wear, moderate wear, and severe wear.

Components A and B experienced wear in four categories and components C and D in three categories. The reason for the differing numbers of categories was related to the wearout properties of the four components and the experts' experience with wear levels for the various components. Not all components may experience the same degrees of wear. For example, some components tended to wear out in a continuous manner, while others tended to fail catastrophically. In addition, it may be that even though certain wear levels are theoretically possible, there is no experience base with such wear.

Similarly, seven levels of degradation were used to describe system performance: nominal, very slight, slight, moderate, large, very large, and excessive. Tables 14.3–14.6 show the rules for relating condition to performance for each of the four components.

In a true fuzzy control systems application, one would completely characterize the performance relationship for combined wear patterns in a fuzzy manner, requiring $4 \times 4 \times 3 \times 3 = 144$ rules. However, in this case, the experts chose to develop fuzzy rules

Table 14.3. *Component* A *performance degradation rules.*

Wear	Performance degradation
negligible	nominal
slight	slight
moderate	very large
severe	excessive

Table 14.4. *Component* B *performance degradation rules.*

Wear	Performance degradation
negligible	nominal
slight	slight
moderate	moderate
severe	large

Table 14.5. *Component* C *performance degradation rules.*

Wear	Performance degradation
negligible	nominal
slight	slight
moderate	moderate

Table 14.6. *Component* D *performance degradation rules.*

Wear	Performance degradation
negligible	nominal
slight	very slight
moderate	moderate

relating only the condition of each individual component to system performance—under the assumption that each of the other components was in a nondegraded condition—and then developed analytical rules for dealing with combined degradation effects.

Figure 14.11 shows the membership functions for the condition of each component using a percentage wear metric. Figure 14.12 shows the seven membership functions for system performance as percent degradation from nominal for each of the four components. Table 14.7 lists the Gaussian parameters for the system performance degradation membership functions.

The analytical rules considered by the experts for combining the wear effects of the four components are a weighted sum of individual degradations and a geometric sum of the individual degradations. Section 14.3.2 describes how uncertainty distributions are obtained from the work described in this section.

14.3.2   Fuzzy development of uncertainty distributions

The method for generating uncertainty distributions will now be illustrated by considering only component A. Suppose component A has wear equal to w and the other three components show no wear.

From Figure 14.11, we can determine the levels at which each of the component A rules are "fired." That is, for a given w, we determine the values of each of the four membership functions. Denote these by $m_{A1}(w), m_{A2}(w), m_{A3}(w)$, and $m_{A4}(w)$. For example, if $w = 50$, then $m_{A2}(w) = 0.25$, $m_{A3}(w) = 0.75$, and $m_{A1}(w) = m_{A4}(w) = 0$.

For a given value of w, by using standard fuzzy systems methodology, we can obtain the corresponding combined membership function for performance. In this case it is a weighted average of the two involved performance membership functions. The combined membership function for $w = 50$ is shown in Figure 14.13.

Departing now from standard fuzzy systems methods, we normalize the trimmed membership function for performance so that it integrates to 1.0. The resulting function, $f_A(x|w)$, is taken to be the performance (x) uncertainty distribution corresponding to the situation where component A wear is equal to w and the other three components show no wear. Figure 14.14 is the CDF form of the uncertainty distribution, $F_A(x|w)$. If performance must exceed some threshold, T, in order for the system to operate successfully, then the reliability of the system for the situation where component A wear is equal to w and the other three components show no wear can be expressed as $R_A(w) = 1 - F_A(T|w)$.

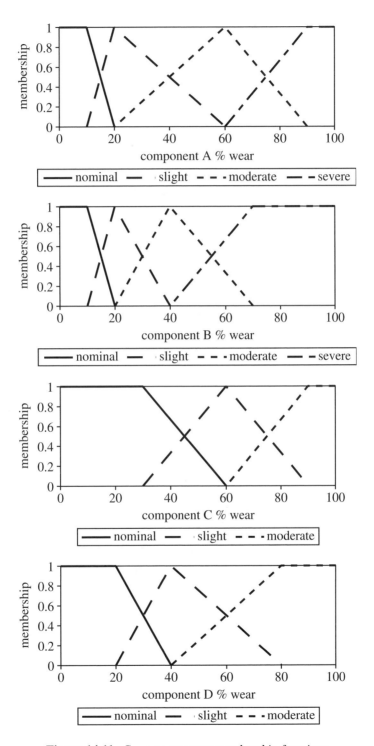

Figure 14.11. *Component wear membership functions.*

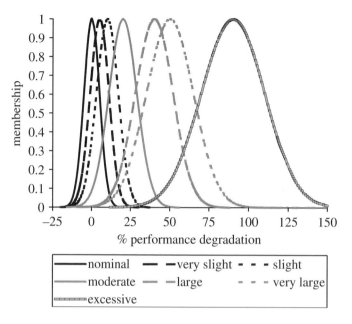

Figure 14.12. *Performance degradation membership functions.*

Table 14.7. *Gaussian parameters for performance degradation membership functions.*

Membership function	Mean	Standard deviation
nominal	0	5
very slight	5	6
slight	10	7
moderate	20	9
large	40	12
very large	50	15
excessive	90	20

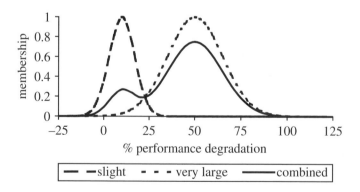

Figure 14.13. *Component A performance degradation membership with 50% wear.*

Clearly, uncertainty or variability in the level of wear for component A induces uncertainty in $R_A(w)$. Suppose that the uncertainty in wear, w, is characterized by some distribution. The distribution of $F_A(x)$, and hence of R_A, then can be obtained. For example,

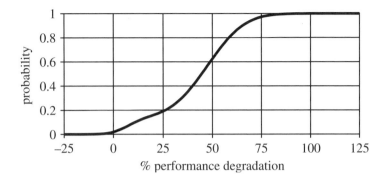

Figure 14.14. *CDF for component* A *performance degradation with 50% wear.*

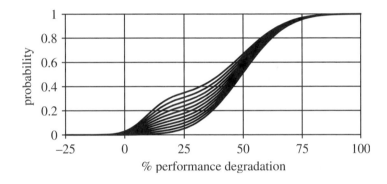

Figure 14.15. *CDFs for sampling from wear distribution for component* A.

suppose the uncertainty in wear for component A is characterized by a uniform distribution on the closed interval [40, 60]. The results of repeatedly sampling w from that distribution and calculating $F_A(x|w)$ are shown in Figure 14.15. The uncertainty in the degradation probability that is due to uncertainty in the level of wear is illustrated by the "envelope" of cumulative distribution functions that is induced by sampling values of w. The median and the fifth and 95th percentiles of the envelope are shown in Figure 14.16. The approximate distribution of R_A can be obtained from such a numerical simulation process.

14.3.3   Probabilistic characterization of the development of uncertainty distributions

A probabilistic alternative to the fuzzy control methodology described above is offered by Laviolette et al. (1995). Following their approach, we describe how the development of uncertainty distributions described above can be approached probabilistically rather than with fuzzy methods.

In the probabilistic case, the membership functions in Figure 14.11 are replaced with wear classification probability functions, conditioned on the wear level, w. For example, in Figure 14.11, the "nominal" membership functions are replaced with functions giving the probability that the wear category is "nominal" given that the wear level is w. The performance membership functions in Figure 14.12 were each replaced with corresponding

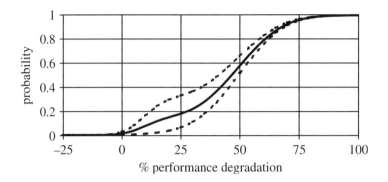

Figure 14.16. *Median, fifth, and* 95*th percentiles of the CDF envelope generated by sampling from component* A *wear distribution.*

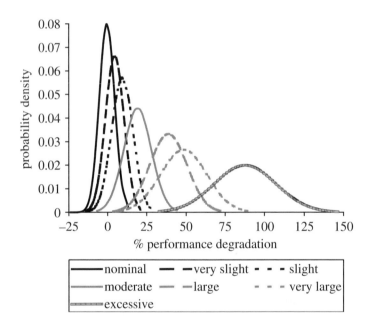

Figure 14.17. *Performance degradation PDFs.*

PDFs that characterize the distribution of performance within each of the seven performance categories (Figure 14.17). For example, the PDF corresponding to "slight" performance degradation will provide the distribution of performance given slight degradation. For a given weight, w, the combined membership function for performance of Figure 14.13 is replaced with a PDF that is a weighted average (mixture) of the appropriate performance distributions (Figure 14.18). The weights are the wear category membership probabilities for wear equal to w.

An alternative method of obtaining the combined performance degradation is to use a linear combination of the individual performance degradation PDFs using the wear category membership probabilities for wear as the weights (Figure 14.19). If, as in this case, the membership functions are Gaussian, $N(\mu, \sigma^2)$, then the linear combination is also Gaussian, $N(w_1\mu_1 + w_2\mu_2, w_1^2\sigma_2^2 + w_2^2\sigma_2^2)$.

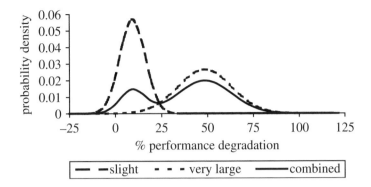

Figure 14.18. *Component* A *performance degradation PDF with* 50% *wear using "mixture" of PDFs.*

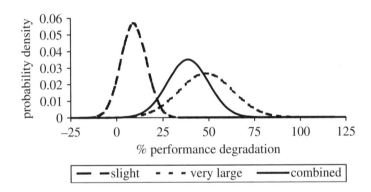

Figure 14.19. *Component* A *performance degradation PDF with* 50% *wear using linear combination of PDFs.*

14.3.4 Extended example

Let us consider an extended example that fully illustrates each approach. Consider the case where component A is worn 50%, component B 15%, component C 40%, component D 70%. Using the methods described in the previous two sections, one can derive the performance degradation PDFs for each component. Figure 14.20 shows the performance degradation PDFs using the fuzzy methodology. Figure 14.21 shows the performance degradation PDFs using the probabilistic methodology "mixture." Figure 14.22 shows the performance degradation PDFs using the probabilistic methodology of linear combination.

Consider one of the analytical rules given by an expert for combining the wear effects of the four components: "The largest effect will dominate, the second largest effect will contribute one half as much as it would independently, the third largest—one fourth—and the smallest—one eighth" (Smith et al. (1997)). In this case, A has the largest effect, followed by D, B, and C. One can then combine the effects shown in Figures 14.20, 14.21, and 14.22 using Monte Carlo simulation techniques (in the probabilistic–linear combination case an analytical solution exists). The final combined performance degradation for this example is shown in Figure 14.23.

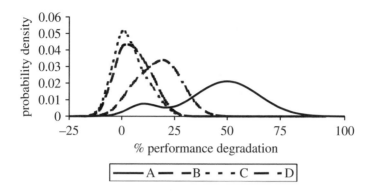

Figure 14.20. *Performance degradation PDFs, extended example, fuzzy methodology.*

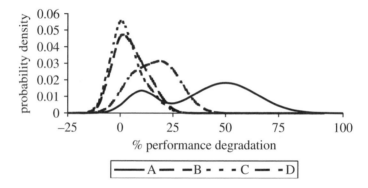

Figure 14.21. *Performance degradation PDFs, extended example, probabilistic methodology, mixture.*

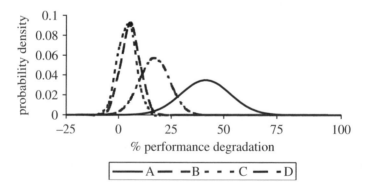

Figure 14.22. *Performance degradation PDFs, extended example, probabilistic methodology, linear combination.*

14.3.5   Concluding remarks

Recall that our interest is in situations where the relationship between wear and performance is not well understood, and where the understanding may be better for some wear ranges than for others. In this type of application, the fuzzy system was used as an approximation to an

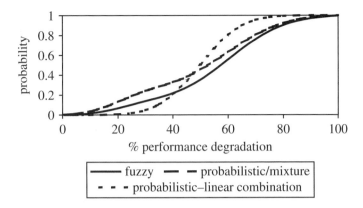

Figure 14.23. *Combined performance degradation CDFs.*

unknown wear–performance function. In Kosko (1994) and elsewhere, the implications of using a fuzzy systems approach to approximate continuous functions are discussed. This sort of approximation has the advantage of being able to combine information from a variety of sources, ranging from highly data driven to highly subjective. Such situations are common when equipment is used outside of its original design envelope, for example.

A potential drawback for this method is that the combined performance distributions for wear values in unexplored ranges will be multimodal, as in Figures 14.13 and 14.18. The multimodality proceeds directly from the fuzzy characterization of the relationship and the fact that the relationship is understood better for some ranges of wear than for others. However, if the multimodality is seriously contrary to the judgment of experts, that fact itself may indicate that additional information (i.e., in addition to the fuzzy system rules and membership functions) is available and should be included in characterizing the wear–performance relationship. The linear combination approach also may solve the multimodal problem.

14.4 Summary

It is evident after reading this chapter and observing the examples, and especially Figures 14.10 and 14.23, that this technique does not produce perfect values. There are several possible CDFs that could be produced by these techniques. If the developer has sufficient numerical data or some mathematical relationship to produce a PDF, then the standard statistical techniques will most likely work better. On the other hand, when data or reliable mathematical descriptions are not available, a semantic description by an expert or knowledgeable person can be very useful. We feel strongly that a PDF generated by a rule-based method as described here is better than one that came directly from an expert's imagination. The quality should be better for no other reason than that the expert has to break down his/her reasoning into small, logical steps in order to solve the problem. The choice of whether to use a fuzzy technique or a probabilistic technique for the rule-based method probably should be based on which technique the expert is more comfortable with, which in turn corresponds to the form of his/her knowledge. The quality of the final solution highly depends on the quality of the expert's description. Finally, if the expert has no strong opinion about which method to use, the developer is free to choose the technique. Remember, the analyst is judging the quality of the results, not the technique used.

References

B. KOSKO (1994), Fuzzy systems as universal approximators, *IEEE Trans. Comput.*, 43, pp. 1329–1333.

M. LAVIOLETTE, J. W. SEAMAN JR., J. D. BARRETT, AND W. H. WOODALL (1995), A probabilistic and statistical view of fuzzy methods, *Technometrics*, 37, pp. 249–261.

M. A. MEYER AND J. M. BOOKER (2001), *Eliciting and Analyzing Expert Judgment: A Practical Guide*, ASA–SIAM Series on Statistics and Applied Probability, SIAM, Philadelphia, ASA, Alexandria, VA.

W. J. PARKINSON, K. W. HENCH, M. R. MILLER, R. E. SMITH, T. R. BEMENT, M. A. MEYER, J. M. BOOKER, AND P. J. WANTUCK (1999), The use of fuzzy expertise to develop uncertainty distributions for cutting tool wear, in *Proceedings of the 3rd International Conference on Engineering Design and Automation*, Vancouver, BC, Canada, Integrated Technology Systems, Inc., CD-ROM.

R. E. SMITH, J. M. BOOKER, T. R. BEMENT, W. J. PARKINSON, AND M. A. MEYER (1997), The use of fuzzy control system methods for characterizing expert judgment uncertainty distributions, in *Proceedings of the Joint Statistical Meeting*, Vol. 1, American Statistical Association, Alexandria, VA, Elsevier, New York, pp. 497–502.

R. E. SMITH, T. R. BEMENT, W. J. PARKINSON, F. N. MORTENSEN, S. A. BECKER, AND M. A. MEYER (1998), The use of fuzzy control system techniques to develop uncertainty distributions, in *Proceedings of the 4th International Conference on Probabilistic Safety Assessment and Management*, Vol. I, Springer-Verlag, New York, pp. 497–502.

Chapter 15

Signal Validation Using Bayesian Belief Networks and Fuzzy Logic[*]

Hrishikesh Aradhye and A. Sharif Heger

Abstract. Safe and reliable control of many processes, simple or complex, relies on sensor measurements. Sophisticated control algorithms only partially satisfy the growing demand for reliability in complex systems because they depend on the accuracy of the sensor input. Noise-ridden or faulty sensors can lead to wrong control decisions, or may even mask a system malfunction and delay critical evasive actions. Thus for optimal and robust operation and control of a process system, correct information about its state in terms of *signal validation* is of vital importance. To this end, process control methods need to be augmented with *sensor fault detection, isolation,* and *accommodation* (SFDIA) to detect and localize faults in instruments. SFDIA can be defined as (a) the detection of sensor faults at the earliest, (b) isolation of the faulty sensor, (c) classification of the type of fault, and (d) providing alternative estimates for the variable under measurement.

In this chapter, we introduce the use of Bayesian belief networks (BBN) for SFDIA. Bayesian belief networks are an effective tool to show flow of information and to represent and propagate uncertainty based on a mathematically sound platform. Using several illustrations, we will present its performance in detecting, isolating, and accommodating sensor faults. A probabilistic representation of sensor errors and faults will be used for the construction of the Bayesian network. We limit sensor fault modes considered in this work to bias, precision degradation, and complete failure. Fuzzy logic forms the basis for our second approach to SFDIA. In decision-making related to fault detection and isolation, fuzzy logic removes the restrictions of hard boundaries set by crisp rules. The main advantage of fuzzy logic is its simplicity without compromising performance. As shown in the results, three fuzzy rules per variable are sufficient for a model-based fuzzy SFDIA scheme. A comparison of these two methods based on the experience of this application provides valuable insights.

15.1 Introduction

This chapter focuses on sensor fault detection, isolation, and accommodation (SFDIA) using Bayesian belief networks (BBN) and fuzzy logic. The intent is to apply these methods to

---

[*]This chapter is based on the first author's thesis (Aradhye (1997)).

this important field and highlight the advantages and limitations of each method. SFDIA is particularly important in the operation of complex systems such as chemical processes, power generation, aviation, and space exploration. In all these systems, any fault can have serious consequences in terms of economics, environmental concerns, or human health effects. Control actions designed to mitigate the effect of faults or maintain the system at its optimal state depend on correct interpretation of its state. The determination of the system state depends on the readings provided by sensors. Hence, noise-ridden or faulty sensors can lead to wrong control decisions or may mask a system malfunction. Thus for reliable and optimal operation and control of a system, sensor validation is of vital importance.

Almost all the SFDIA schemes use redundancies in available information to detect inconsistencies between various sources, which lead to diagnoses of probable faults. One of the common approaches is the use of *hardware redundancy*, i.e., the use of multiple instruments to measure the same quantity. The *functional* or *analytical redundancy* approach is based on the fact that readings of instruments measuring different quantities for the same plant or device are correlated and are governed by the laws of known or unknown process dynamics. Each of the two approaches has its benefits and drawbacks. Hardware redundancy suffers from the obvious drawbacks of higher instrumentation and maintenance cost, higher information processing load, and higher weight requirements. On the other hand, the model predictions in an analytical redundancy approach may be wrong in cases of unforeseen process faults and/or process conditions outside the operating limits of the model. The hardware redundancy approach holds even in the case of abnormal process conditions and may be necessary in safety-critical variables. To be effective, an SFDIA scheme must take advantage of both these approaches and

1. must handle both dynamic and steady state sensor validation;

2. should allow for continuous as well as discrete-valued representation;

3. should incorporate analytical as well as knowledge-based, system-specific information;

4. must learn and adapt from experience.

Current approaches to SFDIA fall into two broad categories of statistical and AI-based methods. Statistical methods can be further subcategorized into the parity-space approach (Potter and Sunman (1977), Desai and Ray (1981)), observer-based approaches (Clark, Fosth, and Walton (1975), Frank (1987a, 1987b, 1994a, 1994b)), parameter-based approaches (Isermann (1984), Isermann and Freyermuth (1991a, 1991b)), and frequency-domain approaches (Ding and Frank (1990), Frank and Ding (1994)). These methods are, in general, quantitative and nonlearning. Examples of AI-based schemes are expert systems or decision support systems (Singh (1987), Watton (1994), Lee (1994)), neural networks–based approaches (Bernieri et al. (1994), Mageed et al. (1993), Napolitano et al. (1995)), fuzzy logic–based approaches (Park and Lee (1993), Heger, Holbert, and Ishaque (1996)), and neurofuzzy approaches (Mourot et al. (1993), Sauter et al. (1994)). AI-based methods provide qualitative reasoning and learning capability. It is desirable to combine the robustness and mathematical soundness of quantitative methods with the simplicity, intuitive appeal, and adaptation capabilities of the qualitative methods. The methods presented in this chapter attempt to achieve this goal. To this end, we will

• develop SFDIA methods using BBNs and fuzzy logic;

• compare the features of each method;

- broaden the basic goal of SFDIA to encompass process as well as sensor fault detection, multisensor fusion, data reconciliation, and analytical as well as knowledge-based modeling.

BBNs are an effective tool to show flow of information and to represent and propagate uncertainty based on a mathematically sound platform (Pearl (1988), Spiegelhalter et al. (1993), Lauritzen and Spiegelhalter (1988)). Using provable schemes of *inference* and *evidence propagation*, a Bayesian network modeling a system can help to detect, isolate, and accommodate sensor faults in a single, unified scheme. It is no longer necessary to construct separate modules to perform each of these functions. Uncertain instantiations and unavailable data are handled implicitly. Addition, removal, or replacement of one or more sensors can be handled by changing accordingly the nodes corresponding to the sensors in question and does not affect the entire network.

Many examples of the use of BBNs for process fault detection can be found in the published literature (Deng (1993), Kirch and Kroschel (1994), Nicholson and Brady (1994), Nicholson (1996)). The message-passing and inference schemes in these networks are based on sound theorems of probability, whereas the network structure has the ability to capture causal relationships. Thus this network is a combination of analytical as well as knowledge-based approaches. The use of BBNs for the detection of sensor faults is introduced in this chapter. This method has the potential to incorporate the four features listed above, as demonstrated through several examples. This scope of the examples is limited to sensor bias, precision degradation, and complete failure.

The relative speed and simplicity of approximate reasoning and calculation have been the basis of the fuzzy logic domain. Traditional sensor validation methods rely on crisp data and produce either a valid or failed decision. A binary-valued decision for the sensor status increases the probability of a missed or false alarm, which may be reduced due to the continuous-valued approach used in fuzzy logic.

A solution to the fault detection problem with this approach has been proposed as application of fuzzy rule bases either on residual readings or along with neural nets using the functional redundancy approach (Dexter (1993), Singer (1991), Mourot, Bousghiri, and Kratz (1993), Goode and Chow (1993), Sauter et al. (1994)). The fuzzy logic–based hardware redundancy scheme (Holbert, Heger, and Ishaque (1995), Heger, Holbert, and Ishaque (1996)) uses two or three redundant sensors. In contrast, the work presented in this chapter makes use of fuzzy rules in an analytical redundancy-based setup.

15.2 Bayesian belief networks

In this section, we start with a brief introduction of BBNs. Following that, we present the use of the networks for SFDIA through a simple example for a single-variable, single-sensor case. Sample calculations demonstrate the mechanisms by which new evidence is propagated through the network, and new inferences are drawn. A good SFDIA system uses the redundancy in information to detect and mitigate the effects of faults and provide reduction in noise. To this end, the incorporation of redundancies in the network structure is discussed in the next section. Application of the method to a continuous stirred tank reactor (CSTR) is presented.

BBNs (probabilistic networks) are causal acyclic graphs which are hypothesized to represent the flow of human reasoning in domains with inherent uncertainty (Pearl (1988), Lauritzen and Spiegelhalter (1988), Spiegelhalter et al. (1993)). The nodes in this network represent propositional variables and the directed link between two nodes represents a causal influence of the parent node on the child node. Associated with each child node is a probability distribution conditional on its parents. The nodes with no parents are called "root nodes" and each has an a priori probability distribution.

At any instant, a certain body of external information is made available to the network. This "evidence" is incorporated by instantiating the corresponding nodes. The directed structure of the network and the conditional probabilities associated with each link store a body of prior knowledge. This allows for the calculation of probability distributions of the uninstantiated nodes in the network, given the evidence. This represents the conclusion inferred by the network in light of the evidence presented and is called "inference." The process of message passing that results in the inference is called "propagation" of evidence.

The inference scheme uses the Bayesian method to calculate updated "beliefs," i.e., the probabilities. A typical inference process consists of one initialization step, a series of updating probabilities, and exchanging π and λ messages among nodes. A λ-message, which is sent by a child node to its parent, represents the *diagnostic information*. A π-message from a parent node to a child node represents the *causal influence*. Once the network stabilizes, the updated probabilities associated with each node represent the belief about the state of the node in light of the evidence.

Consider a simple system that consists of a temperature sensor/indicator (Figure 15.1). If the sensor is operational, then there is a correlation between its value and that of the temperature. The lack of correlation in a faulty sensor is represented as a uniform probability distribution in the conditional probability table of the indicator node; i.e., both *high* and *low* are equally probable values for sensor reading, irrespective of the *high* or *low* value of the temperature. When the temperature is above a certain threshold (high), the indicator is red; otherwise it is blue. The network representation of this system uses three discrete-valued nodes: one for the indicator reading (node I), one for the status of the sensor (node S), and another (node T) for temperature, whose actual value is unknown at any given time. The causal relationships and dependencies between the nodes are shown in Figure 15.1. Node I has two possible values of "red" and "blue." Node S can be either "functional" or "inoperative" representing the corresponding status of the sensor. Node T represents the actual temperature that can be "high" if it is above the threshold or "low" otherwise. The actual states of nodes S and T are not known at any given time. The network helps to form a probabilistic inference about these states based on the indicator reading.

Based on the manufacturer's specifications, let the a priori probability of the *functional* state of node S be 0.65. Since each node is binary-valued, a total of four conditional probability statements are needed to uniquely define the dependence of I on T and S. These values are shown in boxes attached to each node in Figure 15.1. At this point, the a priori distribution for the node T needs to be provided for the network to be completely defined. As a starting point, assume a noninformative prior, assigning 0.5 for the two possible states. An observation about the indicator reading is presented to the network via instantiation of the node I. Given this evidence, the network updates its beliefs about the other two nodes, S and T. For example, if the indicator is observed to be *red*, the probability of the temperature being *high* increases from 50% to 82.5%. Beliefs about node S remain unchanged, due to the noninformative prior. The calculations showing the propagation of probabilities in the network and the resultant inference about the status of the sensor are shown in detail below.

Initialization.   The objective of this stage is to initialize the BBN such that the network becomes ready to accept evidence. Probabilities for all nodes and all π and λ messages are initialized for this purpose.

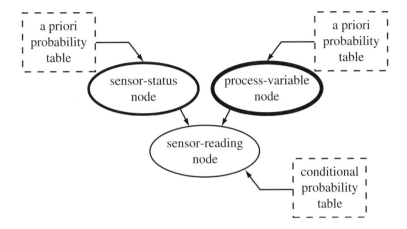

Figure 15.1. *The basic unit of a BBN for SFDIA. The sensor-reading node is dependent on the sensor-status node and the process-variable node. The links and the conditional probability tables define this causal relationship. The sensor-reading node is shown with a single line, the sensor-status node with double lines, and the process-variable node is shown with triple lines. This representation scheme is used for all BBNs presented in this chapter.*

I. Set all π and λ messages and λ values to 1:

$$\pi(S) = P(S) = (0.65 \quad 0.35),$$
$$\pi(T) = P(T) = (0.5 \quad 0.5).$$

II. Calculate π messages, exchange messages between S and I, and calculate updated probabilities for I.

A. Calculate π messages from I to S:

$$\pi_I(S = \text{functional}) = \frac{P'(S = \text{functional})}{\lambda_I(S = \text{functional})} = \frac{0.65}{1} = 0.65,$$

where P' denotes the updated (posterior) probability, which at this point is equal to the a priori probability P for the root nodes:

$$\pi_I(S = \text{inoperative}) = \frac{P'(S = \text{inoperative})}{\lambda_I(S = \text{inoperative})} = \frac{0.35}{1} = 0.35.$$

B. Calculate π values for node I:

$$\pi(I = \text{red}) = P(\text{red}|\text{high, functional}) \cdot \pi_I(S = \text{functional}) \cdot \pi_I(T = \text{high})$$
$$+ \cdots + P(\text{red}|\text{low, inoperative}) \cdot \pi_I(S = \text{inoperative})$$
$$\cdot \pi_I(T = \text{low})$$
$$= 1.0 \times 0.65 \times 1.0 + 0.0 + 0.5 \times 0.35 \times 1.0 + 0.5 \times 0.35 \times 1$$
$$= 1.0,$$
$$\pi(I = \text{blue}) = 0.0 + 1.0 \times 0.65 \times 1.0 + 0.5 \times 0.35 \times 1.0 + 0.5 \times 0.35 \times 1$$
$$= 1.0.$$

C. Update probabilities for I based on its π and λ values and a proportionality constant α:

$$P'(I = \text{red}) = \alpha \cdot \lambda(I = \text{red}) \cdot \pi(I = \text{red})$$
$$= \alpha \times 1.0 \times 1.0$$
$$= \alpha,$$
$$P'(I = \text{blue}) = \alpha \cdot \lambda(I = \text{blue}) \cdot \pi(I = \text{blue})$$
$$= \alpha \times 1.0 \times 1.0$$
$$= \alpha,$$
$$P'(I = \text{red}) + P'(I = \text{blue}) = 1.0,$$
$$\therefore \alpha + \alpha = 1.0,$$
$$\Rightarrow \alpha = 0.5,$$
$$\therefore P'(I = \text{red}) = P'(I = \text{blue}) = 0.5.$$

I has no child to send any π messages. Hence this branch of the propagation algorithm terminates.

III. Repeat step II for node T; i.e., calculate π messages, exchange messages between T and I, and calculate updated probabilities for I.

A. Calculate π messages from I to T:

$$\pi_I(T = \text{high}) = \frac{P'(T = \text{high})}{\lambda_I(T = \text{high})} = \frac{0.5}{1} = 0.5,$$
$$\pi_I(T = \text{low}) = \frac{P'(T = \text{low})}{\lambda_I(T = \text{low})} = \frac{0.5}{1} = 0.5.$$

B. Calculate π values for node I:

$$(I = \text{red}) = P(\text{red}|\text{high, functional}) \cdot \pi_I(S = \text{functional})$$
$$\cdot \pi_I(T = \text{high})$$
$$+ \cdots + P(\text{red}|\text{low, inoperative}) \cdot \pi_I(S = \text{inoperative})$$
$$\cdot \pi_I(T = \text{low})$$
$$= 1.0 \times 0.65 \times 0.5 + 0.0 + 0.5 \times 0.35 \times 0.5 + 0.5 \times 0.35 \times 0.5$$
$$= 0.5,$$
$$\pi(I = \text{blue}) = 0.0 + 1.0 \times 0.65 \times 0.5 + 0.5 \times 0.35 \times 0.5 + 0.5 \times 0.35 \times 0.5$$
$$= 0.5.$$

C. Update probabilities for I based on its π and λ values and a proportionality constant α:

$$P'(I = \text{red}) = \alpha \cdot \lambda(I = \text{red}) \cdot \pi(I = \text{red})$$
$$= \alpha \times 1.0 \times 0.5$$
$$= 0.5\alpha,$$
$$P'(I = \text{blue}) = \alpha \cdot \lambda(I = \text{blue}) \cdot \pi(I = \text{blue})$$
$$= \alpha \times 1.0 \times 0.5$$

$$= 0.5\alpha,$$
$$\therefore \alpha = 1.0,$$
$$\therefore P'(I = \text{red}) = P'(I = \text{blue}) = 0.5.$$

I has no child to send any π messages. Hence this branch of the propagation terminates.

Since all branches of the propagation scheme have terminated, the computation stops. The network now is completely initialized and is ready to accept new evidence.

Instantiation.   The indicator is observed to be red and hence I is instantiated to *red*. This instantiation causes node I to send λ messages to its parents and π messages to its children.

I. Change the posterior probabilities and λ values of I to reflect the evidence:

A.
$$P'(I = \text{red}) = 1.0,$$
$$P'(I = \text{blue}) = 0.0,$$
$$\lambda(I = \text{red}) = 1.0,$$
$$\lambda(I = \text{blue}) = 0.0.$$

B. Calculate λ messages from I to its parents:

$$\begin{aligned}
\lambda_I(S = \text{functional}) &= \pi_I(T = \text{high}) \\
&\quad \cdot (P(\text{red}|\text{high, functional}) \cdot \lambda(I = \text{red}) \\
&\quad\quad + P(\text{blue}|\text{high, functional}) \cdot \lambda(I = \text{blue})) \\
&\quad + \pi_I(T = \text{low}) \\
&\quad \cdot (P(\text{red}|\text{low, functional}) \cdot \lambda(I = \text{red}) \\
&\quad\quad + P(\text{blue}|\text{low, functional}) \cdot \lambda(I = \text{blue})) \\
&= 0.5(1.0 \times 1.0 + 0.0) + 0.5(0.0 + 0.0) \\
&= 0.5, \\
\lambda_I(S = \text{inoperative}) &= 0.5(0.5 + 0.0) + 0.5(0.5 + 0.0) \\
&= 0.5, \\
\lambda_I(T = \text{high}) &= 0.65(1.0 + 0.0) + 0.35(0.5 + 0.0) \\
&= 0.825, \\
\lambda_I(T = \text{low}) &= 0.65(0.0 + 0.0) + 0.35(0.5 + 0.0) \\
&= 0.175.
\end{aligned}$$

I has no child to send any π messages. Hence this branch of the propagation terminates.

Propagation.   Receiving λ messages from node I triggers nodes S and T into the appropriate updating procedure.

I. S receives a λ message from I.

A. Update λ values for S:

$$\lambda(S = \text{functional}) = \lambda_I(S = \text{functional}) = 0.5,$$
$$\lambda(S = \text{inoperative}) = \lambda_I(S = \text{inoperative}) = 0.5.$$

B. Update probabilities for S:

$$P'(S = \text{functional}) = \alpha \cdot \lambda(S = \text{functional}) \cdot \pi(S = \text{functional})$$
$$= \alpha \times 0.5 \times 0.65$$
$$= 0.325\alpha,$$
$$P'(S = \text{functional}) = \alpha \cdot \lambda(S = \text{functional}) \cdot \pi(S = \text{functional})$$
$$= \alpha \times 0.5 \times 0.35$$
$$= 0.175\alpha,$$
$$0.325\alpha + 0.175\alpha = 1.0,$$
$$\therefore \alpha = 2,$$
$$\therefore P'(S = \text{functional}) = 0.65,$$
$$P'(S = \text{functional}) = 0.35.$$

S has no parent to send λ messages. Hence this branch of the propagation terminates. Also, S has no other children to send new π messages; hence this branch of the propagation terminates.

II. T receives a λ message from I:

A. Update λ values for T:

$$\lambda(T = \text{high}) = \lambda_I(T = \text{high}) = 0.825,$$
$$\lambda(T = \text{low}) = \lambda_I(T = \text{low}) = 0.175.$$

B. Update probabilities for S:

$$P'(T = \text{high}) = \alpha \cdot \lambda(T = \text{high}) \cdot \pi(T = \text{high})$$
$$= \alpha \times 0.825 \times 0.5$$
$$= 0.4125\alpha,$$
$$P'(T = \text{low}) = \alpha \cdot \lambda(T = \text{low}) \cdot \pi(T = \text{low})$$
$$= \alpha \times 0.175 \times 0.5$$
$$= 0.0875\alpha,$$
$$0.4125\alpha + 0.0875\alpha = 1,$$
$$\therefore \alpha = 2,$$
$$\therefore P'(T = \text{high}) = 0.825,$$
$$P'(T = \text{low}) = 0.175.$$

T has no parent to send λ messages. Hence this branch of the propagation terminates. Also, T has no other child to send new π messages. Hence this branch of the propagation terminates.

Since all branches of the propagation scheme have terminated, the computation stops. The arrival of evidence results in updating the probabilities that are stored as posterior probabilities with each node.

15.2.1   BBNs for SFDIA

Using the inference and evidence propagation methods from the previous section, a BBN can be used to detect, isolate, and accommodate sensor faults in a single, unified scheme. With this approach it is no longer necessary to construct separate modules to perform each of these functions. In addition, the network can handle uncertain or missing data implicitly. Further, the network can accommodate the addition, removal, or replacement of one or more sensors by changing the corresponding nodes without affecting the entire network. This approach is relatively new for SFDIA. Its use in the past has been limited to discrete domain, where the network structure encodes the rules defining the system. The message-passing and inference schemes in these networks are based on sound theorems of probability, whereas the network structure has the ability to capture causal relationships. Thus this network is a combination of analytical as well as knowledge-based approaches. These networks also can be dynamic, leading to online and localized learning (Deng (1993), Kirch and Kroschel (1994), Nicholson and Brady (1994), Nicholson (1996)). Unlike neural networks, these learning schemes do not aim at minimizing a measure for training error, but update themselves according to what is probabilistically correct.

15.2.2   Construction of nodes, links, and associated functions

For a BBN representation of SFDIA, three basic types of nodes are associated with each sensor:

1. "sensor-reading" nodes that represent the mechanisms by which the information is communicated to the BBN;

2. "sensor-status" nodes that convey the status of the corresponding sensors at any given time;

3. "process-variable" nodes that are a conceptual representation of the actual values of the process variables, which are unknown.

To define the communication of information among the basic types of nodes, specification of the links among them is necessary. In a BBN, there exists a link between each cause and its direct effect, directed from the former to the latter. A sensor reading is a reflection of the value of the process variable that it is measuring. Hence a parent process-variable node exists for each sensor-reading node. Also, a normal or faulty mode of operation of a sensor affects the sensor reading, and hence the sensor-status node forms the other parent of a sensor-reading node.

In addition to the directed links, its conditional probability table specifies the dependence of each child on its parents. Distributions of the root variables are defined in terms of the a priori probabilities. The network structure and the probability tables completely define the knowledge about how the basic unit models an SFDIA system.

Figure 15.2 shows a BBN representing the basic SFDIA unit. It contains three nodes of the three basic types—sensor-reading, sensor-status, and process-variable. The sensor-status node, being a root node, has an a priori probability table associated with it. The same is true for the process-variable node. The sensor-reading node is a common child of the nodes sensor-status and process-variable, and it has a conditional probability table.

After identification of the basic types of nodes, further model developments are based on the following assumptions:

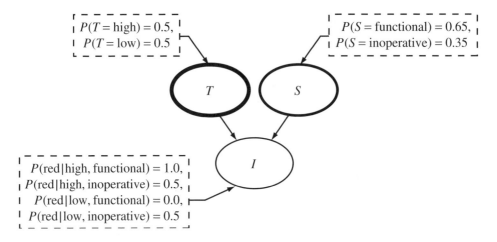

Figure 15.2. *Bayesian network representation of the single-sensor single-variable problem. It consists of one SFDIA basic unit. All three nodes are binary-valued. Nodes S and T are root nodes, and both have their a priori probability tables in attached boxes. Node I has a conditional probability table attached to it. For example, when T = low and S = functional, the probability of I = red is 0.0.*

1. Given the parent process variable, the sensors are independent of each other.

2. The status of one sensor is independent of the status of any other sensor.

3. The status of a sensor is independent of the state of the process, i.e., the values of the process variables.

A complete BBN for an SFDIA problem will involve several sensor-reading, sensor-status, and process-variable nodes. The primary functions associated with the task of SFDIA are divided between these nodes. The functions are providing input to the network and detecting, classifying, and accommodationing faults. In SFDIA, evidence consists of the sensor readings available at that instant, achieved via instantiations of the sensor-reading nodes. Given the sensor readings, it is desired to infer the status of each sensor and also estimates of actual values of the process variables that the sensors are measuring. Updated beliefs (i.e., probabilities) about the sensor-status and process-variable nodes represent the inferred knowledge. After evidence propagation, monitoring of the updated beliefs of the sensor-status nodes leads to fault detection and classification. The updated beliefs about the process-variable nodes are a result of multisensor fusion and fault accommodation functions. Thus although the functions associated with each node are different, they are achieved via a single unified evidence propagation and inference scheme.

15.2.3   A single-variable, single-sensor system

Earlier in this section, we demonstrated the use of BBNs for representing a single temperature sensor/indicator and updating its state as new data become available. Any statement about the value of temperature is a logical consequence of our knowledge about the sensor. The higher the reliability of the sensor, the stronger the effect of the indicator reading on the belief about temperature. The values of belief about the actual value of temperature as a function of the a priori sensor fault probability are summarized in Table 15.1. As the a priori sensor fault probability increases, the sensor becomes less reliable and hence the actual value of the temperature has a lower contribution from the sensor reading. Hence the updated probability of $P(T = \text{high})$ is lower, showing lower belief in the sensor reading. For example, as shown in Table 15.1, as the a priori sensor fault probability goes from 0.45 to 0.05, the updated probability of $T = \text{high}$ increases from 0.775 to 0.995. More definitive statements about the a priori temperature distribution result in a stronger statement about the sensor status, as shown in Table 15.2. When the a priori $P(T = \text{high})$ is low, the network concludes that the sensor is inoperative. When $P(T = \text{high})$ is high, it is more probable that the sensor is functional. For example, as shown in Table 15.2, as the a priori $P(T = \text{high})$ is increased from 0.02 to 0.09, updated $P(S = \text{functional})$ rises from 0.07 to 0.77.

Table 15.1. *Updated belief about node T as a function of the a priori fault probability of the sensor, given that a sensor reading of high has been observed. A priori temperature distribution remains constant at equal values of* 0.5.

A priori sensor fault probability	Updated probability of $T = \text{high}$
0.45	0.775
0.35	0.825
0.15	0.925
0.05	0.975

Table 15.2. *The effect of prior knowledge on the inference of the network, given that a sensor reading of high has been observed. A priori* $P(S = functional)$ *is kept constant. Note that when the a priori* $P(T = high)$ *is low, the network concludes that the sensor is inoperative and thus the sensor reading is invalid. When* $P(T = high)$ *is high, there is a high degree of belief in the sensor being operational and thus the sensor reading being valid.*

Prior probability distribution	Posterior probability distribution
$P(T = \text{high}) = 0.02,$ $P(S = \text{functional}) = 0.65$	$P(T = \text{high}) = 0.09,$ $P(S = \text{functional}) = 0.07$
$P(T = \text{high}) = 0.20,$ $P(S = \text{functional}) = 0.65$	$P(T = \text{high}) = 0.54,$ $P(S = \text{functional}) = 0.46$
$P(T = \text{high}) = 0.50,$ $P(S = \text{functional}) = 0.65$	$P(T = \text{high}) = 0.825,$ $P(S = \text{functional}) = 0.650$
$P(T = \text{high}) = 0.90,$ $P(S = \text{functional}) = 0.65$	$P(T = \text{high}) = 0.98,$ $P(S = \text{functional}) = 0.77$

15.2.4   Incorporation of redundancies in the network structure

The basic unit developed in the previous section is devoid of any redundant information and its function as an SFDIA unit is limited. As was discussed earlier in the chapter, hardware and analytical redundancies are typically used for SFDIA. This section builds on the basic unit to construct BBNs that incorporate these redundancies. To this end, we will progressively develop a general network structure for a system with hardware and analytical redundancy and discuss it with respect to an example system.

15.2.4.1   Hardware redundancy

Hardware redundancy is a direct means of sensor fault accommodation. The BBN representation of a system with multiple sensors measuring the same process variable must contain multiple sensor-reading nodes for the same process-variable node. Each sensor-reading node will have its own sensor-status node as one of its parents. The other parent for all sensor-reading nodes will be common, viz., the process-variable node representing the process variable that the sensors are measuring.

Consider an ensemble of five indicators, I_1 through I_5, assigned to measure and report the temperature of a given process (Smets (1992)). The description of operational characteristics of the sensors are similar to those for the single-sensor case, i.e., a priori fault probabilities of all sensors are available. Given the sensor readings, the objective is to make a statement about the actual value of the process variable. The BBN model of the system is given in Figure 15.3, where node T represents the temperature being measured. The actual temperature influences the sensor reading as reported by its indictor. This influence is shown in form of the directed link from node T to any indicator node I. If the sensor is operational, there is a correlation between its value and that of the temperature. The lack of correlation in a faulty sensor is represented as a uniform probability distribution in the conditional probability table of the indicator node; i.e., both *high* and *low* are equally probable values for sensor reading, irrespective of the *high* or *low* value of the temperature. Obviously, the status of the sensor, as represented by node S, influences the corresponding indicator node. This pattern of influence is represented as a directed link from nodes S to nodes I.

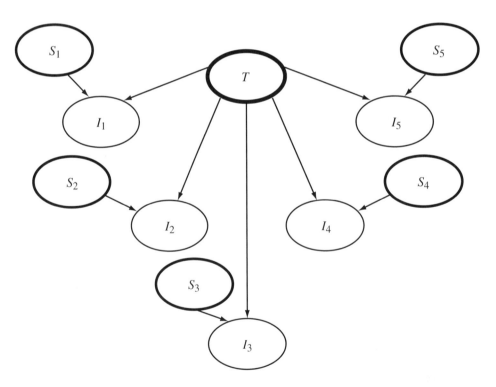

Figure 15.3. *The Bayesian network model of a system consisting of one process variable and five sensors, measuring its value. Nodes I_1 through I_5 are the sensor-reading nodes. Nodes S_1 through S_5 are the corresponding sensor-status nodes. Node T represents the temperature.*

15.2.4.2 Analytical redundancy

Correlations exist among process variables and these can be taken advantage of for the goal of sensor fault detection and accommodation. A process model attempts to encompass these relationships that exist between different process variables. In a BBN representation, each variable is represented as a process-variable node. Thus a model has to be represented as conditional dependence of one or more process-variable nodes on the rest of the process-variable nodes.

A BBN representation would require separation of the process variables into parent, intermediate, and terminal nodes. Advantage can be taken of the fact that a process model is often represented as differential and/or algebraic equations. A state-space type reorganization of the characteristic equation(s) will allow distinction into independent and dependent variables, thus facilitating a causal representation of the dependency.

After constructing the part of the BBN representing the process model, the next requirement is formulation of the conditional probability tables for the dependent variables and the a priori probability tables for the independent variables. As a first step, a discrete representation for the process variables, which are continuous-valued in real life, is necessary. Each parent variable is assigned one of its possible values. The value of the child process variable is then calculated and discretized. This process is then repeated until all possible combinations of parent variables are covered. This data is then used to construct the conditional probability table for the child variable.

15.2.4.3 Construction of a discrete, steady-state knowledge base

Consider a simple example of an adiabatic CSTR (Karjala and Himmelblau (1994)). Description of the process, its definition in terms of the characteristic equations, and a progressive development of a BBN model are discussed. The system consists of a CSTR undergoing a first order, exothermic reaction. The reactor is provided with a cooling jacket or coil. The process variables considered here are the temperature and concentration of the inlet stream, and the temperature and concentration prevailing inside the CSTR. A schematic diagram is shown in Figure 15.4, which shows the symbols used to denote the variables and the corresponding sensors.

The steady-state equations that model this system are as follows:

$$\frac{dT}{dt} = \frac{q}{V}(T_0 - T) - \frac{\Delta H_r}{\rho C_p} k_0 C \exp\left(\frac{-E_a}{T}\right) - \frac{U A_r}{\rho C_p V}(T - T_c) = 0, \quad (15.1)$$

$$\frac{dC}{dt} = \frac{q}{V}(C_0 - C) - k_0 C \exp\left(\frac{-E_a}{T}\right) = 0. \quad (15.2)$$

All variables except T, C, T_0, and C_0 are assumed to be constant. The physical meanings of the variables involved and their typical values and units are provided in Table 15.3. For simulation, the equations are made dimensionless using the following dimensionless variables:

$$T' = \frac{T}{T_r}, \quad C' = \frac{C}{C_r}, \quad T_0' = \frac{T_0}{T_r}, \quad \text{and} \quad C_0' = \frac{C_0}{C_r}. \quad (15.3)$$

Substituting into (15.1) and (15.2),

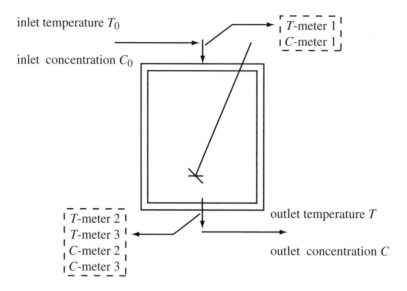

inlet temperature T_0

inlet concentration C_0

T-meter 1
C-meter 1

T-meter 2
T-meter 3
C-meter 2
C-meter 3

outlet temperature T

outlet concentration C

Figure 15.4. *A schematic diagram of the CSTR. Partial hardware redundancy is assumed. The sensors T-meter 1 and C-meter 1 measure inlet temperature and concentration, respectively. Outlet temperature is reported by two sensors, viz. T-meter 2 and T-meter 3. Similarly, the sensors C-meter 2 and C-meter 3 report on the value of outlet concentration.*

Table 15.3. *CSTR model parameters (Karjala and Himmelblau (1994)).*

Parameter	Value	Units
flow rate (q)	10.0	$cm^3 \cdot s^{-1}$
volume (V)	1,000.0	cm^3
heat of reaction (ΔH_r)	$-27,000.0$	$cal \cdot mol^{-1}$
density (ρ)	0.001	$g \cdot cm^{-3}$
specific heat (C_p)	1.0	$cal(g \cdot K)^{-1}$
heat transfer coefficient (U)	5.0×10^{-4}	$cal(cm^2 \cdot s \cdot K)^{-1}$
heat transfer area (A_r)	10.0	cm^2
coolant temperature (T_c)	340.0	K
arrhenius constant (k_0)	7.86×10^{12}	s^{-1}
activation energy (E_a)	14,090.0	K
reference concentration (C_r)	1.0×10^{-6}	$mol \cdot cm^{-3}$
reference temperature (T_r)	100.0	K

$$\frac{dT'}{dt} = \frac{q}{V}(T'_0 - T') - \frac{\Delta H_r}{\rho C_p T_r} k_0 C' C_r \exp\left(\frac{-E_a}{T'T_r}\right) - \frac{U A_r}{\rho C_p V}(T' - T'_c) = 0, \qquad (15.4)$$

$$\frac{dC'}{dt} = \frac{q}{V}(C'_0 - C') - k_0 C' \exp\left(\frac{-E_a}{T'T_r}\right) = 0. \qquad (15.5)$$

Using (15.4) and (15.5), C' and T' may be treated as variables dependent on the values of C'_0 and T'_0. As shown in Figure 15.5, the central structure of the BBN representing the CSTR contains these four process-variable nodes, with the independent variables being the parents of the dependent variables. For the requirements of stability, the random perturbations in the input variables are restricted to be within $\pm 3\%$ of the steady-state values. Fractional deviation about the steady state for each variable is discretized into five equal slots: very

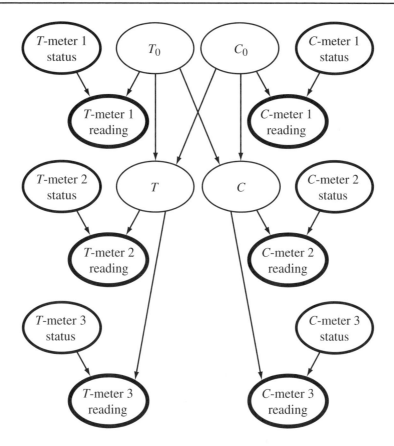

Figure 15.5. *Discretized Bayesian network for the CSTR. Corresponding to the schematic diagram of the CSTR in Figure* 15.4, *four process-variable nodes* (C, T, C_0, T_0) *are involved. Each process variable is linked to sensor-reading nodes corresponding to the sensors measuring that variable (Figure* 15.4)*. Sensor-status nodes represent the status of these sensors.*

low (VL), low (L), medium (M), high (H), and very high (VH). This discretization enables the approximation of the steady-state correlation as conditional probability tables for the dependent variables. These values are provided as Tables 15.4(a) and (b).

Each of the process-variable nodes is linked, as a parent, to a number of sensor-reading nodes equal to the number of sensors measuring that variable. Each sensor-reading node in turn will have its own sensor-status node as the other parent. Specifications about a priori fault probability are available for each sensor, enabling the definition of the conditional dependence of the corresponding sensor-status node. A uniform distribution for a priori probabilities for all process-variable nodes is assumed. Readings of faulty sensors are assumed to be uniformly distributed random values. They are thus completely independent of the process variable under measurement. Table 15.5 shows three simulated instances of steady-state data, each involving a sensor fault.

Each entry in Table 15.5 contains simulated sensor readings for a steady state. These are provided as input to the network via instantiations of the corresponding sensor-reading nodes. The BBN responds to this evidence by updating the probabilities of all the uninstantiated nodes, i.e., the process-variable and sensor-status nodes. Figures 15.6, 15.7, and 15.8 show

Table 15.4. *Conditional probability tables for the discrete CSTR BBN as shown in Figure 15.5. (a) Representation of conditional probability table for node T (outlet tempera-ture). This table shows various values taken by T corresponding to different combinations of the values of the parent nodes. For example, when $C_0 = M$ and $T_0 = H$, $T = H$ with a probability of 1.0. (b) Representation of conditional probability table for node C (outlet concentration). Similar to part (a), this table shows various values taken by C corresponding to different combinations of the values of the parent nodes.*

(a)

$\dfrac{C_0}{T_0}$	VL	L	M	H	VH
VL	VL	VL	L	L	L
L	L	L	L	M	M
M	L	M	M	H	H
H	M	H	H	H	VH
VH	H	H	VH	VH	VH

(b)

$\dfrac{C_0}{T_0}$	VL	L	M	H	VH
VL	VH	H	H	M	M
L	H	M	M	L	L
M	M	L	L	VL	VL
H	L	VL	VL	VL	VL
VH	VL	VL	VL	VL	VL

Table 15.5. *Simulated steady-state data and sensor readings for the CSTR example. Three sets of data are presented in this table. Values of the independent variables (i.e., T_0 and C_0) are generated randomly. Values of variables T and C are then calculated. Sensor readings are then generated for each variable. These sensor readings are used to instantiate the BBN in Figure 15.5. Each experiment involves one faulty sensor. The rest of the sensors are assumed to be working in the normal mode. For example, T-meter 1 reading in experiment 1 shows a faulty reading H (high). Its response in the normal mode would have been L (low), which is the actual value of C_0.*

Experiment number	T_0		C_0		T			C		
	Actual value	T-meter 1 reading	Actual value	C-meter 1 reading	Actual value	T-meter 2 reading	T-meter 3 reading	Actual value	C-meter 2 reading	C-meter 3 reading
1	L	H	M	M	L	L	L	M	M	M
2	H	H	L	L	H	H	H	VL	H	VL
3	VH	VH	VL	H	VH	VH	VH	VL	VL	VL

the results corresponding to the input from Table 15.5. Each figure shows the updated probability distributions and the actual values of these nodes. As can be seen, the network is capable of correctly inferring the status of each sensor and the actual value of the process variable. Thus the network can model a system with hardware and analytical redundancies together. Sensor faults are correctly detected and isolated. Therefore, in a discrete, steady-state domain, the network is capable of fault detection, isolation, and accommodation with a single, unified scheme.

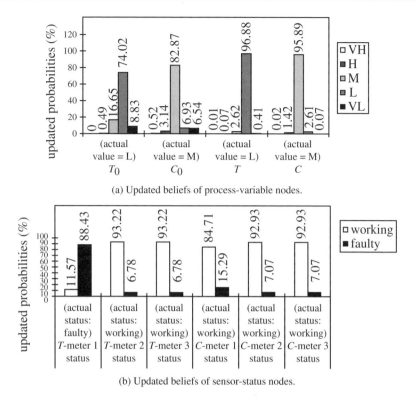

(a) Updated beliefs of process-variable nodes.

(b) Updated beliefs of sensor-status nodes.

Figure 15.6. *Results corresponding to experiment* 1 *in Table* 15.5. *Sensor T -meter* 1 *reports a faulty reading H corresponding to the actual value L of the inlet temperature. This fault is detected as seen from the high probability of its status being faulty (part (b)). This implies a successful fault detection and isolation. The updated beliefs of process variable nodes show that updated beliefs corresponding to actual values of the variables are high (part (a)). Thus correct inference about all process variables, including inlet temperature, is drawn. This is a result of multisensor fusion and fault accommodation.*

15.3   Fuzzy logic

In terms of the crisp set theory, a point either is or is not a member of a set. This enforces hard boundaries on the definitions of sets and their membership functions. Fuzzy or multiple-valued logic on the other hand allows infinite possible truth-values between the interval 0 to 1. Any proposition thus can be partly true and partly false. Sets defined in terms of this concept are termed "fuzzy sets." Data thus can have different degrees of memberships of these sets.

Consider a fuzzy set defined over a domain consisting of data points. Each data point can be said to be a member of the fuzzy set with the degree of membership ranging from 0 (no membership) to 1 (full membership). A fuzzy set thus can be said to be a collection of elements in a universe of information where the boundary of the set is ambiguous or imprecise. When the data points are real-valued, this relationship between the point and its membership in the fuzzy set can be defined in terms of a function. Such a function is called a "membership function."

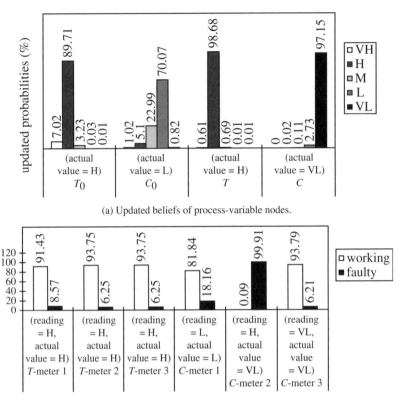

(a) Updated beliefs of process-variable nodes.

(b) Updated beliefs of sensor-status nodes.

Figure 15.7. *Results corresponding to experiment* 2 *in Table* 15.5. *A fault in C-meter* 2 *causes it to report an erroneous value of H, whereas the actual value of outlet concentration is VL. The fault is correctly detected, isolated, and accommodated, as can be seen from the high updated probability of C-meter* 2 *being faulty (part* (b)). *Also, probabilities corresponding to actual values of the process variables are high (part* (a)).

15.3.1 Using fuzzy logic for fault detection

The fuzzy logic–based hardware redundancy SFDIA scheme (Holbert, Heger, and Ishaque (1995), Heger, Holbert, and Ishaque (1996)) uses two or three redundant sensors. The deviation (i.e., the residual) between each sensor pairing is computed and classified into three fuzzy sets. A fuzzy rule base, which is created allowing human perception of the situation to be represented mathematically, is made to operate on these fuzzy input variables. Finally, a defuzzification scheme is used to find the centroid location and hence the sensor status. The potential for error is reduced with the fuzzy logic approach of sensor validation by introducing an intermediate state of "suspect" between the failed and the valid status.

The approach due to Park and Lee (1993) applies fuzzy logic to instrument fault detection by dividing the process into three different stages—multisensor fusion process, temporal fusion process, and diagnostic fusion process—using the uncertainty reductive fusion technique (URFT). It uses redundant hardware and does not use analytical redundancy. Each sensor reading is represented as a fuzzy number with a mean value and a triangular spread with uncertainty values. The multisensor fusion process consists of fusion of mean values and fusion of uncertainty levels. The temporal fusion process is used for time

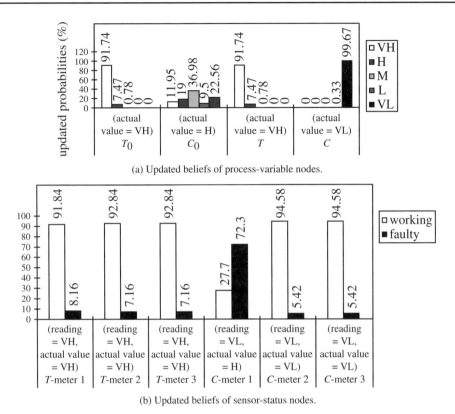

(a) Updated beliefs of process-variable nodes.

(b) Updated beliefs of sensor-status nodes.

Figure 15.8. *Results corresponding to experiment 3 in Table 15.5. C-meter 1 reports a value of VL when the actual value of inlet concentration is H. This fault is correctly detected and isolated, as seen from the high probability of C-meter 1 being in its faulty mode of operation (part (b)). Due to a higher number of possible combinations, the network was unable to accurately accommodate the fault as seen by the distribution of C_0 (part (a)). This implies unsuccessful fault accommodation. This is caused by a large number of possible hypotheses to explain the observed sensor readings.*

consistency with the prior information stored in the system database. Temporal fusion detects the change in the state of the system, whereas the multisensor fusion combines the readings from the redundant hardware. The diagnostic fusion stage uses the information about the nominal state of the system, which is the last known state without a fault. A multiplexer is used to choose between the output of the temporal fusion process and the diagnostic fusion process. This reduces the possibility of propagation of the fault in the system. The data flow is designed to be sensitive to sensor faults with a minimum number of false alarms.

The use of fuzzy logic for (a) adaptive thresholding for residual evaluation and (b) isolation of a faulty sensor is proposed by Sauter et al. (1994). This work uses the quantitative approach of a Kalman filter for residual generation. Residuals may be nonzero even under fault-free conditions due to measurement noise and model uncertainties, and hence decisions are based on the difference between a residual and a certain threshold residual. Adaptive threshold selection chooses a threshold based on the distance of the current state of the system from the nominal condition. Fuzzy logic is used for this purpose. Fuzzy rules and a statistical dedicated observer scheme are combined for fault isolation as well. Calculated residuals are separated into clusters using the fuzzy c-means clustering algorithm. The allocation of a new data point to one of these clusters leads to fault isolation.

In this section we present a possible approach to SFDIA that is mainly based on fuzzy processing of deviations from model predictions. It makes use of a mathematical predictive model (i.e., an observer) for the system. The model uses the sensor readings at the current time step to predict system state at the next time step. In case of a sensor fault, the deviation between the sensor reading and the model prediction rises, leading to fault detection and isolation. The fault is accommodated by replacing the reading of the faulty sensor by the model prediction.

15.3.1.1   Fuzzy flow control

Consider a simple example frequently found in control engineering literature. A system consists of a reactor vessel with an intake with dimensionless volumetric flow rate q_i and a dimensionless drainage q_o through an orifice at the bottom of the tank. The equation used for simulation is

$$\frac{dh}{dt} = \psi q_i - \xi q_o, \quad \text{where } q_o = \sqrt{h} \quad \text{and} \quad h(k = 0) = 0.5. \tag{15.6}$$

Parameter h is the dimensionless height of liquid level, and ψ, ξ are constants equal to 0.1. The input disturbance q_i is simulated as random step changes. The values of h and q_o can be found by numerically integrating the system, given the initial condition. Random Gaussian noise with zero mean was added to each variable to obtain simulated sensor readings $q_{i_s}(k)$, $h_s(k)$, and $q_{o_s}(k)$.

In this application, the inlet flow rate is the dependent variable and can be used to control the level of the tank. A simple heuristic may be to increase the inflow if the level is dropping below the desired value and decrease it if the level is too high. Two simple fuzzy rules encompass this control algorithm:

1. IF the level is high and the rate of change of level is positive, THEN change the inflow to low.

2. IF the level is low and the rate of change of level is negative, THEN change the inflow to high.

15.3.1.2   Flow of data

The steps for fuzzy logic–based SFDIA can be outlined as given below and in Figure 15.9.

1. Actual values of level and outlet flow-rate at time $k + 1$ (i.e., $h(k + 1)$ and $q_o(k + 1)$) are simulated using $h(k)$, $q_o(k)$, and $q_i(k)$.

2. Random Gaussian noise is added to obtain simulated normal or faulty sensor readings $h_s(k + 1)$ and $q_{o_s}(k + 1)$.

3. In parallel to step 2, two independent observers (observers 1 and 2 in Figure 15.9) are used to calculate estimates $\hat{h}(k + 1)$ and $\hat{q}_o(k + 1)$ in terms of $h_s(k)$, $q_{o_s}(k)$, $q_i(k)$, $h^{\text{SFDIA}}(k)$, and $q_o^{\text{SFDIA}}(k)$. The values $h^{\text{SFDIA}}(k)$ and $q_o^{\text{SFDIA}}(k)$ are the estimates provided by the fuzzy SFDIA scheme at the previous time step.

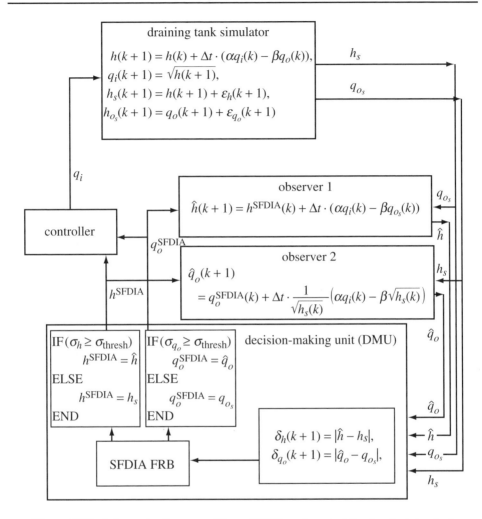

Figure 15.9. *Schematic diagram of fuzzy SFDIA and control for the draining tank.*

4. In the decision-making unit (DMU), the deviations $\delta_i(k+1)$ between the actual sensor reading and observer prediction are calculated as a crisp number. Each $\delta_i(k+1)$ is then mapped by the fuzzy rule builder (FRB) onto three fuzzy membership functions: small, medium, and large. It is then subjected to fuzzy rules that evaluate the sensor status (σ_i) in terms of three fuzzy sets: valid, suspect, and failed (Figures 15.10 and 15.11). The status is then subjected to defuzzification. Thus we have a crisp number for σ_i, which ranges from 0 to 1. This value is then compared to a threshold, and a fault is declared to be present if σ_i is greater than the threshold. Appropriate values (sensor reading or observer estimate) are assigned to $h^{\text{SFDIA}}(k+1)$ and $q_o^{\text{SFDIA}}(k+1)$. For example, when a fault in the level sensor is detected, $h^{\text{SFDIA}} = \hat{h}$. In the absence of level sensor fault, $h^{\text{SFDIA}} = h_s$.

5. The output estimates by the DMU, $h^{\text{SFDIA}}(k+1)$, and $q_o^{\text{SFDIA}}(k+1)$ are used by the controller for providing the control action $q_i(k+1)$. Step 1 is then repeated.

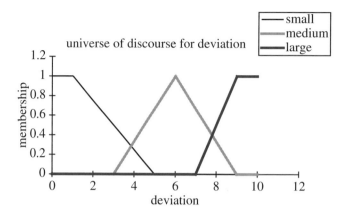

Figure 15.10. *Universe of discourse for deviation. The deviation is fuzzified (i.e., converted from crisp number to a fuzzy number) into three fuzzy sets:* small, medium, *and* large. *The membership functions of these three fuzzy sets have been defined in the above chart.*

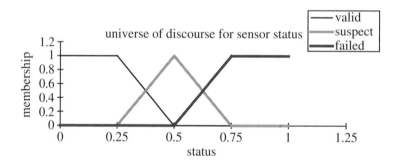

Figure 15.11. *Universe of discourse for sensor status. The fuzzy rule base determines the status of a sensor as three fuzzy sets:* valid, suspect, *and* failed. *The membership functions of these sets are defined above. These membership functions are used to defuzzify the sensor status, i.e., to convert from a fuzzy value to a crisp number.*

15.3.1.3   Simulation results

To simulate the dynamic control system, the fuzzy rule base for SFDIA was constructed in FuLDeK® (Bell Helicopter Textron (1995)) using the membership functions and rules described above. It was then converted to C using the FuLDeK FULCODE-C utility. Thus the simulator, controller, and FRB could communicate with one another. Initially, a precision degradation fault is introduced in the sensor measuring the liquid level. This is reflected in the sensor error plotted in Figure 15.12(a). As a result, the calculated deviation (i.e., residual) also increases as shown in Figure 15.12(b). The FRB for SFDIA detects this increase, which is reflected as the stepping up of a "sensor fault detection flag" used internally by the program. A value of 0 for this flag is intended to represent the absence of any detected faults, and a value of 1 represents the detection of a fault. As can be seen from Figure 15.12(a), the change in the fault detection flag corresponds to the occurrence of the fault in the sensor.

    This method has several limitations. The most important is that the observers must assume a mathematical model and, at any given step, the sensor readings at the previous

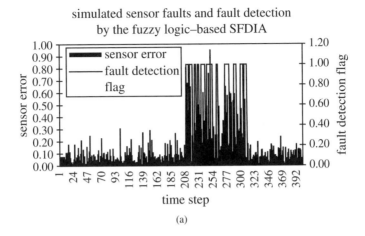

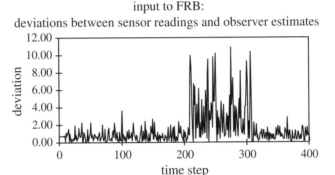

Figure 15.12. *Fuzzy logic–based SFDIA scheme. The graph with solid lines (left ordinate) in* (a) *depicts the simulated sensor fault by a large increase in the magnitude of its error between time steps* 206 *and* 306. *The deviation between predicted and observed sensor readings is calculated and provided as an input to the FRB* (b). *As can be seen, these deviations increase during the period of sensor fault. The output of the FRB is used to raise a flag for sensor fault detection. The value of* 1 *of this flag means a detected sensor fault, while a value of* 0 *means the absence of a detected fault. The value of this flag is plotted on the right ordinate of part* (a). *It can be seen that the flag remains* 0 *throughout the normal operation period and is raised to the maximum value more often during faulty sensor operation. It shows that both the onset and removal of the sensor fault is promptly detected.*

time step are used as its input. Since these readings are noisy, the observer estimate itself could be further away from the actual value. The system is nonlearning: it does not try to improve upon the model. There is no attempt at data reconciliation and the system, at best, can mimic normal sensor readings.

15.3.1.4 Neurofuzzy SFDIA

A combination of an internally recurrent neural network (IRNN) and fuzzy logic for SFDIA can reduce the limitations that were discussed in the previous section. For this combined approach, the overall structure and data flow remains the same as those in section 15.3.1.2, except that the IRNN is used as an observer for residual generation. The IRNN has an input layer, an output layer, and a hidden layer with full interconnections onto itself but with a unit delay. This type of simple recurrence can be expressed in terms of a regular feed-forward network with a set of "context nodes" as shown in Figure 15.13. The advantage of this recurrent structure for dynamic modeling is due to its implicit representation of time via these context nodes. For predictive purposes, the use of any moving window of input is no longer necessary. Thus the network size is trimmed and hence the training period. Similar to the fuzzy scheme, the fuzzy DMU then evaluates the residuals. This arrangement combines the respective strengths of both the approaches and hence is more effective.

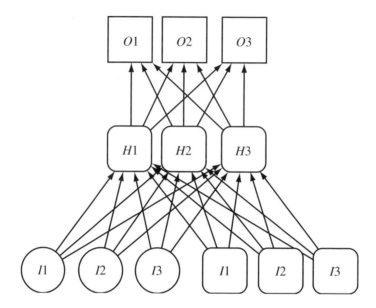

Figure 15.13. *Schematic diagram of an IRNN. The nodes I1, I2, and I3 represent input nodes; nodes H1, H2, and H3 are hidden layer nodes; and nodes O1, O2, and O3 are output nodes. Nodes C1, C2, and C3 are "context layer" nodes, which store the outputs of the hidden, layer nodes at the previous time step.*

The IRNN is an n-input, n-output predictive network. For a process with n variables, the input vector to the IRNN consists of the readings of the corresponding n sensors. In other words, the IRNN does not need redundant sensors. The network can be trained using standard backpropagation with the sensor readings at the next time step as the desired output, thus learning to predict its own input at the next time step. Let k be the time step index such that $t = k\Delta t$, where t is time and Δt is the duration of a unit time step. Let $\vec{X}(k)$ be an n-dimensional vector representing the values of the n process variables at time step k, $\{x_1(k), x_2(k), \ldots, x_i(k), \ldots, x_n(k)\}$. Let the measurement vector $\vec{X}_s(k)$ be comprised of the corresponding sensor readings $\{x_{1_s}(k), x_{2_s}(k), \ldots, x_{i_s}(k), \ldots, x_{n_s}(k)\}$. At time step k, the IRNN takes the previous measurement vector $\vec{X}_s(k-1)$ as its input and provides an estimate of $\vec{X}(k)$ as its output vector, $\hat{\vec{X}}_{\text{IRNN}}(k)$. It is, however, incapable of totally eliminating bias in the sensor reading. Since the output of IRNN is dependent on the readings of all sensors, its accuracy is affected by faults in any of these. Hence, although it is a good data reconciliation tool, a single IRNN cannot be used as a fully functional SFDIA tool.

Figure 15.14 shows the relative performance of these three cases—ideal, in presence of faults without SFDIA, and with the neurofuzzy SFDIA in conjunction with a fuzzy control algorithm. The draining tank system is simulated first without any sensor noise and later with a sensor failure of the precision degradation type introduced at time step 200, followed by a bias introduced at time step 700. The objective of the fuzzy controller is to maintain the fluid level in the tank at 0.5 using the inlet flow rate as the manipulated variable. In the ideal case, the fuzzy control works very well as it succeeds in maintaining the desired tank fluid level. On introduction of sensor failures, however, the controller performs poorly, as the input sensor readings used for control decision-making are not accurate due to sensor faults. On the other hand, both bias and precision degradation types of sensor faults are promptly detected and successfully accommodated using our neurofuzzy scheme. As a result, the controller can maintain the level closer to the desired level of 0.5.

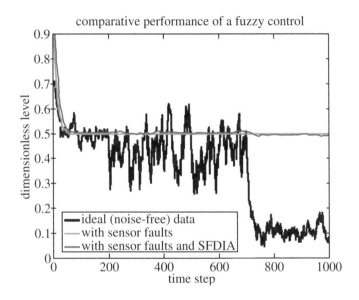

Figure 15.14. *Fuzzy control with neurofuzzy SFDIA scheme. This figure shows that the neurofuzzy SFDIA scheme assists the controller to maintain the liquid level close to the desired value of 0.5.*

15.4   Summary

Each of the methods discussed in this chapter has its own advantages and disadvantages. In general, it is desirable to combine the mathematical soundness and robustness of quantitative model-based methods with the knowledge encoding, reasoning, and learning capabilities of the qualitative, AI-based methods. Some researchers have achieved this task by providing an outer qualitative shell over a quantitative residual generation scheme. On the other hand, Bayesian networks have the unique capability of representation of quantitative as well as qualitative information in a single network structure and inference scheme via a mechanism which is both mathematically sound and close to intuitive human reasoning. Hence Bayesian networks form an effective powerful framework for a fault diagnosis.

The results presented in this work demonstrate a simple scheme that can carry out sensor fault detection, isolation, and accommodation (SFDIA) using Bayesian belief networks (BBNs) over a discrete problem domain. It inherently integrates both analytical and hardware redundancy approaches for instrument fault diagnosis into an elegant unified scheme. Process systems are dynamic environments: new sensors are frequently added and removed. Unforeseen uncertainties may need to be represented such as weather conditions or changes in demands on the system. The BBN mechanism can be incrementally updated in light of these possibilities. New process-variable and sensor-reading nodes can be added, or existing nodes can be modified or removed.

On the other hand, fuzzy rule-based systems provide a means of encoding the experience of an operator into heuristics. The fuzzy logic–based approach presented in this chapter is both simple and efficient. The method, however, lacks the mathematical basis that BBNs provide. Based on the needs of a given SFDIA application domain, one of the two approaches can be selected and implemented. Making use of internally recursive neural networks (IRNNs), the system can learn a dynamic model of the process and thus improves on the basic fuzzy logic–based SFDIA framework.

Sensor fault detection is often considered the single most important task that plant operators would like automated. A missed sensor fault can cause severe damage to the process equipment and is a hazard from the point of view of human safety. On the other hand, a false alarm can prove costly due to the plant downtime and involved human labor. The schemes discussed in this chapter apply the principles of machine learning for this important problem and help alleviate the information load on the human operator.

References

H. B. ARADHYE (1997), *Sensor Fault Detection, Isolation, and Accommodation Using Neural Networks, Fuzzy Logic, and Bayesian Belief Networks*, Master's thesis, University of New Mexico, Albuquerque, NM.

BELL HELICOPTER TEXTRON (1995), *FuLDeK Version 4.0 Reference Manual*, Fort Worth, TX.

A. BERNIERI, G. BETTA, A. PIETROSANTO, AND C. SANSONE (1994), A neural network approach to instrument fault detection and isolation, in *Proceedings of the 1994 IEEE Instrumentation and Measurement Conference*, IEEE Computer Society Press, Los Alamitos, CA, pp. 139–144.

R. N. CLARK, D. C. FOSTH, AND W. M. WALTON (1975), Detecting instrument malfunctions in control systems, *IEEE Trans. Aeronautical and Electrical Systems*, 465.

R. H. DENG, A. A. LAZAR, AND W. WANG (1993), A probabilistic approach to fault diagnosis in linear lightwave networks, *IEEE J. Selected Areas Comm.*, 11, pp. 1438–1448.

M. DESAI AND A. RAY (1981), A fault detection and isolation methodology, in *Proceedings of the 20th IEEE Conference on Decision and Control*, IEEE Computer Society Press, Los Alamitos, CA, pp. 1363–1369.

A. L. DEXTER (1993), *Fault Detection in HVAC Systems Using Fuzzy Models*, Report AN25/UK/151092/1.1, University of Oxford, Oxford, UK.

X. DING AND P. M. FRANK (1990), Fault detection via factorization approach, *Systems Control Lett.*, 14, pp. 431–436.

P. M. FRANK (1987A), Advanced fault detection and isolation schemes using nonlinear and robust observers, in *Proceedings of the 10th IFAC World Congress*, Munich, Vol. 3, Pergamon Press, Oxford, UK, pp. 63–68.

P. M. Frank(1987b), Fault diagnosis in dynamic systems via state-estimation: A survey, in *System Fault Diagnosis, Reliability and Related Knowledge-Based Approaches*, Vol. 1, S. Tsafestas, M. Singh, and G. Schmidt, eds., D. Reidel, Dordrecht, The Netherlands, pp. 35–98.

P. M. Frank (1994a), On-line fault detection in uncertain nonlinear systems using diagnostic observers: A survey, *Internat. J. Systems Sci.*, 25, pp. 2129–2154.

P. M. Frank (1994b), Enhancement of robustness in observer-based fault detection, *Internat. J. Control*, 59, pp. 955–981.

P. M. Frank and X. Ding (1994), Frequency domain approach to optimally robust residual generation and evaluation for model-based fault diagnosis, *Automatica*, 30, pp. 789–804.

P. V. Goode and M. Chow (1993), Neural/fuzzy systems for incipient fault detection in induction motors, in *Proceedings of the International Conference on Industrial Economics, Control and Instrumentation (IECON'93)*, Vol. 1, IEEE Industrial Electronics Society, New York, pp. 332–337.

A. S. Heger, K. E. Holbert, and A. M. Ishaque (1996), Fuzzy associative memories for instrument fault detection, *Ann. Nuclear Energy*, 23, pp. 739–756.

K. E. Holbert, A. S. Heger, and A. M. Ishaque (1995), Fuzzy logic for power plant signal validation, in *Proceedings of the 9th Power Plant Dynamics, Control and Testing Symposium*, Knoxville, TN, pp. 20.01–20.15.

R. Isermann (1984), Process fault detection based on modeling and estimation methods: A survey, *Automatica*, 20, pp. 387–404.

R. Isermann and B. Freyermuth (1991a), Process fault diagnosis based on process model knowledge, part I: Principles for fault diagnosis with parameter estimation, *Trans. ASME J. Dynam. Systems Measurement Control*, 113, pp. 620–626.

R. Isermann and B. Freyermuth (1991b), Process fault diagnosis based on process model knowledge, part II: Case study experiments, *Trans. ASME J. Dynam. Systems Measurement Control*, 113, pp. 627–633.

T. W. Karjala and D. M. Himmelblau (1994), Dynamic data rectification using the extended Kalman filter and recurrent neural networks, in *Proceedings of the IEEE-ICNN*, Vol. 5, IEEE, Piscataway, NJ, pp. 3244–3249.

H. Kirch and K. Kroschel (1994), Applying Bayesian networks to fault diagnosis, in *Proceedings of the IEEE Conference on Control Applications*, IEEE, Piscataway, NJ, p. 895.

S. L. Lauritzen and D. J. Spiegelhalter (1988), Local computations with probabilities on graphical structures and their application to expert systems, *J. Roy. Statist. Soc. Ser. B*, 50, pp. 157–224.

S. C. Lee (1994), Sensor value validation based on systematic exploration of the sensor redundancy for fault diagnosis KBS, *IEEE Trans. Systems Man Cybernet.*, 24, pp. 594–605.

M. F. Mageed, A. F. S. Abdel, and A. Bahgat (1993), Fault detection and identification using a hierarchical neural network-based system, in *Proceedings of the International Conference on Industrial Electronics, Control and Instrumentation (IECON'93)*, IEEE Industrial Electronics Society, New York p. 338.

G. Mourot, S. Bousghiri, and F. Kratz (1993), Sensor fault detection using fuzzy logic and neural networks, in *Proceedings of the 1993 IEEE International Conference on Systems, Man and Cybernetics*, IEEE, Piscataway, NJ, pp. 369–374.

M. R. Napolitano, C. Neppach, V. Casdorph, and S. Naylor (1995), Neural-network-based scheme for sensor failure detection, identification, and accommodation, *J. Guidance Control Dynam.*, 18, pp. 1280–1286.

A. E. Nicholson and J. M. Brady (1994), Dynamic belief networks for discrete monitoring, *IEEE Trans. Systems Man and Cybernetics*, 24, pp. 1593–1610.

A. E. Nicholson (1996), Fall diagnosis using dynamic belief networks, in *Proceedings of the 4th Pacific Rim International Conference on Artificial Intelligence (PRICAI-96)*, Lecture Notes in Computer Science 1114, Springer-Verlag, Berlin, New York, Heidelberg, pp. 206–217.

S. Park and C. S. G. Lee (1993), Fusion-based sensor fault detection, in *Proceedings of the 1993 IEEE International Symposium on Intelligent Control*, IEEE, Piscataway, NJ, pp. 156–161.

J. Pearl (1988), *Probabilistic Reasoning in Intelligent Systems*, Morgan Kaufmann, San Mateo, CA.

I. E. Potter and M. C. Sunman (1977), Thresholdless redundancy management with arrays of skewed instruments, in *Integrity in Electronic Flight Control Systems*, AGARDO-GRAPH 224, Advisory Group for Aerospace Research and Development, Neuilly-sur-Seine, France, pp. 15.11–15.25.

D. Sauter, N. Mary, F. Sirou, and A. Thieltgen (1994), Fault diagnosis in systems using fuzzy logic, in *Proceedings of the 1994 IEEE Conference on Control Applications*, Vol. 2, IEEE, Piscataway, NJ, pp. 883–888.

R. M. Singer (1991), *Pumping System Fault Detection and Diagnosis Utilizing Pattern Recognition and Fuzzy Inference Techniques*, Report ANL/CP-71255, Argonne National Laboratory, Argonne, IL.

M. G. Singh, K. S. Hindi, G. Schmidt, and S. G. Tzafestas (1987), *Fault Detection and Reliability: Knowledge Based and Other Approaches*, Proceedings of the Second European Workshop on Fault Diagnostics, Reliability and Related Knowledge Based Approaches, April 6–8, 1987, Manchester, UK, Pergamon Press, Oxford, UK.

P. Smets (1992), The nature of the unnormalized beliefs encountered in the transferable belief model, in *Proceedings of the 8th Conference on Artificial Intelligence (AAAI-90)*, American Association for Artificial Intelligence, pp. 292–297.

D. J. Spiegelhalter, A. P. Dawid, S. L. Lauritzen, and R. G. Cowell (1993), Bayesian analysis in expert systems, *Statist. Sci.*, 8, pp. 219–283.

J. Watton, O. Lucca-Negro, and J. C. Stewart (1994), An on-line approach to fault diagnosis of fluid power cylinder drive systems, *J. Systems Control Engrg.*, 208, pp. 249–262.

Index